Prentice Hall Advanced Reference Series

Physical and Life Sciences

BINKLEY *The Pineal: Endocrine and Nonendocrine Function*
CAROZZI *Carbonate Rock Depositional Models: A Microfacies Approach*
EISEN *Mathematical Methods and Models in the Biological Sciences: Linear and One-Dimensional Theory*
EISEN *Mathematical Methods and Models in the Biological Sciences: Nonlinear and Multidimensional Theory*
FRASER *Stratigraphic Evolution of Clastic Depositional Sequences*
JEGER, ED. *Spatial Components of Plant Disease Epidemics*
McLENNAN *Introduction to Nonequilibrium Statistical Mechanics*
PLISCHKE AND BERGERSEN *Equilibrium Statistical Physics*
VALENZUELA AND MYERS *Adsorption Equilibrium Data Handbook*
WARREN *Evaporite Sedimentology: Importance in Hydrocarbon Accumulation*

INTRODUCTION TO NONEQUILIBRIUM STATISTICAL MECHANICS

James A. McLennan

Lehigh University

Prentice Hall, Englewood Cliffs, New Jersey 07632

Library of Congress Cataloging-in-Publication Data

McLennan, James A. (date)
Introduction to nonequilibrium statistical mechanics.

Bibliography: p.
Includes index.
1. Nonequilibrium thermodynamics. 2. Statistical thermodynamics. I. Title.
QC318.I7M35 1989 530.1′3 88-15108
ISBN 0-13-490962-3

Editorial/production supervision
and interior design: *Nancy Havas Farrell*
Cover design: *Ben Santora*
Manufacturing buyer: *Mary Ann Gloriande*

Prentice Hall Advanced Reference Series

A Division of Simon & Schuster
Englewood Cliffs, New Jersey 07632

Printed in the United States of America
10 9 8 7 6 5 4 3 2 1

ISBN 0-13-490962-3

Prentice-Hall International (UK) Limited, *London*
Prentice-Hall of Australia Pty. Limited, *Sydney*
Prentice-Hall Canada Inc., *Toronto*
Prentice-Hall Hispanoamericana, S.A., *Mexico*
Prentice-Hall of India Private Limited, *New Delhi*
Prentice-Hall of Japan, Inc., *Tokyo*
Simon & Schuster Asia Pte. Ltd., *Singapore*
Editora Prentice-Hall do Brasil, Ltda., *Rio de Janeiro*

To

Betsy, Ross, and Mimi

CONTENTS

PREFACE

This book has developed from a course on nonequilibrium statistical mechanics that I have taught at Lehigh University for a number of years. The course is intended for graduate students in physics, but has also been attended by students from chemistry and some engineering fields. It is assumed that the students have had a one-semester course in equilibrium statistical mechanics. More material is included than can be covered in one semester; I usually cover only somewhat more than half (but not the first half).

Statistical mechanics is a broad field, and the methods of nonequilibrium statistical mechanics have been fruitfully applied to a large variety of phenomena and materials. The scope of the book is limited to the fundamental subjects of Brownian motion, kinetic theory of gases, and the method of Green-Kubo formulas and time-dependent correlation functions. In the latter case only gases are considered in detail, although a few references are provided to indicate the widespread use of time-correlation functions. Even within this limitation many important references have been omitted. The bibliography contains over 400 references but should only be treated as a guide to the literature. Fortunately there are many excellent reviews, and their citation has been used to indirectly provide a more complete picture of the literature.

Thanks are due to many students and colleagues for their feedback and support. I am particularly grateful to Professor James W. Dufty for constant help and encouragement over an extended period of time. I am also indebted to Louise McCormack for her patient and careful typing of early versions of the manuscript.

James A. McLennan
Bitnet LJM 3 @ Lehigh

1

DYNAMICS OF FLUIDS

1.1. INTRODUCTION

The purpose of statistical mechanics is to provide a description of the macroscopic behavior of matter in bulk, starting from the microscopic equations which govern the motion of the atoms or molecules from which the matter is constituted. For states of thermal equilibrium this purpose is realized by the derivation of the laws of thermodynamics and of formulas which relate thermodynamic functions such as specific heat and compressibility to the properties and interactions of the constituent atoms or molecules.

The kinetic theory of gases describes the processes which occur in a gas not in equilibrium. The Boltzmann equation enables one to derive the hydrodynamic equations for a low-density monatomic gas and to calculate transport coefficients such as thermal conductivity and viscosity. Nonequilibrium statistical mechanics serves somewhat the same role as kinetic theory but for systems more general than gases; it also provides a basis for the derivation of the equations of kinetic theory.

This book covers kinetic theory including generalizations of the Boltzmann equation to deal with density-dependent effects and internal molecular degrees of freedom. Nonequilibrium statistical mechanics is covered with emphasis on the relation between irreversible processes and spontaneous fluctuations. Brownian motion is discussed in Chapter 2; its description has features in common with both kinetic theory and the theory of fluctuations.

In this chapter the elements of the macroscopic theory of fluids will be presented as background for the statistical treatment to follow. The local conservation laws are introduced, and thermodynamic variables defined for nonequilibrium states. The constitutive equations for slowly varying processes define the transport coefficients and complete the hydrodynamic equations. Much of the later material will be concerned with the derivation, from statistical mechanics or kinetic theory, of the hydrodynamic equations, but this emphasis should not obscure the more general applicability of the statistical approach to phenomena which lie beyond the range of the hydrodynamic equations.

1.2. CONSERVATION LAWS

The variables basic to the description of a fluid are the mass density ρ, the energy density ϵ, and the momentum density $\mathbf{g}$. At this point these are considered to be macroscopic variables, which will later be interpreted as statistical averages of the corresponding microscopic quantities. Initially we suppose there to be only a single chemical constituent; otherwise the concentrations of the various components are needed for a complete description.

In general ρ, ϵ, and $\mathbf{g}$ depend on the position $\mathbf{x}$ and time t. This dependence must be in accordance with conservation of mass, energy, and momentum. Let $\mathbf{j}_m$ be the mass flux, defined so that $\mathbf{j}_m \cdot d\mathbf{A}$ is the rate at which mass crosses the element of area $d\mathbf{A}$. Local conservation of mass is expressed by the equation of continuity,

$$\frac{\partial \rho}{\partial t} + \nabla \cdot \mathbf{j}_m = 0. \tag{1.1}$$

Similarly the local conservation law for energy is

$$\frac{\partial \epsilon}{\partial t} + \nabla \cdot \mathbf{s} = 0 \tag{1.2}$$

where $\mathbf{s}$ is the energy flux, and for momentum

$$\frac{\partial g_i}{\partial t} + \frac{\partial t_{ji}}{\partial x_j} = 0. \tag{1.3}$$

Here t_{ji} is the flux of the i-component of momentum, across an element of area with its normal in the j-direction, and is also called the pressure tensor. We will sometimes denote the pressure tensor by $\mathbf{t}$ and write the preceding equation as

$$\frac{\partial \mathbf{g}}{\partial t} + \nabla \cdot \mathbf{t} = 0. \tag{1.4}$$

For simplicity we assume the pressure tensor to be symmetric, $t_{ij} = t_{ji}$, and defer discussion of this point until Chapter 8.

The total mass, energy, and momentum are obtained by integrating their densities over the volume of the system. They are conserved (that is, time-independent) if their net fluxes across the surface bounding the system vanish. For brevity ρ, ϵ and $\mathbf{g}$ will be referred to as conserved densities, although strictly speaking they are not conserved.

In nonrelativistic mechanics the mass flux $\mathbf{j}_m$ can be identified with the momentum density $\mathbf{g}$. The effects of relativity will not be taken into account in this book and so henceforth $\mathbf{g}$ will be used for the mass flux. The two quantities are nevertheless logically distinct, and indeed in relativity the dependence of mass on velocity prevents their identification.

As written, the local conservation laws apply to an isolated system. If there are external interactions, the conservation laws are augmented by source terms. An external force $\mathbf{F}$ (per unit volume) changes Equation (1.4) to

$$\frac{\partial \mathbf{g}}{\partial t} + \nabla \cdot \mathbf{t} = \mathbf{F}.$$

Similarly, Equation (1.2) is replaced by

$$\frac{\partial \epsilon}{\partial t} + \nabla \cdot \mathbf{s} = w$$

where w is the rate at which work is done, per unit volume, by the external force.

The preceding conservation laws are sufficient for a fluid with only a single chemical component, which is usually called a simple fluid. For other systems there may be more, or different, conservation laws. For example, if there are several nonreacting species, then the number density (or mole density, or mass density) is separately conserved for each species. This case will be considered later.

The local conservation laws both express the fact of conservation and provide definitions of the fluxes. The fluxes, so defined, are not determined uniquely, since the addition to a flux of any quantity with vanishing divergence does not change the conservation law. Furthermore, the fluxes are not measurable in a way which would provide separate definitions, independently of the conservation laws. It is sometimes helpful to visualize $\mathbf{t}$ as the force acting on a surface element but in fact forces act on physical bodies; the net force on an element of matter is measurable but its distribution over the bounding surface is not. This arbitrariness in the fluxes can be reduced or eliminated by special considerations. For example, as already mentioned, we may choose the mass flux to be the same as the momentum density (but it remains to be shown that they have the same divergence). In addition we will require that the fluxes transform in a definite way under transformations of the coordinate system.

We have assumed the conserved densities ρ, ϵ, and $\mathbf{g}$ to be uniquely defined but actually even this is open to question; the total energy of a closed physical system has a clear meaning but its spatial distribution throughout the system is not determined by the total value. There is less ambiguity, with regard to both the conserved densities and their fluxes, in the statistical approach. From the microscopic point of view the mass, energy, and momentum in a volume element can be determined, at least in principle, by adding up contributions from all the atoms contained in the element. In Chapter 8 the microscopic conserved densities and their fluxes will be determined as definite functions of the atomic variables, and there is then no arbitrariness in the macroscopic quantities obtained through averaging.

The flow velocity $\mathbf{u}$ is defined by

$$\mathbf{g} = \rho\mathbf{u}. \tag{1.5}$$

The equation of continuity can then be written

$$\frac{\partial\rho}{\partial t} + \nabla\cdot\rho\mathbf{u} = 0.$$

Equation (1.3) can be manipulated into

$$\rho\frac{\partial\mathbf{u}}{\partial t} = \mathbf{u}\nabla\cdot\rho\mathbf{u} - \nabla\cdot\mathbf{t}$$

which can also be written as

$$\rho\left(\frac{\partial u_i}{\partial t} + \mathbf{u}\cdot\nabla u_i\right) = \frac{\partial\sigma_{ji}}{\partial x_j} \tag{1.6}$$

where

$$\sigma_{ji} = \rho u_i u_j - t_{ji}.$$

Usually σ_{ji} is called the stress tensor, and Equation (1.6) is called the equation of motion.

1.3. GALILEAN TRANSFORMATIONS

The conservation laws have been written in a certain set of coordinates, or frame of reference. This frame must be an inertial one, but the conservation laws should be invariant under a transformation to a different inertial frame. The transformation laws for the conserved densities and their fluxes will be derived from the requirement that they leave the conservation laws invariant.

By "invariant" we mean that, with transformation laws which still have to be determined, the conservation laws in one frame of reference imply those in the other.

Let $\mathbf{x}, \mathbf{x}'$ be the coordinates of a point as seen from two different frames. If the transformation is a translation, $\mathbf{x}' = \mathbf{x} + \mathbf{a}$, $\mathbf{a} =$ constant, then the densities and fluxes have the same values in the two frames, for example

$$\rho'(\mathbf{x}', t) = \rho(\mathbf{x}, t). \tag{1.7}$$

It is immediately verified that the conservation laws are then invariant. The same applies to a translation of the time axis, $t' = t + s$, $s =$ constant, with $\rho'(\mathbf{x}, t') = \rho(\mathbf{x}, t)$, and so forth.

If the transformation is a rotation,

$$x_i' = a_{ij} x_j, \; a_{ij} a_{ik} = \delta_{jk},$$

then ρ and ϵ transform as scalars, that is, as in Equation (1.7). As the notation indicates, g_i and s_i transform as vectors and t_{ij} as a second-order tensor:

$$g_i'(\mathbf{x}', t) = a_{ij} g_j(\mathbf{x}, t), \; s_i'(\mathbf{x}', t) = a_{ij} s_j(\mathbf{x}, t), \; t_{ij}'(\mathbf{x}', t) = a_{ik} a_{jl} t_{kl}(\mathbf{x}, t).$$

That these transformations leave the conservation laws invariant under rotations is shown by a straightforward calculation.

The behavior under a Galilean velocity transformation is more complicated and also more interesting for the later discussion. Let $\mathbf{w}$ be the velocity of the primed system with respect to the unprimed. If the two frames are taken to coincide at time $t = 0$, then

$$\mathbf{x}' = \mathbf{x} - \mathbf{w}t.$$

The density ρ is the same in both frames while the fluid velocity changes according to

$$\mathbf{u}'(\mathbf{x}', t) = \mathbf{u}(\mathbf{x}, t) - \mathbf{w}$$

and consequently

$$\mathbf{g}'(\mathbf{x}', t) = \mathbf{g}(\mathbf{x}, t) - \rho(\mathbf{x}, t)\mathbf{w}. \tag{1.8}$$

The derivative $(\partial\rho/\partial t)'$ in the primed frame is taken with $\mathbf{x}'$ held constant, while $\partial\rho/\partial t$ is taken with $\mathbf{x}$ held constant. We have $(\partial\mathbf{x}/\partial t)' = \mathbf{w}$ and so

$$\left(\frac{\partial\rho}{\partial t}\right)' = \frac{\partial\rho}{\partial t} + \mathbf{w}\cdot\nabla\rho.$$

The spatial derivatives are the same in the two frames, $\nabla' = \nabla$, so we get

$$\left(\frac{\partial\rho}{\partial t} + \nabla\cdot\mathbf{g}\right)' = \frac{\partial\rho}{\partial t} + \nabla\cdot\mathbf{g}$$

which demonstrates the invariance of the equation of continuity to Galilean velocity transformations.

We next use the transformation law (1.8) for $\mathbf{g}$ to deduce the transformation law for $\mathbf{t}$. Arguing in the same way as before, we obtain

$$\left(\frac{\partial \mathbf{g}}{\partial t}\right) = \left(\frac{\partial}{\partial t} + \mathbf{w}\cdot\nabla\right)(\mathbf{g} - \rho\mathbf{w}) = \frac{\partial \mathbf{g}}{\partial t} + \mathbf{w}\cdot\nabla\mathbf{g} + \mathbf{w}(\nabla\cdot\mathbf{g} - \mathbf{w}\cdot\nabla\rho).$$

From this we get

$$t'_{ij} = t_{ij} - w_i g_j - w_j g_i - \rho w_i w_j$$

The transformation law for ϵ is

$$\epsilon' = \epsilon - \mathbf{g}\cdot\mathbf{w} + \frac{1}{2}\rho w^2.$$

(This can be deduced from the transformation law for the energy $E = \frac{1}{2}mv^2$ of a particle.) This yields

$$\left(\frac{\partial \epsilon}{\partial t}\right)' = \frac{\partial \epsilon}{\partial t} + \frac{\partial}{\partial x_i}\left[w_i\left(\epsilon - \mathbf{w}\cdot\mathbf{g} + \frac{1}{2}\rho w^2\right) + w_j t_{ij} - \frac{1}{2}w^2 g_i\right]$$

and the transformation law for $\mathbf{s}$ is found to be

$$s'_i = s_i - \left(\epsilon - \mathbf{w}\cdot\mathbf{g} + \frac{1}{2}\rho w^2\right)w_i - t_{ij}w_j + \frac{1}{2}w^2 g_i.$$

The transformation laws can be solved for the unprimed quantities in terms of the primed. The equations have the same form, but $\mathbf{w}$ is replaced by $-\mathbf{w}$.

1.4. THERMODYNAMICS

Consider an element of fluid with position $\mathbf{x}$, having the velocity $\mathbf{u}$ at time t. Let a transformation be made to a frame of reference moving with velocity $\mathbf{w} = \mathbf{u}(\mathbf{x}, t)$. Of course $\mathbf{w}$ is constant and is equal to the fluid velocity only at the particular position and time. With respect to the new frame of reference, the fluid element is instantaneously at rest, and the new frame will be referred to as the local rest frame. There is a local rest frame for each point in the fluid and for each value of the time. Denoting quantities in the local rest frame by a subscript 0, we obtain from the transformation equations of the previous section

$$\rho_0 = \rho,\ \mathbf{g}_0 = 0,\ \epsilon_0 = \epsilon - \frac{1}{2}\rho u^2,$$
$$t_{0ij} = t_{ij} - \rho u_i u_j,\ s_{0i} = s_i - \left(\epsilon_0 + \frac{1}{2}\rho u^2\right)u_i - t_{0ij}u_j. \tag{1.9}$$

Note that $t_{0ij} = -\sigma_{ij}$ where σ_{ij} is the stress tensor.

Quantities referred to the local rest frame depend only on the internal state of the fluid element. If the element were in thermodynamic equilibrium, its intensive state would be specified by two intensive variables, say ρ and ϵ_0. (At this point we still consider a simple fluid.) Other thermodynamic variables can then be determined as functions of these. For example, the Kelvin temperature T and entropy (per unit mass) s are certain functions of ρ and ϵ_0,

$$T = T(\rho, \epsilon_0), \; s = s(\rho, \epsilon_0). \tag{1.10}$$

The actual forms of the functions depend on the material but can be determined by equilibrium methods and will be assumed to be known.

Suppose now that the fluid element is not in equilibrium. The mass and energy densities are mechanical quantities which may be defined independently of the thermodynamic variables. Hence Equations (1.10) may be used to define T and s for the nonequilibrium state. Since ρ and ϵ_0 in general depend on $\mathbf{x}$ and t, the same will be the case for T and s:

$$T(\mathbf{x}, t) = T[\rho(\mathbf{x}, t), \epsilon_0(\mathbf{x}, t)], \; s(\mathbf{x}, t) = s[\rho(\mathbf{x}, t), \epsilon_0(\mathbf{x}, t)].$$

Other intensive variables of thermodynamics, such as pressure, chemical potential, specific heat, and so on, may be defined in the same way, that is, as the same functions of ρ and ϵ_0 as they are in equilibrium. The various relations of thermodynamics are then valid locally, since they are based on functional relationships which are maintained by the definitions. (At a later point it will be found advantageous to define the thermodynamic variables in a slightly different way which takes into account the statistical correlations between neighboring volume elements.)

The total entropy S is defined as the integral of its density,

$$S = \int d^3x\, \rho\, s.$$

This does not depend on the total mass and energy in the same way as in equilibrium unless the system is in fact uniform.

The definition of thermodynamic variables for nonequilibrium states is straightforward theoretically, but their measurement is problematic. The basis of the measurement of temperature is the zero'th law of thermodynamics: two systems in equilibrium with a third are in equilibrium with each other. This assures that a properly calibrated thermometer will give meaningful readings, regardless of the details of its construction or of the system being measured. Clearly there is no zero'th law for a nonequilibrium situation. The reading of a "thermometer" may depend on its orientation, shape, surface properties, and so on; other variables in the system being observed, such as the gradient of the "temperature" or its rate of change, may also influence the reading. Nonequilibrium temperature is introduced for theoretical convenience rather than to take advantage of a basic principle.

In systems with long-range order such as solids, order parameters provide additional "thermodynamic" variables. These do not correspond, at the microscopic level, to strict conservation laws; an order parameter exists only if the temperature is below a transition point. The incorporation of order parameters into thermodynamics will not be considered here.

In thermodynamics the entropy is used to measure the deviation from equilibrium and to track the approach to equilibrium. According to the entropy principle, the entropy of an isolated system increases monotonically to its maximum as the system approaches equilibrium. However, the statement of the entropy principle must be qualified to account for spontaneous fluctuations. Furthermore, the thermodynamic entropy, as previously defined, is sensitive only to deviations from equilibrium which are associated with spatial variations. A spatially homogeneous system need not be in equilibrium; the thermalization of atomic velocities and the relaxation of internal molecular degrees of freedom are examples of irreversible processes which do not show up in the thermodynamic entropy. Statistical mechanics provides greater discrimination in measuring the deviation from equilibrium, for example by Boltzmann's H-function in kinetic theory or the Gibbs entropy for statistical ensembles. Frequently entropy refers to one of the statistical versions, but the thermodynamic definition is customary in most of the nonstatistical literature. Later we will encounter some of the statistical entropies, but at present the term will be given its thermodynamic definition.

We list here a few thermodynamic equations which will be needed. The first law of thermodynamics, $T\,ds = d(\epsilon_0/\rho) + p\,d(1/\rho)$, is equivalent to

$$T\,d(\rho s) = d\epsilon_0 - \mu\,d\rho \tag{1.11}$$

where the chemical potential (per unit mass) μ is defined by

$$\mu\rho = \epsilon_0 + p - T\rho s.$$

In addition we have

$$dp = \rho(d\mu + s\,dT) \tag{1.12}$$

which is called the Gibbs-Duhem equation. The specific heat (per unit mass) at constant volume is given by

$$C_v = \rho^{-1}(\partial\epsilon_0/\partial T)_\rho$$

and the isothermal compressibility is

$$K_T = \rho^{-1}(\partial\rho/\partial p)_T$$

A standard formula for the difference in specific heats is

$$C_p - C_v = (T/\rho)\, K_T[(\partial p/\partial T)_\rho]^2 \tag{1.13}$$

where C_p is the specific heat at constant pressure.

For states near equilibrium there is a convenient expansion of the total entropy S. Let s be considered as a function of ϵ, ρ and $\mathbf{g}$ rather than of ϵ_0 and ρ. In view of the relation between ϵ and ϵ_0, we have

$$s = s\left(\epsilon - \frac{1}{2\rho}g^2, \rho\right). \tag{1.14}$$

Let $\bar{\rho}$, $\bar{\epsilon}$, and $\bar{\mathbf{g}}$ denote the average values,

$$V\bar{\rho} = \int d^3x\, \rho,\ V\bar{\epsilon} = \int d^3x\, \epsilon,\ V\bar{\mathbf{g}} = \int d^3x\, \mathbf{g},$$

where V is the volume of the system. Let $\delta\rho$, $\delta\epsilon$, and $\delta\mathbf{g}$ denote the deviations from the average values,

$$\delta\rho = \rho - \bar{\rho},\ \delta\epsilon = \epsilon - \bar{\epsilon},\ \delta\mathbf{g} = \mathbf{g} - \bar{\mathbf{g}}. \tag{1.15}$$

We wish to expand S about an equilibrium reference state determined by the values of $\bar{\rho}$, $\bar{\epsilon}$, and $\bar{\mathbf{g}}$. The terms in the expansion which are linear in the deviations (1.15) vanish (since the volume integrals of the small variations vanish), which expresses the fact that the entropy is an extremum in equilibrium. Through second-order terms we get

$$\begin{aligned}\delta S &= S - \bar{S} \\ &= \frac{1}{2}\int d^3x \left\{\rho\frac{\partial^2 s}{\partial\epsilon^2}\delta\epsilon^2 + \frac{\partial^2(\rho s)}{\partial\rho^2}\delta\rho^2 + \rho\frac{\partial^2 s}{\partial g_i\partial g_j}\delta g_i\,\delta g_j\right. \\ &\quad \left. + 2\frac{\partial^2(\rho s)}{\partial\epsilon\,\partial\rho}\delta\epsilon\,\delta\rho + 2\rho\frac{\partial^2 s}{\partial\epsilon\,\partial g_i}\delta\epsilon\,\delta g_i + 2\frac{\partial^2(\rho s)}{\partial\rho\,\partial g_i}\delta\rho\,\delta g_i + \ldots\right\}\end{aligned}$$

where the derivatives are evaluated at the reference state.

The first derivatives of the entropy are given by

$$(\partial s/\partial\epsilon_0)_\rho = 1/(\rho T),\ (\partial s/\partial\rho)_{\epsilon 0} = -(\epsilon_0 + p)/(\rho^2 T).$$

Since the entropy is a Galilean invariant, we may transform to a frame of reference in which $\bar{\mathbf{g}} = 0$. In this case Equation (1.14) yields

$$\partial^2 s/\partial g_i\,\partial g_j = -\delta_{ij}(\rho T)^{-1},\ \partial^2 s/\partial g_i\,\partial\rho = \partial^2 s/\partial g_i\,\partial\epsilon = 0. \tag{1.16}$$

Consequently, we have at this point

$$\delta S = \frac{1}{2}\int d^3x \left\{\left(\frac{\partial(1/T)}{\partial\epsilon_0}\right)_\rho \delta\epsilon^2 - \left(\frac{\partial(\mu/T)}{\partial\rho}\right)_{\epsilon 0}\delta\rho^2 - 2\left(\frac{\partial(\mu/T)}{\partial\epsilon}\right)_\rho \delta\epsilon\,\delta\rho - \frac{1}{\rho T}\delta g^2\right\}.$$

This can be brought into diagonal form by using δT as an independent variable in place of $\delta\epsilon$. After making the substitution $\delta\epsilon = (\partial\epsilon/\partial T)_\rho \delta T + (\partial\epsilon/\partial\rho)_T \delta\rho$ and carrying out some thermodynamic manipulations, we find

$$\delta S = -\frac{1}{2}\int d^3x[(\rho^2 T K_T)^{-1}\,\delta\rho^2 + (\rho C_v/T^2)\,\delta T^2 + (\rho T)^{-1}\,\delta g^2].$$

This can also be written in terms of the velocity by substituting $\delta\mathbf{g} = \rho\,\delta\mathbf{u}$. The coefficients of δT^2, $\delta\rho^2$, and δg^2 are non-negative, so $\delta S \le 0$.

Thermodynamic normal coordinates can be defined on the basis of the preceding quadratic form. Let ξ_α, $\alpha = 1,..,5$ denote the quantities

$$\xi_\alpha = \left\{\frac{1}{\sqrt{TK_T}}\frac{\delta\rho}{\rho}, \sqrt{\rho C_v}\frac{\delta T}{T}, \frac{1}{\sqrt{\rho T}}\delta\mathbf{g}\right\} \tag{1.17}$$

In terms of the ξ_α, δS has the normal form:

$$\delta S = -\frac{1}{2}\int d^3x \sum_{\alpha=1}^{5} \xi_\alpha^2. \tag{1.18}$$

1.5. IRREVERSIBLE FLUXES AND ENTROPY PRODUCTION

The fluxes can be separated into reversible and irreversible parts on the basis of Equation (1.9), which we rewrite as

$$t_{ij} = t_{0ij} + \rho u_i u_j,\; s_i = s_{0i} + \left(\epsilon_0 + \frac{1}{2}\rho\, u^2\right)u_i + t_{0ij}u_j. \tag{1.19}$$

If the volume element is in equilibrium, then for a fluid $\mathbf{t}_0$ is diagonal with diagonal elements equal to the pressure p: $t_{0ij} = p\,\delta_{ij}$. Furthermore, there is no energy flux in the local rest frame, so $\mathbf{s}_0 = 0$. Hence, we have

$$t_{ij} = p\,\delta_{ij} + \rho u_i u_j,\; s_i = \left(h + \frac{1}{2}\rho u^2\right)u_i, \tag{1.20}$$

where $h = \epsilon_0 + p$ is the enthalpy per unit volume.

If the fluid element is not in equilibrium, there will in general be additional contributions to $\mathbf{t}$ which we denote by $\mathbf{t}^*$:

$$t_{ij} = p\,\delta_{ij} + \rho u_i u_j + t_{ij}^*. \tag{1.21}$$

The quantity p is the thermodynamic pressure, defined to be the same function of ρ and ϵ_0 as in equilibrium; it then has no immediate connection with the force per unit area which is determined by $\mathbf{t}$. As is seen from Equation (1.19), t_{ij}^* gives a contribution to s_i of the form $t_{ij}^* u_j$. There will be an additional contribution due to processes such as heat conduction, so we write

$$s_i = \left(h + \frac{1}{2}\rho u^2\right)u_i + t^*_{ij}u_j + s^*_i. \tag{1.22}$$

The preceding two equations define the irreversible fluxes $\mathbf{t}^*$ and $\mathbf{s}^*$. The fluxes in the local rest frame are given by

$$t_{0ij} = p\,\delta_{ij} + t^*_{ij},\ \mathbf{s}_0 = \mathbf{s}^*. \tag{1.23}$$

Equations (1.21) and (1.22) can be used to write the conservation laws in a different way which is sometimes more convenient. Let D/Dt denote the substantial derivative:

$$\frac{D}{Dt} = \frac{\partial}{\partial t} + \mathbf{u}\cdot\nabla. \tag{1.24}$$

Then the equation of motion (1.6) becomes

$$\rho\frac{D\mathbf{u}}{Dt} = -\nabla p - \nabla\cdot\mathbf{t}^* \tag{1.25}$$

and the conservation law for energy can be converted into

$$\frac{D\epsilon_0}{Dt} = -h\nabla\cdot\mathbf{u} - t^*_{ji}\frac{\partial u_i}{\partial x_j} - \nabla\cdot\mathbf{s}^*, \tag{1.26}$$

which is called the energy equation. In addition the equation of continuity takes the form

$$\frac{D\rho}{Dt} = -\rho\nabla\cdot\mathbf{u}. \tag{1.27}$$

The energy equation can be replaced by an equation for the temperature as follows. We have

$$\frac{DT}{Dt} = \left(\frac{\partial T}{\partial\rho}\right)_{\epsilon_0}\frac{D\rho}{Dt} + \left(\frac{\partial T}{\partial\epsilon_0}\right)_\rho\frac{D\epsilon_0}{Dt}$$
$$= -\left[\rho\left(\frac{\partial T}{\partial\rho}\right)_{\epsilon_0} + h\left(\frac{\partial T}{\partial\epsilon_0}\right)_\rho\right]\nabla\cdot\mathbf{u} - \left(\frac{\partial T}{\partial\epsilon_0}\right)_\rho\left[t^*_{ji}\frac{\partial u_i}{\partial x_j} + \nabla\cdot\mathbf{s}^*\right].$$

To reduce the coefficient of $\nabla\cdot\mathbf{u}$ we use Equation (1.11) to express h as

$$h = -\rho^2 T\left(\frac{\partial s}{\partial\rho}\right)_T + \rho\left(\frac{\partial\epsilon_0}{\partial\rho}\right)_T.$$

The relation

$$d[(\epsilon_0/\rho) - T\,s] = (p/\rho^2)d\rho - s\,dT$$

shows that

$$(\partial s/\partial \rho)_T = -(1/\rho^2)(\partial p/\partial T)_\rho \tag{1.28}$$

so h can be written as

$$h = T\left(\frac{\partial p}{\partial T}\right)_\rho + \rho\left(\frac{\partial \epsilon_0}{\partial \rho}\right)_T.$$

Using this we find the coefficient of $\nabla \cdot \mathbf{u}$ to reduce to $T(\partial p/\partial \epsilon_0)_\rho$. The factor $(\partial \epsilon_0/\partial T)_\rho$ can be expressed in terms of the specific heat, and the desired equation for the temperature is

$$\frac{DT}{Dt} = -T\left(\frac{\partial p}{\partial \epsilon_0}\right)_\rho \nabla \cdot \mathbf{u} - (\rho C_v)^{-1}\left(t_{ji}^*\frac{\partial u_i}{\partial x_j} + \nabla \cdot \mathbf{s}^*\right). \tag{1.29}$$

The quantity $(\partial p/\partial \epsilon_0)_\rho$ can be obtained from standard thermodynamic functions by

$$(\partial p/\partial \epsilon_0)_\rho = \alpha/(\rho C_v K_T)$$

where α is the coefficient of expansion, $\alpha = -\rho^{-1}(\partial \rho/\partial T)_p$. This can also be written

$$(\partial p/\partial \epsilon_0)_\rho = (\gamma - 1)/\alpha T$$

where $\gamma = C_p/C_v$.

The irreversible nature of $\mathbf{t}^*$ and $\mathbf{s}^*$ can be exhibited by calculating their effect on the entropy. Using Equations (1.27) and (1.29) we find

$$\begin{aligned}\frac{Ds}{Dt} &= \left(\frac{\partial s}{\partial T}\right)_\rho \frac{DT}{Dt} + \left(\frac{\partial s}{\partial \rho}\right)_T \frac{D\rho}{Dt} \\ &= -\left[T\left(\frac{\partial s}{\partial T}\right)_\rho \left(\frac{\partial p}{\partial \epsilon_0}\right)_\rho + \rho\left(\frac{\partial s}{\partial \rho}\right)_T\right]\nabla \cdot \mathbf{u} - (\rho C_v)^{-1}\left(\frac{\partial s}{\partial T}\right)_\rho \left[t_{ji}^*\frac{\partial u_i}{\partial x_j} + \nabla \cdot \mathbf{s}^*\right].\end{aligned}$$

Using Equation (1.28) we find the coefficient of $\nabla \cdot \mathbf{u}$ to vanish, while the coefficient of the remaining term is $1/\rho T$. Hence, we get

$$\frac{Ds}{Dt} = -(\rho T)^{-1}\left[t_{ji}^*\frac{\partial u_i}{\partial x_j} + \nabla \cdot \mathbf{s}^*\right].$$

This can be written in terms of the entropy per unit volume as

$$\frac{\partial(\rho s)}{\partial t} + \nabla \cdot (\rho s \mathbf{u}) = -T^{-1}\left[t_{ji}^*\frac{\partial u_i}{\partial x_j} + \nabla \cdot \mathbf{s}^*\right]. \tag{1.30}$$

The quantity $\rho s \mathbf{u}$ is the flux of entropy, so if the right-hand side vanished we would have a conservation law for entropy. The source terms containing $\mathbf{t}^*$ and $\mathbf{s}^*$ describe the irreversible contributions to the entropy.

If Equation (1.30) is integrated over the volume, the result is an equation for the rate of change of the total entropy S which we put in the form

$$\frac{dS}{dt} = \int d^3x \left[\mathbf{s}^* \cdot \nabla T^{-1} - T^{-1} t_{ji}^* \frac{\partial u_i}{\partial x_j} \right] - \int d\mathbf{A} \cdot (\rho s\, \mathbf{u} + T^{-1} \mathbf{s}^*)$$

where $d\mathbf{A}$ is an element of area on the bounding surface. The surface integral describes the rate of change of entropy due to flow across the surface of entropy-bearing fluid, and to energy crossing the surface. The integrand of the volume integral,

$$\sigma = \mathbf{s}^* \cdot \nabla T^{-1} - T^{-1} t_{ji}^* \frac{\partial u_i}{\partial x_j}, \tag{1.31}$$

is called the entropy production. This can also be written in the form

$$\sigma = \rho\, Ds/Dt + \nabla \cdot (\mathbf{s}^*/T).$$

In some cases the entropy production can be shown to be non-negative, $\sigma \geq 0$, with equality holding only in equilibrium. Then if the system is isolated (so the normal components of $\mathbf{s}^*$ and $\mathbf{u}$ vanish at the boundary), S increases monotonically with time until it reaches its maximum in equilibrium. However, the approach to equilibrium cannot always be attributed to a non-negative entropy production. For example, in materials with slow internal relaxation, σ does not have a definite sign.

1.6. CONSTITUTIVE EQUATIONS

From the microscopic point of view, the conservation laws are consequences of the atomic equations of motion. When the system is treated macroscopically, the point of view is taken that the conservation laws themselves are the equations of motion for ρ, ϵ, and $\mathbf{g}$. In order for these to form a closed set of equations, it is necessary that $\mathbf{t}^*$ and $\mathbf{s}^*$ be functionally related to ρ, ϵ, and $\mathbf{g}$, as otherwise there are more variables than equations.

The simplest possibility is to assume that $\mathbf{t}^*$ and $\mathbf{s}^*$ vanish. The equation of motion and temperature equation reduce to

$$\rho \frac{D\mathbf{u}}{Dt} = -\nabla p, \quad \frac{DT}{Dt} = -T \left(\frac{\partial p}{\partial \epsilon} \right)_\rho \nabla \cdot \mathbf{u}. \tag{1.32}$$

These, together with the equation of continuity, are Euler's equations for an ideal fluid.

To discuss $\mathbf{t}^*$ and $\mathbf{s}^*$ more generally, we suppose the deviation from equilibrium to be completely characterized by spatial variations in T, ρ and $\mathbf{u}$.

This means in particular that $\mathbf{t}^*$ and $\mathbf{s}^*$ are functions of $\mathbf{x}$ and t only through their dependence on $T(\mathbf{x},t)$, $\rho(\mathbf{x},t)$ and $\mathbf{u}(\mathbf{x},t)$,

$$\mathbf{t}^* = F[T(\mathbf{x},t),\rho(\mathbf{x},t),\mathbf{u}(\mathbf{x},t)], \mathbf{s}^* = G[T(\mathbf{x},t),\rho(\mathbf{x},t),\mathbf{u}(\mathbf{x},t)]. \tag{1.33}$$

Here F and G are functionals which vanish when T, ρ and $\mathbf{u}$ are spatially uniform. They may involve spatial derivatives of all orders and may be nonlinear.

The Equations (1.33) are called constitutive equations, or transport equations. When combined with the conservation laws, they yield the hydrodynamic equations, which are the equations of motion for T, ρ, and $\mathbf{u}$.

The constitutive equations could be written in terms of ρ, ϵ and $\mathbf{g}$ rather than T, ρ and $\mathbf{u}$, but the form given is more convenient. To distinguish them from the conserved densities, T, ρ and $\mathbf{u}$ will be referred to as the thermodynamic variables.

The discussion of the previous sections was quite general, being based only on the conservation laws and on thermodynamics. However, the constitutive equations are special in nature and the conditions for their application need discussion. The first point to notice is that the constitutive equations in effect make the fluxes definite functionals of the conserved densities, but from the microscopic point of view it is easily seen that such a dependence does not hold in general. There is sufficient freedom to adjust the atomic coordinates and momenta to give the fluxes almost any values independently of the densities.

This point is easily understood on the basis of the Boltzmann equation, to be discussed in later chapters. Let $f(\mathbf{x}, \mathbf{v}, t)$ be the distribution function, defined so that fd^3xd^3v is the number of atoms in the volume element d^3x and with their velocity $\mathbf{v}$ in d^3v, at time t. The conserved densities and their fluxes are given in terms of f by

$$\begin{aligned} \rho &= \int d^3v\, mf,\ \mathbf{g} = \int d^3v\, m\mathbf{v}f,\ \epsilon = \int d^3v\, \frac{1}{2}mv^2 f, \\ \mathbf{s} &= \int d^3v \frac{1}{2}mv^2\mathbf{v}f,\ t_{ij} = \int d^3v\, mv_iv_jf \end{aligned} \tag{1.34}$$

where m is the mass of an atom. Clearly f can be given a form which, within limits, will assign almost any values to the densities and fluxes independently of one another. (The limits are set by certain inequalities which follow from the preceding equations, such as $\epsilon \geq g^2/2\rho$.)

In the theory of the Boltzmann equation it is found that there is a special class of solutions, called normal solutions, with the property that the dependence on $\mathbf{x}, t$ of f occurs entirely through its functional dependence on the thermodynamic variables. It follows immediately that the fluxes calculated from Equations (1.34) are also determined as functionals of the thermodynamic variables, so constitutive equations hold when f has the normal form.

The Boltzmann equation is limited to a low-density gas, but the concept of normal states or processes which are completely characterized by the thermodynamic variables is more general.

The special nature of normal processes is due to the following. Consider first a gas which is spatially uniform but whose atomic velocities are not distributed according to equilibrium. Collisions cause the distribution function to approach its equilibrium form, the particular equilibrium state being fixed by the values of the conserved variables such as the total energy. Collisions are very frequent and so the approach to equilibrium takes place rapidly. (This velocity relaxation cannot be described by constitutive equations, since $\mathbf{s}^*$ and $\mathbf{t}^*$ may be nonvanishing even if there are no spatial variations.) Now suppose the system to be put into a state which is not spatially uniform. The atoms within a small volume element (macroscopically small, that is, but large enough to contain many atoms) will, through collisions, quickly adjust to the local values of the thermodynamic variables. The latter are restricted by the conservation laws; the energy, say, must be redistributed throughout the entire system and this redistribution is relatively slow. Collisions force the distribution function to be synchronized with the slowly varying local values of the conserved densities. There is an initial period during which this synchronization takes place, and then a slower hydrodynamic stage during which the distribution function has the normal form.

The initial period is called the ageing period, and one speaks of ageing to hydrodynamics. The length of the ageing period depends on the initial state, but a typical order of magnitude is about the same as the mean time between collisions. At standard conditions this is about 10^{-9} seconds, and in dense systems it is even shorter. The duration of the hydrodynamic stage is much longer since it is determined by macroscopic parameters including the size of the system. The normal solution has a special role because it lasts throughout the hydrodynamic stage.

There are some materials for which the ageing period is long because of slow structural relaxation, slow chemical reactions, and so on. One can then modify the constitutive equations so that they depend on time derivatives, or even the entire past history, of the thermodynamic variables. Such a procedure is also indicated for a material with a nominally short ageing period if the "hydrodynamics" is very rapid, as in a shock or high-frequency sound. On the other hand, in some cases a long internal relaxation time is best handled by the introduction of extra "conserved" quantities, an example being the slow conversion of the ortho and para states of hydrogen.

A process analogous to ageing also takes place near a physical boundary or wall. There is a thin boundary layer, called the Knudsen layer in gases, within which the state of the fluid is affected by the nature of the wall and its interaction with the fluid. As one goes into the body of the fluid, the fluxes again become synchronized with the thermodynamic variables and the constitutive equations hold. The Knudsen layer leads to "slip"; boundary values used in the solution of the hydrodynamic equations must be altered from the actual boundary values to correct for the layer in which the hydrodynamic equations do not hold. It was noticed by Maxwell that this leads to a reduction in the apparent viscosity of gases, which however is negligible unless the Knudsen layer occupies a significant fraction of the volume of the gas. There is a similar effect in initial-value problems, which has been called "initial slip" by Grad: If the hydrodynamic equations are solved subject to assigned initial values it may be necessary to adjust the initial values to extrapolate properly through the ageing period.

The approach to hydrodynamics is, however, certainly more complex than we have indicated. One can be confident that the linear constitutive equations introduced in the next section are valid in the situations which are standard for the measurement of viscosity and thermal conductivity, that is, a steady state with small uniform gradients of temperature or velocity. More generally, the conditions for existence of a hydrodynamic stage, sharply distinguished from initial or boundary effects, may be severely limited.

1.7. NAVIER-STOKES EQUATIONS

Consider now a particular form for the constitutive equations. If the spatial variations are slow, $\mathbf{s}^*$ and $\mathbf{t}^*$ may be assumed to be linear in the gradients of the thermodynamic variables. For a fluid with spherical symmetry (including inversion symmetry), the coefficients in such a relation must be scalars. The only possibilities with the correct transformation properties are

$$\mathbf{s}^* = -\lambda \nabla T,\ t_{ij}^* = -2\eta\, \tilde{D}_{ij} - \eta'\, \delta_{ij} D \tag{1.35}$$

where we use the abbreviations

$$D_{ij} = \frac{1}{2}\left(\frac{\partial u_i}{\partial x_j} + \frac{\partial u_j}{\partial x_i}\right),\ D = D_{ii} = \nabla \cdot \mathbf{u},\ \tilde{D} = D_{ij} - \frac{1}{3}\delta_{ij} D. \tag{1.36}$$

The coefficient λ is the thermal conductivity, while η and η' are the shear and bulk viscosities. Actually symmetry would allow a term in $\mathbf{s}^*$ proportional to $\nabla\rho$ but the microscopic theory shows it to be absent. (This can also be shown by Onsager's theorem, which will be discussed later.) The first preceding equation is Fourier's law of heat conduction, and the second is Newton's viscosity law. The set of equations obtained when these constitutive equations are combined with the conservation laws is called the Navier-Stokes equations.

The entropy production (1.31) can be put in the form

$$\sigma = \lambda T^{-2}(\nabla T)^2 + T^{-1}[2\eta \tilde{D}_{ij}\tilde{D}_{ij} + \eta'\, D^2]. \tag{1.37}$$

The transport coefficients λ, η, and η' are non-negative and so σ is also.

Equations (1.35) are the lowest-order terms in an expansion. In higher orders there occur higher derivatives, and products of derivatives, of the thermodynamic variables. (Derivatives of ρ occur in higher orders since the argument referred to previously prevents their appearance only in the lowest order.) In the case of gases, the hydrodynamic equations in the second approximation [the next beyond (1.35)] are called the Burnett equations. For some processes such as high-frequency sound, it is necessary to use the Burnett equations or even a higher approximation.

1.8. DIFFUSION

If the system is a mixture of nonreacting chemical species, then there is a conservation law for the number of particles of each kind. Let n_r denote the number density of the r'th species and $\mathbf{j}_r$ its flux. The local conservation laws for particle number are

$$\frac{\partial n_r}{\partial t} + \nabla \cdot \mathbf{j}_r = 0. \tag{1.38}$$

The mass density ρ can be written in terms of n_r as

$$\rho = \sum_{r=1}^{R} m_r n_r \tag{1.39}$$

where m_r is the mass of a particle of the r'th species and R is the total number of species. Similarly, the mass flux, or momentum density, is

$$\mathbf{g} = \sum_{r=1}^{R} m_r \mathbf{j}_r. \tag{1.40}$$

Under a Galilean velocity transformation, n_r and $\mathbf{j}_r$ transform according to

$$n_r' = n_r, \; \mathbf{j}_r' = \mathbf{j} - n_r \mathbf{w}. \tag{1.41}$$

It is readily verified that this leaves the conservation laws invariant. In the local rest frame we have

$$n_{r0} = n_r, \; \mathbf{j}_{r0} = \mathbf{j}_r - n_r \mathbf{u}. \tag{1.42}$$

In equilibrium $\mathbf{j}_{r0}$ vanishes. We denote its value for a nonequilibrium state by $\mathbf{j}_r^*$ so

$$\mathbf{j}_r = n_r \mathbf{u} + \mathbf{j}_r^*. \tag{1.43}$$

The $\mathbf{j}_r^*$ are not independent since Equations (1.5), (1.39) and (1.40) imply

$$\sum_{r=1}^{R} m_r \mathbf{j}_r^* = 0. \tag{1.44}$$

The chemical potentials μ_r are defined by the Gibbs equation

$$\sum_{r=1}^{R} \mu_r n_r = \epsilon_0 + p - T\rho s.$$

Here it is convenient to use the chemical potential per particle rather than per unit mass. In terms of μ_r, the first law of thermodynamics is

$$Td(\rho s) = d\epsilon_0 - \sum_{r=1}^{R} \mu_r dn_r$$

and the Gibbs-Duhem equation is

$$\sum_{r=1}^{R} n_r d\mu_r = -\rho s dT + dp.$$

These equations hold for the nonequilibrium state since μ_r is defined to be the same function of ϵ_0 and n_r as in equilibrium.

Equations (1.27), (1.25), and (1.26) hold without modification but in addition

$$Dn_r/Dt = -n_r \nabla \cdot \mathbf{u} - \nabla \cdot \mathbf{j}_r^*.$$

After some manipulation, one finds the temperature equation (1.29) to be replaced by

$$\frac{DT}{Dt} = -T\left(\frac{\partial p}{\partial \epsilon_0}\right)_\rho \nabla \cdot \mathbf{u} - (\rho C_v)^{-1}\left\{t_{ji}^* \frac{\partial u_i}{\partial x_j} + \nabla \cdot \mathbf{s}^*\right\} - \sum_{r=1}^{R}\left(\frac{\partial T}{\partial n_r}\right)_{\epsilon_0} \nabla \cdot \mathbf{j}_r.$$

The entropy per unit mass satisfies

$$\frac{Ds}{Dt} = -(\rho T)^{-1}\left\{t_{ji}^* \frac{\partial u_i}{\partial x_j} + \nabla \cdot \mathbf{s}^* - \sum_{r=1}^{R} \mu_r \nabla \cdot \mathbf{j}_r^*\right\}.$$

In terms of the entropy per unit volume, this can be written

$$\frac{\partial(\rho s)}{\partial t} + \nabla \cdot \rho s \mathbf{u} = -T^{-1}\left\{t_{ji}^* \frac{\partial u_i}{\partial x_j} + \nabla \cdot \mathbf{s}^* - \sum_{r=1}^{R} \mu_r \nabla \cdot \mathbf{j}_r^*\right\}.$$

From this equation the entropy production is found to be

$$\sigma = -T^{-2}\, \mathbf{s}^* \cdot \nabla T - T^{-1} t_{ji}^* \frac{\partial u_i}{\partial x_j} - \sum_{r=1}^{R} \mathbf{j}_r^* \cdot \nabla(\mu_r/T).$$

In the approximation corresponding to the Navier-Stokes equations, the fluxes are linear in the gradients of the thermodynamic variables. Equation (1.35) for $\mathbf{t}^*$ is unchanged while

$$\mathbf{s}^* = -\lambda\nabla T - \sum_{r=1}^{R} L_r \nabla\nu_r,\ \mathbf{j}_r^* = -L_r'\nabla T - \sum_{p=1}^{R} L_{rp}\,\nabla\nu_p. \tag{1.45}$$

Here we have introduced the abbreviation $\nu_r = \mu_r/T$; of course μ_r, or the concentrations, could be used as independent variables rather than ν_r but the latter are more convenient, particularly in the statistical theory. The linearity of the particle fluxes in the gradients of the chemical potentials or concentrations is referred to as Fick's law.

The particle flux due to a gradient of ν_r (or a gradient of concentration) is ordinary diffusion. A temperature gradient can set up a particle flux; this is called thermal diffusion or the Soret effect. The coefficients L_r describe a heat flux due to concentration gradients, which is called the diffusion thermoeffect or the Dufour effect. If the particles are electrically charged, there are similar phenomena which form the subject of thermoelectricity; an electric current can be driven by a temperature gradient as well as by an electric field, and there are contributions to the heat flux due to the electric field.

The constants L_r, L_r', and L_{rp} are not all independent. It follows from Equation (1.44) that

$$\sum_{r=1}^{R} m_r L_r' = 0,\ \sum_{r=1}^{R} m_r L_{rp} = 0. \tag{1.46}$$

There are additional relations which follow from Onsager's theorem, to be discussed in the following section. These are

$$L_{rp} = L_{pr},\ L_r = T^2 L_r'. \tag{1.47}$$

Consequently, we have

$$\sum_{r=1}^{R} m_r L_r' = \sum_{r=1}^{R} m_r L_{pr} = 0. \tag{1.48}$$

When the constitutive equations are substituted into the entropy production, it becomes

$$\sigma = (\lambda/T^2)(\nabla T)^2 + T^{-1}(2\eta\tilde{D}_{ij}\tilde{D}_{ij} + \eta'\,D^2) + (2/T^2)\sum_r L_r\nabla T\cdot\nabla\nu_r + \sum_{p,r} L_{pr}\,\nabla\nu_p\cdot\nabla\nu_r.$$

In the statistical theory it will be shown that the coefficients of this quadratic form satisfy conditions which insure that it be non-negative.

Because of Equations (1.46) and (1.48) not all of the ν_r occur independently in the constitutive equations. The latter can be rewritten as, for example,

$$\mathbf{s}^* = -\lambda\nabla T - \sum_{r=1}^{R-1} L_r \nabla \nu_r',$$

$$\mathbf{j}_r^* = -L_r'\nabla T - \sum_{p=1}^{R-1} L_{rp}\,\nabla\nu_p',\ r = 1,.., R-1$$

where $\nu_r' = \nu_r - (m_r/m_R)\nu_R$.

The simplest case is the binary mixture, with $R = 2$. There is only one independent particle flux and Equations (1.45) can be written

$$\mathbf{s}^* = -\lambda\nabla T - L\nabla\nu,\ \mathbf{j}_1^* = -L'\,\nabla T - L_{11}\,\nabla\nu \tag{1.49}$$

where $\nu = \nu_1 - (m_1/m_2)\nu_2$. At constant temperature the heat and diffusion fluxes are related by

$$\mathbf{s}^* = (L/L_{11})\mathbf{j}_1^*.$$

If there is a temperature gradient but no diffusion, that is, $\mathbf{j}_1^* = 0$, then $\nabla\nu$ can be eliminated to give

$$\mathbf{s}^* = -[\lambda - (LL'/L_{11})]\nabla T.$$

Normally the thermal conductivity is measured under conditions of vanishing $\mathbf{j}_r^*$ rather than constant ν_r. Hence, the preceding coefficient, rather than λ, is the thermal conductivity as it is usually defined.

Self-diffusion is a special case when the particles are identical except that some are distinguished by being labeled (for example, by being radioactive). Clearly a labeling cannot give rise to an energy flux nor can a temperature gradient set up a flux of labeled particles. Hence, for self-diffusion the coefficients L and L' in Equation (1.49) must vanish. The labeling changes the entropy by a combinatorial, or mixing, term given by

$$\Delta(\rho\, s) = -\kappa\sum_r n_r \ln(n_r/n)$$

where $n = \Sigma n_r$ and κ is Boltzmann's constant. It follows that

$$\mu_r = -T\partial(\rho s)/\partial n_r = \mu_0 + \kappa T\ln(n_r/n)$$

where μ_0 is the chemical potential in the absence of labeling. The quantity ν defined previously is $\nu = (\mu_1 - \mu_2)/T = \kappa\ln(n_1/n_2)$ where n_1, n_2 denote the densities of labeled and unlabeled particles, respectively. We assume the fraction of labeled particles to be small so $\nabla\nu = \kappa\nabla\ln n_1$ and the constitutive equation for $\mathbf{j}_1^*$ is

$$\mathbf{j}_1^* = -D_s\,\nabla n_1$$

where $D_s = (\kappa/n_1)L_{11}$ is called the coefficient of self-diffusion. The continuity equation for n_1 yields

$$\frac{\partial n_1}{\partial t} = \nabla \cdot (D_s \nabla n_1)$$

which if nonlinear terms are neglected reduces to the diffusion equation

$$\frac{\partial n_1}{\partial t} = D_s \nabla^2 n_1.$$

1.9. ONSAGER'S THEOREM

It is evident that the transport coefficients are restricted by spatial symmetries of the material; spherical symmetry, for example, was used in writing down the Navier-Stokes equations for a fluid. Time-reversal invariance also imposes a restriction but in a way which is not so evident.

Time-reversal invariance of an equation means that if t is changed to $-t$, then there is a transformation of the dependent variables such that the equation maintains its form. Time-reversal invariance is also called reversibility, or microscopic reversibility when it refers to the microscopic equations of motion. The equations of motion at the microscopic level are in fact reversible (neglecting effects in particle physics which are irrelevant for our purposes), although the reversibility can be destroyed by an external magnetic field.

The macroscopic equations are normally written for a particular sign of the time and so are not reversible. With a given initial (microscopic) state, the Navier-Stokes equations, for example, are valid only *after* an ageing period. Microscopic reversibility implies the validity of "anti-hydrodynamic" equations, which apply prior to a specified "final" state. The augmented set of equations comprising the "normal" (for $t > 0$) and "anti" (for $t < 0$) hydrodynamic equations is then reversible. Our natural bias is such that the normal case is usually the one of interest, but the initial growth of a spontaneous fluctuation is described by the anti-hydrodynamic equations.

It was shown by Onsager that microscopic reversibility implies a symmetry of the transport coefficients in those constitutive equations which are linear in gradients of the thermodynamic variables. This is referred to as Onsager's theorem, or also as the Onsager-Casimir theorem since Casimir provided a generalization of the theorem. In this section a discussion of the theorem will be given. Since the proof is based on the microscopic equations of motion, it will be given in a later chapter.

The theorem is applicable to a more general class of systems than previously treated, and so we first reconsider some of the previous discussion from a more abstract and general point of view. Let ξ_α, $\alpha = 1,2..$, denote a set of independent quantities satisfying local conservation laws, and let $\boldsymbol{\psi}_\alpha$ be their fluxes. The local conservation laws are

$$\partial\xi_\alpha/\partial t + \nabla\cdot\boldsymbol{\psi}_\alpha = 0.$$

The quantity

$$\chi_\alpha = -\partial(\rho s)/\partial\xi_\alpha \tag{1.50}$$

is thermodynamically conjugate to ξ_α. A potential, or free energy, q is defined by the Legendre transformation

$$q = \rho s + \sum_\alpha \chi_\alpha\xi_\alpha.$$

Considering q as a function of the χ_α, we obtain

$$dq = \sum_\alpha \xi_\alpha d\chi_\alpha$$

and

$$\xi_\alpha = \partial q/\partial\chi_\alpha. \tag{1.51}$$

The symmetric matrix of second derivatives of q will be abbreviated as g:

$$\mathsf{g}_{\alpha\beta} = \partial^2 q/\partial\chi_\alpha\partial\chi_\beta = \partial\xi_\alpha/\partial\chi_\beta = \partial\xi_\beta/\partial\chi_\alpha. \tag{1.52}$$

The matrices with elements $\partial\xi_\alpha/\partial\chi_\beta$ and $\partial\chi_\alpha/\partial\xi_\beta$ are inverse, and it follows that

$$\mathsf{g}^{-1}_{\alpha\beta} = -\partial^2(\rho s)/\partial\xi_\alpha\partial\xi_\beta.$$

For states near equilibrium the deviation of the entropy from its maximum defines a quadratic form

$$\delta S = -\frac{1}{2}\int d^3x \sum_{\alpha\beta} \mathsf{g}^{-1}_{\alpha\beta}\delta\xi_\alpha\delta\xi_\beta.$$

If the $\delta\xi_\alpha$ are arranged as the elements of a column matrix, this can be written in matrix notation as

$$\delta S = -\frac{1}{2}\int d^3x\, \delta\xi^T\mathsf{g}^{-1}\delta\xi \tag{1.53}$$

where the superscript T denotes the transpose. It follows from Equation (1.51) that, in matrix form, $\delta\xi = \mathsf{g}\delta\chi$ so Equation (1.53) has the equivalent forms

$$\delta S = -\frac{1}{2}\int d^3x\delta\xi^T\delta\chi = -\frac{1}{2}\int d^3x\delta\chi^T\mathsf{g}\delta\chi.$$

Since δS cannot be positive, g must be a non-negative matrix.

Let $\boldsymbol{\psi}_\alpha^0$ denote the fluxes in the Euler, or reversible, approximation. In this approximation we have

$$\frac{\partial(\rho s)}{\partial t} = \sum_\alpha \frac{\partial(\rho s)}{\partial \xi_\alpha}\frac{\partial \xi_\alpha}{\partial t} = \sum_\alpha \chi_\alpha \nabla \cdot \boldsymbol{\psi}_\alpha^0. \tag{1.54}$$

For the simple fluid treated earlier, this is the "conservation law" for entropy

$$\frac{\partial(\rho s)}{\partial t} + \nabla \cdot (\rho s \mathbf{u}) = 0.$$

More generally the right-hand side of Equation (1.54) will be a divergence, which can be expressed by the statement that $\Sigma\chi_\alpha\delta\boldsymbol{\psi}_\alpha^0$ is a perfect differential; there exists a function **a** such that

$$\sum_\alpha \chi_\alpha \delta \boldsymbol{\psi}_\alpha^0 = -\delta \mathbf{a}. \tag{1.55}$$

Equation (1.54) takes the form

$$\frac{\partial(\rho s)}{\partial t} + \nabla \cdot \mathbf{a} = 0. \tag{1.56}$$

Equation (1.55) implies that

$$\partial \boldsymbol{\psi}_\alpha^0 / \partial \chi_\beta = \partial \boldsymbol{\psi}_\beta^0 / \partial \chi_\alpha. \tag{1.57}$$

This is a condition on the fluxes imposed by the reversibility of the Euler approximation. It follows that there exists a potential **b** for the Euler fluxes

$$\boldsymbol{\psi}_\alpha^0 = \partial \mathbf{b} / \partial \chi_\alpha.$$

To avoid terms like $t_{ij}^* u_j$ in Equation (1.22) we work in the local rest frame and write

$$\boldsymbol{\psi}_\alpha = \boldsymbol{\psi}_\alpha^0 + \boldsymbol{\psi}_\alpha^*. \tag{1.58}$$

In place of Equation (1.56) we obtain

$$\partial(\rho s)/\partial t + \nabla \cdot \mathbf{a} = \sum_\alpha \chi_\alpha \nabla \cdot \boldsymbol{\psi}_\alpha^*$$

and the entropy production is given by

$$\sigma = -\sum_\alpha \boldsymbol{\psi}_\alpha^* \cdot \nabla \chi_\alpha.$$

In the approximation corresponding to the Navier-Stokes equations the fluxes can be written as

$$\boldsymbol{\psi}_\alpha^* = -\sum_\beta L_{\alpha\beta} \nabla \chi_\beta. \tag{1.59}$$

Microscopic reversibility defines a real linear transformation τ of ξ,

$$(\tau\xi)_\alpha = \sum_\beta \tau_{\alpha\beta}\xi_\beta,$$

which is an involution, $\tau^2 = 1$. Then Onsager's theorem states that τL is symmetric,

$$(\tau L)^T = \tau L. \tag{1.60}$$

It is also a consequence of microscopic reversibility that

$$\mathsf{g} = \tau \mathsf{g} \tau^T.$$

(From its definition, g is real and symmetric.)

If an external magnetic field is present, the statement of Onsager's theorem must be modified in a way which will be described later when the proof is provided.

The choice of the variables ξ is not unique. Let a linear transformation be made to a new set $\xi' = R\xi$. It is readily verified that g and L transform according to

$$\mathsf{g}' = R\mathsf{g}R^T,\ L' = R\ L\ R^T.$$

The transformation for τ is $\tau' = R\tau R^{-1}$ and it follows that Equation (1.60) still holds after the transformation.

We may choose R to bring τ into diagonal form, with diagonal elements τ_α equal to ± 1. The ξ_α then separate into two classes with $\tau_\alpha = \pm 1$. Those with $\tau_\alpha = 1$ are even under a time reversal, and those with $\tau_\alpha = -1$ are odd. In fact the natural choice of the ξ_α has a definite parity with respect to time reversal.

To apply Onsager's theorem to a simple fluid, let the ξ_α be ρ, ϵ, and $\mathbf{g}$. The χ_α can be determined from their definition (1.50) and

$$Td(\rho s) = d\epsilon - \left(\mu - \frac{1}{2}mu^2\right)d\rho/m - \mathbf{u}\cdot d\mathbf{g}.$$

(Here again μ denotes chemical potential per particle rather than per unit mass.) Using a notation $\chi(\xi)$ rather than χ_α, we find

$$\chi(\rho) = \left(\mu - \frac{1}{2}mu^2\right)/mT,\ \chi(\epsilon) = -1/T,\ \chi(\mathbf{g}) = -\mathbf{u}/T.$$

In the local rest frame we have $\nabla\chi(\rho) = \nabla(\mu/mT)$, $\nabla\chi(\mathbf{g}) = -\nabla\mathbf{u}/T$. The variables ρ and ϵ are even under a time reversal, and $\mathbf{g}$ is odd. With these assignments the Euler equations are invariant to a time reversal.

Consider the constitutive equations (1.59). There are no gradient terms in the mass flux since it is the same as the momentum density. It follows from Onsager's theorem that none of the fluxes $*_\alpha$ can contain contributions depending on $\nabla\chi(\rho)$. This is equivalent to the absence of a term in $\nabla\rho$, which confirms the statement to that effect following Equation (1.36). Equations (1.35) contain no cross-coefficients mixing the heat flux with velocity gradients, or the momentum flux with the temperature gradient, and so are trivially consistent with Onsager's theorem.

For a mixture, the variable ρ is replaced by n_r. The χ's are the same as before except that

$$\chi(n_r) = \left(\mu_r - \frac{1}{2}m_r u^2\right)/T.$$

The coefficients in Equation (1.59) are related to those of Equation (1.45) by

$$L(\epsilon,\epsilon) = T^2\lambda,\ L(\epsilon,n_r) = L_r,\ L(n_r,\epsilon) = T^2L'_r,\ L(n_r,n_p) = L_{rp}.$$

Onsager's theorem then gives the restrictions stated in Equation (1.47).

Onsager's theorem applies only to coefficients which relate the fluxes linearly to the first spatial derivatives of the thermodynamic variables. Microscopic reversibility imposes in any case a restriction on the constitutive equations, but it cannot always be expressed in a simple form.

The linearity in gradients is only local and does not imply a limitation of Onsager's theorem to linear processes. The L-coefficients are in general functions of the thermodynamic variables, and this introduces nonlinearity into the hydrodynamic equations. In addition, higher-order terms (such as occur in the Burnett equations) may be present and even large.

Notes on Chapter 1

General references are Landau and Lifshitz [1959], de Groot and Mazur [1962], and Prigogine [1967].

1.3 The Galilean transformation of the conservation laws is worked out in Green [1954].

1.5 A more complete discussion of ageing to hydrodynamics is given in Uhlenbeck and Ford [1963].

1.6 Initial slip is discussed in Grad [1963a].

1.9 Original references to Onsager's theorem are Onsager [1931a, b] and Casimir [1945]. A survey of the comparison with experiment is given in Miller [1973]. There is an extensive literature; a few selected references are de Groot and Mazur [1962], Felderhof [1966], McLennan [1974], Grmela [1977], Grmela and Iscoe [1978], Geigenmuller, Titulaer, and Felderhof [1983], and Dufty and Rubi [1987].

2

BROWNIAN MOTION

2.1. INTRODUCTION

If a powder of small particles is mixed with a fluid and observed under a microscope, the particles can be seen to undergo a perpetual fluctuating motion called Brownian motion. Brownian motion may have been observed by the earliest microscopists, but the first reports of a systematic investigation were by the botanist Robert Brown in 1828. First noticing the phenomena with particles of pollen, Brown went on to investigate a variety of materials, including even fragments from the Sphinx. He concluded that the motion is a general phenomenon, occurring independently of the constitution of the particles.

The earliest observations were on particles suspended in water but Brownian motion can also be seen in other liquids and in gases such as air. The particles themselves may be solid or liquid. Typically, the size of a particle is about 1 micron, say in the range 0.2 to 1.0 microns. For larger particles the magnitude of the Brownian motion, which is inversely proportional to the size, is negligible, while smaller particles cannot be seen with a microscope.

Brownian motion is a direct manifestation of the atomic structure of matter: The particle follows an irregular trajectory because of irregular bombardment by the atoms of the fluid. A detailed explanation based on the atomic theory was published by Einstein in 1905. His main result was a formula for the mean-square displacement of a particle in a given interval of time; this result was also obtained by Smoluchowski using different arguments.

Experimental confirmation was provided by Perrin in 1908 and subsequently by others; a notable example is the extensive series of measurements by Millikan and Fletcher on Brownian motion in air.

Brownian motion is just one example of the general phenomena of fluctuations due to atomic structure. Gases exhibit fluctuations in density: The number of molecules in a small volume element varies erratically with time. The motion of a molecule is similar to that of a particle undergoing Brownian motion, and so the theory of density fluctuations, developed particularly by Smoluchowski, resembles that of Brownian motion. Density fluctuations can be observed by their scattering of light, which is particularly intense near the critical point.

Historically, the theories of Brownian motion and density fluctuations and their experimental confirmation had the utmost importance. During the latter half of the nineteenth century, evidence had accumulated for the existence of atoms which obey the laws of mechanics. However, the same period also saw many successes of thermodynamics. Superficially the atomic and thermodynamic points of view are contradictory and a dispute developed between the "atomicists" and the "energeticists." Arguments were provided, particularly by Boltzmann, that the atomic theory provided an explanation of the laws of thermodynamics but these arguments, although sound, were not universally accepted. In the cases of Brownian motion and density fluctuations, the connection between observation and the underlying atomic structure was too direct to be dismissed. The theoretical understanding of these phenomena helped to resolve the apparent conflict between thermodynamics and atomic theory.

2.2. EINSTEIN'S THEORY

Consider a cell containing a number of particles suspended in a fluid and undergoing Brownian motion. To distinguish them from the molecules of the fluid, the larger particles will be called B-particles.

Instead of focusing on the motion of a single B-particle, we may count the numbers in small volume elements. The density of B-particles, as a function of position and time, becomes the interesting variable. From this point of view the gas of B-particles behaves much as an ordinary molecular gas.

The ideal gas law is $p = n\kappa T$ where p is the pressure, n the density, κ Boltzmann's constant and T the Kelvin temperature. In a mixture of gases each component contributes a partial pressure given by the ideal gas law. Since the size of the particles does not appear, the B-particles should also contribute a partial pressure even though they are far larger than a molecule. Such a pressure can occur only if they continually move about. This argument is Einstein's existence proof for Brownian motion.

If the density of B-particles is nonuniform there will be a tendency for them to diffuse from regions of high density to regions of low density. We describe this by Fick's law

$$\mathbf{j} = -D\nabla n \tag{2.1}$$

where $\mathbf{j}$ is the flux of B-particles and D their coefficient of diffusion. The equation of continuity is

$$\partial n/\partial t + \nabla\cdot\mathbf{j} = 0$$

and this yields the diffusion equation

$$\partial n/\partial t = D\nabla^2 n.$$

It is convenient to use the quantity P defined by $n = NP$ where N is the total number of particles in the cell. Then P can be interpreted as the probability density, that is, $P(\mathbf{x},t)d^3x$ is the probability of finding a given B-particle in the volume element d^3x at time t. Clearly P also satisfies the diffusion equation.

Let a particle be located initially at $\mathbf{x}_0$, so $P(\mathbf{x},0) = \delta(\mathbf{x} - \mathbf{x}_0)$. The solution to the diffusion equation with this initial condition is

$$P(\mathbf{x},t) = (4\pi Dt)^{-3/2}\exp[-(\mathbf{x} - \mathbf{x}_0)^2/4Dt]. \tag{2.2}$$

(We have assumed the volume of the cell to be large enough that boundary effects can be neglected.) From this we can calculate the mean-square displacement in time t:

$$\langle(\Delta\mathbf{x})^2\rangle = \int d^3xP(\mathbf{x},t)(\mathbf{x} - \mathbf{x}_0)^2 = 6Dt. \tag{2.3}$$

For the displacement along a single axis, say the x-axis, we get $\langle(\Delta x)^2\rangle = 2Dt$.

The average can be calculated from observations on a large number of particles, or from measurements on a single particle over a long interval of time. We assume the two methods to give the same results.

The coefficient of diffusion was calculated by Einstein in the following way. Consider a spherical particle moving slowly through the fluid. The fluid will exert a drag force on the particle which is proportional, but in the opposite direction, to the velocity $\mathbf{v}$ of the particle:

$$\mathbf{F}_D = -m\xi\mathbf{v}.$$

The mass m of the particle has been factored out for later convenience. The quantity ξ is called the friction constant, and its inverse is called the relaxation time. The velocity per unit force, or $1/m\xi$, is referred to as the mobility. (For charged particles, the mobility is usually taken to be the velocity per unit electric field, or $q/m\xi$ where q is the particle's charge.) It can be shown from fluid dynamics that a spherical particle moving with constant velocity in an incompressible fluid experiences a drag force given by Stokes' law:

$$\mathbf{F}_D = -6\pi\eta a\mathbf{v}.$$

Here a is the radius of the particle and η the viscosity of the fluid. (Brownian motion is so slow that little error is involved in assuming the fluid to be incompressible, even for Brownian motion in gases.) The value given by Stokes' law for the friction constant is $\xi = 6\pi\eta a/m$. The derivation of Stokes' law requires that the particle be large in comparison to the mean free path in the gas. In fact the mean free path is typically about 10 microns, which is larger than the B-particle. Corrections to Stokes' law to take into account mean-free-path effects were calculated by Cunningham and determined experimentally by Millikan.

Now suppose there is an external force $\mathbf{F}$ acting on the B-particle. Then a terminal velocity $\mathbf{v}_T$ will be reached such that $\mathbf{F} + \mathbf{F}_D = 0$ which gives

$$\mathbf{v}_T = (1/m\xi)\mathbf{F}.$$

If there are n particles per unit volume moving with this velocity, their flux is

$$\mathbf{j}_\mathbf{F} = n\mathbf{v}_T = (n/m\xi)\mathbf{F}.$$

This is to be added to the diffusion flux (2.1) to obtain the total flux $\mathbf{j}$:

$$\mathbf{j} = -D\nabla n + (n/m\xi)\mathbf{F}. \tag{2.4}$$

When this is combined with the equation of continuity, the result is

$$\partial n/\partial t = D\nabla^2 n - (1/m\xi)\nabla\cdot n\mathbf{F}. \tag{2.5}$$

This is known as Smoluchowski's equation and describes diffusion in an external field.

Suppose the external field is that of gravity. Then in equilibrium the particles distribute themselves according to the law of atmospheres:

$$n(z) = n(0)\exp(-mgz/\kappa T).$$

Here g is the acceleration of gravity and the z-axis has been taken to be upward. The external force is $F_z = -mg$. In equilibrium the flux (2.4) must vanish, and this yields

$$D = \kappa T/m\xi \tag{2.6}$$

which is known as Einstein's relation between diffusion and mobility.

If Stokes' law applies, we have $D = \kappa T/(6\pi\eta a)$ so the mean-square displacement can be written as

$$\langle(\Delta\mathbf{x})^2\rangle = (\kappa T/\pi\eta a)t. \tag{2.7}$$

Perrin's experimental results confirmed this formula over a range of values for the temperature and particle size; he also confirmed that the law of atmospheres is valid for B-particles.

To illustrate orders of magnitude, consider a numerical example of Brownian motion in air with $a = 0.2 \times 10^{-6}$ m, $\eta = 1.8 \times 10^{-5}$ kg/m-sec. Taking the density of the particle to be the same as that of water and neglecting corrections to Stokes' law, we get $m = 3 \times 10^{-17}$ kg, $1/\xi = 5 \times 10^{-7}$ sec, $D = 6 \times 10^{-11}$ m^2/sec. For a time interval of 30 sec, the root-mean-square displacement works out to about 60 microns.

Another rough calculation will bring out an important feature of Brownian motion. The flux of gas molecules can be estimated as nv where at standard conditions $n = 2.5 \times 10^{25}$ m^{-3} and v is about 500 m/sec. This gives a flux of about $7.5 \times 10^{27}/(\text{m}^2-\text{sec})$. Multiplying by the surface area of a sphere with radius 0.2 microns, we find that the B-particle is bombarded by molecules at a rate of about 10^{15} collisions/sec. On the other hand a B-particle is some 10^9 times as massive as an atom, and so any single collision has a very weak effect. Furthermore, the impulses received from two successive collisions have little correlation between them. Thus, Brownian motion is the cumulative effect of a rapid and random sequence of weak impulses.

The mean-square velocity of a B-particle should satisfy the principle of equipartition of energy, $\frac{1}{2}m\langle v^2\rangle = \frac{3}{2}\,\kappa T$. However, it is nearly impossible to obtain meaningful measurements of instantaneous velocity since the trajectory of a B-particle is so erratic. As will be seen, confirmation of equipartition by direct velocity measurement would require observations over times small in comparison to the relaxation time and this is difficult even with fast electronic instrumentation. The importance of Einstein's result (2.7) lies not only in its detailed form, but also in the fact that it focused attention on the mean-square displacement as the quantity most easily accessible to experiment.

2.3. LANGEVIN EQUATION

A different approach due to Langevin provides a more detailed picture of Brownian motion. Langevin's method is to treat the velocity as a random variable and to write the equation of motion in the form

$$\frac{d\mathbf{v}}{dt} = -\xi\mathbf{v} + \mathbf{A}. \tag{2.8}$$

The first term corresponds to the drag force (say to Stokes' law) and $\mathbf{A}$ is a random force (actually acceleration) which takes into account the random bombardment by the atoms of the surrounding fluid. The average of $\mathbf{A}$ is assumed to vanish,

$$\langle\mathbf{A}\rangle = 0 \tag{2.9}$$

so that on the average the velocity satisfies $d\langle \mathbf{v}\rangle/dt = -\xi\langle \mathbf{v}\rangle$. It is assumed that $\mathbf{A}$ is independent (statistically and functionally) of the position and velocity of the B-particle, and further assumptions on its statistical properties follow.

Let $\mathbf{q}$ denote the displacement in time t of the B-particle, $\mathbf{q} = \Delta\mathbf{x} = \mathbf{x} - \mathbf{x}_0$. The velocity is $\mathbf{v} = d\mathbf{q}/dt$ and Equation (2.8) yields

$$\mathbf{q}\cdot\frac{d^2\mathbf{q}}{dt^2} = -\xi\mathbf{q}\cdot\frac{d\mathbf{q}}{dt} + \mathbf{q}\cdot\mathbf{A}.$$

We average this equation and use Equation (2.9) plus the identity

$$\mathbf{q}\cdot\frac{d^2\mathbf{q}}{dt^2} = \frac{1}{2}\frac{d^2(q^2)}{dt^2} - \left(\frac{d\mathbf{q}}{dt}\right)^2.$$

The result is

$$\frac{d^2}{dt^2}\langle q^2\rangle + \xi\frac{d}{dt}\langle q^2\rangle = \langle v^2\rangle.$$

The right-hand side is determined from equipartition, $\langle v^2\rangle = 3\kappa T/m$. The equation can be solved for $\langle q^2\rangle$ and the result is

$$\langle q^2\rangle = 6D[t - \xi^{-1}(1 - e^{-\xi t})]. \tag{2.10}$$

(The constants of integration are determined by noting that, from its definition, both $\langle q^2\rangle$ and its first derivative vanish at $t = 0$.) This result, which generalizes Einstein's formula (2.7), is due to Ornstein and Furth.

For times large in comparison to the relaxation time, (2.10) reduces to Einstein's formula. However for small times, motion with constant velocity is obtained:

$$\langle q^2\rangle = (3\kappa T/m)t^2 = \langle v^2\rangle t^2,\ \xi t << 1.$$

The inertia of the B-particle prevents changes in its velocity over times which are much less than the relaxation time.

The solution to Equation (2.8) is

$$\mathbf{v}(t) = e^{-\xi t}\left[\mathbf{v}_0 + \int_0^t ds\, e^{\xi s}\mathbf{A}(s)\right] \tag{2.11}$$

where $\mathbf{v}_0$ is the velocity at time $t = 0$. Taking the average, we get

$$\langle \mathbf{v}(t)\rangle_0 = e^{-\xi t}\mathbf{v}_0 \tag{2.12}$$

where $\langle\ \rangle_0$ denotes an average with the initial velocity $\mathbf{v}_0$ held fixed. Next we multiply by $\mathbf{v}_0$ and average over $\mathbf{v}_0$, assigning the equipartition value to $\langle \mathbf{v}_0^2\rangle$. This average converts the restricted average $\langle\ \rangle_0$ to the full average $\langle\ \rangle$ and we get

$$\langle \mathbf{v}(t)\cdot\mathbf{v}(0)\rangle = (3\kappa T/m)e^{-\xi t}. \tag{2.13}$$

The quantity on the left is referred to as the velocity autocorrelation. The normalized velocity autocorrelation $\chi(t)$ is defined by

$$\chi(t) = \langle \mathbf{v}(t) \cdot \mathbf{v}(0) \rangle / \langle v^2 \rangle$$

and is given by

$$\chi(t) = e^{-\xi t}. \tag{2.14}$$

With an initial time other than 0, the velocity autocorrelation is a function only of the time difference because the random process is stationary in time. Thus in general

$$\langle \mathbf{v}(t_1) \cdot \mathbf{v}(t_2) \rangle = (3\kappa T/m) \exp(-\xi |t_1 - t_2|). \tag{2.15}$$

The mean-square displacement can be obtained directly from the velocity autocorrelation. We have

$$\mathbf{q}(t) = \int_0^t ds \mathbf{v}(s)$$

and so

$$\langle q^2(t) \rangle = \int_0^t ds_1 \int_0^t ds_2 \langle \mathbf{v}(s_1) \cdot \mathbf{v}(s_2) \rangle.$$

This is equivalent to

$$\langle q^2(t) \rangle = 2\langle v^2 \rangle \int_0^t ds(t - s)\chi(s) \tag{2.16}$$

as can be verified by differentiating both expressions with respect to t. If (2.14) is used for χ, the previous result (2.10) is obtained.

The velocity autocorrelation can be calculated in a different way which provides additional information on the random force. We write Equation (2.11) twice for times t_1 and t_2, multiply the two equations together, and average. This yields

$$\langle v_i(t_1) v_j(t_2) \rangle_0 = e^{-\xi(t_1 + t_2)} \left\{ v_{0i} v_{0j} + \int_0^{t_1} ds_1 \int_0^{t_2} ds_2 e^{\xi(s_1 + s_2)} \langle A_i(s_1) A_j(s_2) \rangle_0 \right\}. \tag{2.17}$$

Since $\mathbf{A}$ is assumed independent of velocity, there is actually no dependence on $\mathbf{v}_0$ in the last term. Our picture of the random force is such that its values at two different times are uncorrelated unless the two times are close together. We idealize this by a delta function,

$$\langle A_i(s_1) A_j(s_2) \rangle = \alpha \delta_{ij} \delta(s_1 - s_2). \tag{2.18}$$

Here α is taken to be time-independent because the statistical character of Brownian motion does not depend on the absolute time. (The formula (2.17) depends on t_1 and t_2 separately, rather than just on their difference, because there is a preferred origin, namely the time when the velocity was $\mathbf{v}_0$.) A random variable with the above property is said to be delta correlated; the term "white noise" is also used since the Fourier transform of the correlation function is independent of frequency.

With Equation (2.18) we can evaluate the integral in Equation (2.17) and the result is

$$\langle v_i(t_1)v_j(t_2)\rangle_0 = \frac{1}{3}\delta_{ij}\left\{e^{-\xi(t_1+t_2)}\left[v_0^2 - (\alpha/2\xi)\right] + (\alpha/2\xi)e^{-\xi|t_1-t_2|}\right\}. \tag{2.19}$$

To evaluate α, set the two times equal,

$$\langle v^2(t)\rangle_0 = e^{-2\xi t}v_0^2 + (\alpha/2\xi)[1 - e^{-2\xi t}]. \tag{2.20}$$

We require that this approach the equipartition value $3\kappa T/m$ for large t, which yields $\alpha = 6\kappa T\xi/m$. If we now average over $\mathbf{v}_0$, we obtain the previous result (2.15). The assumption of white noise for the random force is essential as otherwise the two different arguments would not give the same result for the velocity autocorrelation. Thus, the Langevin equation itself determines the statistical properties of the random force.

2.4. FOKKER-PLANCK EQUATION

A more complete description of Brownian motion is provided by the probability distribution function from which averages such as those of the previous section can be calculated. Actually there are many such functions, depending on the variables of interest. One example is $P(\mathbf{x},t)$; the function (2.2) depends on the initial position $\mathbf{x}_0$ and will be written as $P(\mathbf{x},t|\mathbf{x}_0)$ or if the initial time is t_0 rather than 0, as $P(\mathbf{x},t|\mathbf{x}_0,t_0)$. It is a conditional probability density, defined so that $P(\mathbf{x},t|\mathbf{x}_0\, t_0)d^3x$ is the probability for finding the particle in d^3x at time t, given that it was at $\mathbf{x}_0$ at time t_0.

Similarly, we define a function $P(\mathbf{v},t|\mathbf{v}_0,t_0)$ such that $P(\mathbf{v},t|\mathbf{v}_0,t_0)d^3v$ is the probability of the particle having its velocity in d^3v at time t, given that its velocity was $\mathbf{v}_0$ at time t_0. Of course $P(\mathbf{v},t|\mathbf{v}_0,t_0)$ and $P(\mathbf{x},t|\mathbf{x}_0,t_0)$ are different functions; we will depend on the arguments to indicate which is meant. To calculate averages of quantities depending on both position and velocity, we need $P(\mathbf{x},\mathbf{v},t|\mathbf{x}_0,\mathbf{v}_0,t_0)$ whose definition is similar to those just given.

For brevity let y denote the random variables $\mathbf{x}$, or $\mathbf{v}$, or $\mathbf{x}$, $\mathbf{v}$, as the case may be. Thus $P(y,t|y_0,t_0)$ may stand for any of the functions just introduced. Much of the discussion to follow applies to more general situations than Brownian motion so y may represent a general class of random variables.

In addition to the conditional probability densities already introduced, there are joint probability densities which will generally be denoted by W. For example, $W(y,t;y_0,t_0)$ has the meaning that $W(y,t;y_0,t_0)dy_0dy$ is the probability for finding the random variable in the range dy at time t *and* in the range dy_0 at time t_0. This is related to $P(y,t|y_0,t_0)$ by

$$W(y,t;y_0,t_0) = W(y_0)\, P(y,t|y_0,t_0). \tag{2.21}$$

Here $W(y)$ is the probability density for y itself and is assumed to be time independent. We may go on to define probability densities of higher order, for example,

$$P(y_2,t_2|y_0,t_0;y_1,t_1)dy_2$$

is the probability of finding the random variable in the range dy_2 at time t_2, given that it had the value y_0 at time t_0 and the value y_1 at time t_1.

It follows from their definitions that P and W must be non-negative. There are additional conditions which are readily obtained from their definitions, such as

$$\int dy P(y,t|y_0,t_0) = 1,$$

$$\int dy_0 P(y,t|y_0,\, t_0) = W(y),$$

$$\int dy_2 W(y_0,t_0;y_1,t_1;y_2,t_2) = W(y_0,t_0;y_1,t_1).$$

The simplest case is that of a purely random process when the conditional probability density for y at any time is independent of its values at other times. Then $P(y,t|y_0,t_0)$ is independent of y_0, and also of t if the random process is stationary. The case of main interest in the theory of Brownian motion is that of a Markov process, for which the probability at time t is determined by the value of y at any earlier time t_0 but then is independent of values of y at times previous to t_0. In this case we have for $t_1 < t_0 < t$

$$P(y,t|y_1,t_1;y_0,t_0) = P(y,t|y_0,t_0),$$

with similar equations for the functions of higher order. Thus the P's of all orders are determined by the single function $P(y,t|y_0,t_0)$. The same function also determines the joint probabilities as in

$$W(y_0,t_0;y_1,t_1;y_2,t_2;\ldots) = W(y_0)\, P(y_1,t_1|y_0,t_0)\, P(y_2,t_2|y_1,t_1)\ldots \tag{2.22}$$

where the times are ordered as $t_0 < t_1 < t_2 \ldots$

For a Markov process, $P(y,t|y_0,t_0)$ satisfies a basic equation which can be derived as follows. We have

$$\int dy_1 W(y_0,t_0;y_1,t_1;y_2,t_2) = W(y_0,t_0;y_2,t_2).$$

Combining this with Equation (2.22) and (2.21) we get for $t_0 \le t_1 \le t_2$

$$P(y_2,t_2|y_0,t_0) = \int dy_1\, P(y_2,t_2|y_1,t_1)\, P(y_1,t_1|y_0,t_0)$$

which is known as the Chapman-Kolmogorov equation. If the Markov process is stationary, we may write $P(y,t|y_0,t_0) = P(y,t - t_0|y_0)$ and the Chapman-Kolmogorov equation becomes

$$P(y,t + \tau|y_0) = \int dy_1 P(y,t|y_1)P(y_1,\tau|y_0),\ t \ge 0,\ \tau \ge 0. \tag{2.23}$$

The Chapman-Kolmogorov equation can be written more compactly as follows. Let $P(t)$, $t \ge 0$ be an operator on functions defined by

$$P(t)F(y) = \int dy_0 P(y,t|y_0)F(y_0).$$

Then Equation (2.23) yields

$$P(t)P(\tau) = P(t + \tau).$$

Clearly $P(0) = 1$ so $P(t), t \ge 0$ is a semi-group. (It is not a group since $P(t)$ has not been defined for negative t.)

An equation for the time derivative of P can be obtained from the Chapman-Kolmogorov equation as follows. Let $R(y)$ be a function with derivatives of all orders which vanishes at the boundary of the region covered by y. We have

$$\begin{aligned}\int dyR(y)\frac{\partial P(y,t|y_0)}{\partial t} &= \lim_{\tau\to 0} \tau^{-1}\int dyR(y)[P(y,t + \tau|y_0) - P(y,t|y_0)]\\ &= \lim_{\tau\to 0} \tau^{-1}\int dyR(y)\left[\int dy'P(y,\tau|y')P(y',t|y_0) - P(y,t|y_0)\right]\\ &= \lim_{\tau\to 0} \tau^{-1}\int dyP(y,t|y_0)\left[\int dy'R(y')P(y',\tau|y) - R(y)\right].\end{aligned}$$

For small τ, $P(y',\tau|y)$ should be small unless y is near to y'. Hence we expand R about y,

$$R(y') = R(y) + (y' - y)\frac{\partial R(y)}{\partial y} + \frac{1}{2}(y' - y)^2\frac{\partial^2 R(y)}{\partial y^2} + \ldots$$

This yields

$$\begin{aligned}\int dyR(y)\frac{\partial}{\partial t}P(y,t|y_0) &= \lim_{\tau\to 0} \tau^{-1}\int dyP(y,t|y_0)\\ &\cdot\int dy'P(y',\tau|y)\left[(y' - y)\frac{\partial R(y)}{\partial y} + \frac{1}{2}(y' - y)^2\frac{\partial^2 R(y)}{\partial y^2} + \ldots\right]\end{aligned}$$

An integration by parts gives

$$\int dy R(y) \frac{\partial P(y,t|y_0)}{\partial t} =$$

$$\lim_{\tau\to 0} \tau^{-1} \int dy R(y) \left[-\frac{\partial}{\partial y} \int dy'(y' - y) P(y',\tau|y) P(y,t|y_0) \right.$$
$$\left. + \frac{1}{2} \frac{\partial^2}{\partial y^2} \int dy'(y' - y)^2 P(y',\tau|y) P(y,t|y_0) + \ldots \right]$$

Since R is arbitrary, this requires P to satisfy

$$\frac{\partial}{\partial t} P(y,t|y_0) = -\frac{\partial}{\partial y} a(y) P(y,t|y_0) + \frac{1}{2} \frac{\partial^2}{\partial y^2} b(y) P(y,t|y_0) + \ldots \tag{2.24}$$

where

$$a(y) = \lim_{\tau\to 0} \tau^{-1} \int dy'(y' - y) P(y',\tau|y),$$

$$b(y) = \lim_{\tau\to 0} \tau^{-1} \int dy'(y' - y)^2 P(y',\tau|y).$$

It will be assumed that terms in the expansion beyond those written out are negligible. Equation (2.24) is known as the Fokker-Planck equation.

The preceding formulas for $a(y)$ and $b(y)$ cannot be used directly since if the limit $\tau \to 0$ is taken, the result involves the unknown time derivative of P. However, the integrals are the mean of the change in y and its square during the time τ,

$$a(y) = \lim_{\tau\to 0} \tau^{-1} \langle \Delta y(\tau) \rangle, \quad b(y) = \lim_{\tau\to 0} \tau^{-1} \langle [\Delta y(\tau)]^2 \rangle. \tag{2.25}$$

These can be evaluated by independent considerations, for example, by using the Langevin equation.

We first apply the Fokker-Planck equation to obtain Einstein's description of Brownian motion by the diffusion equation. In this case y stands for the three components of position and Equation (2.24) takes the form

$$\frac{\partial}{\partial t} P(\mathbf{x},t|\mathbf{x}_0) = -\frac{\partial}{\partial x_i} a_i P(\mathbf{x},t|\mathbf{x}_0) + \frac{1}{2} \frac{\partial^2}{\partial x_i \partial x_j} b_{ij} P(\mathbf{x},t|\mathbf{x}_0) \tag{2.26}$$

where

$$\mathbf{a} = \lim_{\tau\to 0} \tau^{-1} \langle \Delta \mathbf{x} \rangle, \quad b_{ij} = \lim_{\tau\to 0} \tau^{-1} \langle \Delta x_i \Delta x_j \rangle.$$

Spherical symmetry implies that **a** vanishes, while

$$b_{ij} = \frac{1}{3}\delta_{ij}\lim_{\tau\to 0}\tau^{-1}\langle(\Delta\mathbf{x})^2\rangle.$$

Einstein's relation (2.3) gives $b_{ij} = 2Dt\delta_{ij}$ and Equation (2.26) then reduces to the diffusion equation.

We next consider a process of diffusion in the phase space with coordinates $\mathbf{x}, \mathbf{v}$. Then P is the function $P(\mathbf{x}, \mathbf{v}, t|\mathbf{x}_0, \mathbf{v}_0)$ and the Fokker-Planck equation contains a sum of first and second derivatives with respect to the components of $\mathbf{x}$ and $\mathbf{v}$. A slight change in notation is needed because the initial value of y in Equation (2.25) is held fixed and the average is that indicated previously by $\langle\ \rangle_0$; however, y itself denotes $\mathbf{x}$, $\mathbf{v}$ rather than $\mathbf{x}_0$, $\mathbf{v}_0$. Certain of the coefficients are readily shown to vanish:

$$b_{ij}(\mathbf{x}) = \lim_{\tau\to 0}\tau^{-1}\langle\Delta x_i\Delta x_j\rangle_{\mathbf{v}} = 0,\ b_{ij}(\mathbf{x},\mathbf{v}) = \lim_{\tau\to 0}\tau^{-1}\langle\Delta x_i\Delta v_j\rangle_{\mathbf{v}} = 0.$$

In addition we have

$$\mathbf{a}(\mathbf{x}) = \lim_{\tau\to 0}\tau^{-1}\langle\Delta\mathbf{x}\rangle_{\mathbf{v}} = \mathbf{v}. \tag{2.27}$$

Using Equation (2.12) we obtain

$$\mathbf{a}(\mathbf{v}) = \lim_{\tau\to 0}\tau^{-1}\langle\mathbf{v}(\tau) - \mathbf{v}\rangle_{\mathbf{v}} = \lim_{\tau\to 0}\tau^{-1}(e^{-\xi\tau} - 1)\mathbf{v} = -\xi\mathbf{v}.$$

Furthermore, we have

$$\begin{aligned} b_{ij}(\mathbf{v}) &= \lim_{\tau\to 0}\tau^{-1}\langle[v_i(\tau) - v_i][v_j(\tau) - v_j]\rangle_{\mathbf{v}} \\ &= \frac{1}{3}\delta_{ij}\lim_{\tau\to 0}\tau^{-1}[\langle v^2(\tau)\rangle_{\mathbf{v}} - 2\mathbf{v}\cdot\langle\mathbf{v}(\tau)\rangle_{\mathbf{v}} + v^2]. \end{aligned}$$

With the help of Equations (2.12) and (2.20) we can reduce this to

$$\begin{aligned} b_{ij}(\mathbf{v}) &= \frac{1}{3}\delta_{ij}\lim_{\tau\to 0}\tau^{-1}[v^2(1 - e^{-\xi\tau})^2 + (3\kappa T/m)(1 - e^{-2\xi\tau})] \\ &= (2\xi\kappa T/m)\delta_{ij}. \end{aligned}$$

When the a and b coefficients are substituted into Equation (2.24), the result is

$$\frac{\partial P}{\partial t} + \mathbf{v}\cdot\nabla P = \xi\nabla_{\mathbf{v}}\cdot[\mathbf{v}P + (\kappa T/m)\nabla_{\mathbf{v}}P]$$

where $\nabla_{\mathbf{v}}$ denotes the gradient in velocity space. In the context of Brownian motion, this is the equation usually meant by the term Fokker-Planck equation, and it will be so referred to in the following.

2.5. SOLUTION TO THE FOKKER-PLANCK EQUATION

Let the Fokker-Planck equation be written in the form

$$\frac{\partial P}{\partial t} + \mathbf{v}\cdot\nabla P = \Omega P$$

where the operator Ω is defined by

$$\Omega P = \xi\nabla_{\mathbf{v}}\cdot[\mathbf{v}P + (\kappa T/m)\nabla_{\mathbf{v}}P] \tag{2.28}$$

We consider an unbounded system with the boundary condition that P vanish as $\mathbf{x}$, $\mathbf{v}$ become large.

The equation of continuity can be derived from the Fokker-Planck equation as follows. The expected value for the density n is

$$n(\mathbf{x},t) = N\int d^3v\, P(\mathbf{x},\mathbf{v},t)$$

where N is the total number of B-particles. Taking the time derivative, we get

$$\frac{\partial n}{\partial t} + \nabla\cdot N\int d^3v\,\mathbf{v}\,P = N\int d^3v\,\Omega\, P.$$

On the right-hand side the integration can be carried out immediately and the result vanishes because of the boundary condition on P. Hence, we get

$$\frac{\partial n}{\partial t} + \nabla\cdot\mathbf{j} = 0$$

where

$$\mathbf{j} = N\int d^3v\,\mathbf{v}\,P$$

is the flux of B-particles.

The drag law can also be obtained from the Fokker-Planck equation. The average velocity of a particle is

$$\langle\mathbf{v}\rangle = \int d^3x\,d^3v\,\mathbf{v}\,P$$

and its time derivative is

$$d\langle\mathbf{v}\rangle/dt = -\int d^3x\,d^3v\,\mathbf{v}[\mathbf{v}\cdot\nabla P + \Omega\, P].$$

The first term is a divergence in $\mathbf{x}$-space and the integral vanishes because of the boundary condition. The last term can be calculated by integrating by parts and the result is the drag law, $d\langle\mathbf{v}\rangle/dt = -\xi\langle\mathbf{v}\rangle$.

In equilibrium P is independent of position and time so the equation reduces to $\Omega P = 0$. It is readily shown that a solution is $P = \phi$ where ϕ denotes the Maxwell-Boltzmann distribution function

$$\phi(\mathbf{v}) = (m/2\pi\kappa T)^{3/2} \exp(-mv^2/2\kappa T). \tag{2.29}$$

For brevity we will frequently refer to this function as the Maxwellian.

If P is independent of position but not of time, the Fokker-Planck equation becomes

$$\frac{\partial P}{\partial t} = \Omega P. \tag{2.30}$$

To solve this equation we first consider the eigenvalue problem for Ω,

$$\Omega F = \lambda F. \tag{2.31}$$

The substitution $F = \phi U$ converts this into

$$(\kappa T/m)\nabla_{\mathbf{v}}^2 U - \mathbf{v}\cdot\nabla_{\mathbf{v}} U = (\lambda/\xi)U. \tag{2.32}$$

The Hermite polynomials $H_n(y)$, $n = 0,1,2\ldots$, satisfy

$$\frac{d^2H_n}{dy^2} - 2y\frac{dH_n}{dy} + 2nH_n = 0.$$

A straightforward calculation shows that Equation (2.32) is solved by the functions

$$U_{n1,n2,n3} = AH_{n1}(y_1)\,H_{n2}(y_2)\,H_{n3}(y_3)$$

where A is a constant, $y_i = \sqrt{m/2\kappa T}\,v_i$, and the eigenvalues are

$$\lambda = \lambda_{n1,n2,n3} = -\xi(n_1 + n_2 + n_3).$$

For brevity we will denote the three integers n_1, n_2, n_3 by $\mathbf{n}$ and write

$$U_{\mathbf{n}}(\mathbf{y}) = A_{\mathbf{n}}H_{\mathbf{n}}(\mathbf{y}).$$

If we choose the $A_{\mathbf{n}}$ to be

$$A_{\mathbf{n}} = [2^{n1+n2+n3}\,n_1!n_2!n_3!]^{-1/2}$$

then it follows from properties of the Hermite polynomials that the $U_{\mathbf{n}}$ are an orthonormal set in the sense that

$$\int d^3v\,\phi\,U_{\mathbf{n}}\,U_{\mathbf{m}} = \delta_{\mathbf{n},\mathbf{m}}.$$

In terms of the original operator Ω we have

$$\Omega F_{\mathbf{n}} = \lambda_{\mathbf{n}}\,F_{\mathbf{n}}$$

where the eigenfunctions are $F_{\mathbf{n}} = \phi U_{\mathbf{n}}$ and satisfy

$$\int d^3v\phi^{-1}F_{\mathbf{n}}F_{\mathbf{m}} = \delta_{\mathbf{n},\mathbf{m}}. \tag{2.33}$$

The solution to Equation (2.31) can also be obtained by the substitution $F = \sqrt{\phi}W$. The equation for W is

$$-(\kappa T/m)\,\nabla_{\mathbf{v}}^2\,W + (mv^2/4\kappa T)W = \left[-(\lambda/\xi) + \frac{3}{2}\right]W. \tag{2.34}$$

This is the same as the eigenvalue equation for the energy of a three-dimensional harmonic oscillator in quantum mechanics, and the solutions are known to be $\sqrt{\phi}$ times the product of Hermite polynomials.

The time-dependent equation (2.30) can be solved by expanding in the eigenfunctions of Ω:

$$P(\mathbf{v},t) = \sum_{\mathbf{n}} C_{\mathbf{n}} F_{\mathbf{n}} \exp(\lambda_{\mathbf{n}} t).$$

The expansion coefficients can be determined from the initial value $P(\mathbf{v},0)$ by using Equation (2.33):

$$C_{\mathbf{n}} = \int d^3v\,\phi^{-1} F_{\mathbf{n}} P(\mathbf{v},0).$$

We apply this to the function $P(\mathbf{v},t|\mathbf{v}_0)$ for which the initial condition $P(\mathbf{v},0|\mathbf{v}_0) = \delta(\mathbf{v} - \mathbf{v}_0)$ yields $C_{\mathbf{n}} = F_{\mathbf{n}}(\mathbf{v}_0)/\phi(\mathbf{v}_0)$ so

$$P(\mathbf{v},t|\mathbf{v}_0) = \phi^{-1}(\mathbf{v}_0) \sum_{\mathbf{n}} F_{\mathbf{n}}(\mathbf{v}) F_{\mathbf{n}}(\mathbf{v}_0) \exp(\lambda_{\mathbf{n}} t). \tag{2.35}$$

To sum the series we use the formula for Hermite polynomials

$$H_n(y) = (2^n/\sqrt{\pi}) \int_{-\infty}^{\infty} ds\, e^{-s^2}(y + is)^n$$

which yields

$$\sum_{n=0}^{\infty} \frac{z^n}{n!} H_n(x) H_n(y) = \frac{1}{\pi} \int_{-\infty}^{\infty} ds \int_{-\infty}^{\infty} dr\, e^{-(s^2+r^2)}\, e^{4z(x+ir)(y+is)} \tag{2.36}$$

$$= (1 - 4z^2) \exp\left\{-\frac{4z^2}{1 - 4z^2}[x^2 + y^2 - (xy/z)]\right\}.$$

Using this we get for Equation (2.35)

$$P(\mathbf{v},t|\mathbf{v}_0) = (\gamma/\pi)^{3/2} \exp[-\gamma(\mathbf{v} - e^{-\xi t}\mathbf{v}_0)^2] \tag{2.37}$$

where $\gamma = (m/2\kappa T)(1 - e^{-2\xi t})^{-1}$.

To solve the equation with a spatial dependence included, we take a Fourier transform:

$$\hat{P}(\mathbf{k},\mathbf{v},t) = \int d^3x\, P(\mathbf{x},\mathbf{v},t) \exp(i\mathbf{k}\cdot\mathbf{x}).$$

Then $\hat{P}$ satisfies

$$\partial\hat{P}/\partial t = (\Omega + i\mathbf{k}\cdot\mathbf{v})\hat{P}. \tag{2.38}$$

Now we consider the eigenvalue equation for the operator on the right,

$$(\Omega + i\mathbf{k}\cdot\mathbf{v})F' = \lambda' F'.$$

Making the substitution $F' = \sqrt{\phi}\, W'$, we get

$$-(\kappa T/m)\,\nabla_{\mathbf{v}}^2 W' + (m/4\kappa T)\,(\mathbf{v} - 2Di\mathbf{k})^2\, W'$$

$$= \xi^{-1}\left[-\lambda' - Dk^2 + \frac{3}{2}\xi\right] W' \qquad (2.39)$$

where D is the diffusion constant (2.6). If W is a solution to Equation (2.34) then $W'(\mathbf{v}) = W(\mathbf{v} - 2Di\mathbf{k})$ is a solution to (2.39) with $\lambda' = \lambda - Dk^2$. It follows that the functions

$$F'_{\mathbf{n}}(\mathbf{v}) = \sqrt{\phi(\mathbf{v})/\phi(\mathbf{v} - 2Di\mathbf{k})}\; F_{\mathbf{n}}(\mathbf{v} - 2Di\mathbf{k})$$

are eigenfunctions of $\Omega + i\mathbf{k}\cdot\mathbf{v}$ with the eigenvalues $\lambda'_{\mathbf{n}} = \lambda_{\mathbf{n}} - Dk^2$. In addition these functions satisfy

$$\int d^3v\, \phi^{-1}\, F'_{\mathbf{n}}\, F'_{\mathbf{m}} = \delta_{\mathbf{n},\mathbf{m}}.$$

A solution to the time-dependent equation (2.38) can be written in the form

$$\hat{P}(\mathbf{k},\mathbf{v},t) = \sum_{\mathbf{n}} C_{\mathbf{n}} F'_{\mathbf{n}}(\mathbf{v}) \exp(\lambda'_{\mathbf{n}} t) \qquad (2.40)$$

where the coefficients $C_{\mathbf{n}}$ are to be determined from the initial conditions.

The function $P(\mathbf{q},\mathbf{v},t|\mathbf{v}_0) = P(\mathbf{x},\mathbf{v},t|\mathbf{x}_0,\mathbf{v}_0)$, where $\mathbf{q} = \mathbf{x} - \mathbf{x}_0$, is a solution to the Fokker-Planck equation with the initial condition

$$P(\mathbf{q},\mathbf{v},0|\mathbf{v}_0) = \delta(\mathbf{q})\delta(\mathbf{v} - \mathbf{v}_0).$$

Its Fourier transform has the initial condition

$$\hat{P}(\mathbf{k},\mathbf{v},0) = \delta(\mathbf{v} - \mathbf{v}_0).$$

From this the coefficients in the expansion (2.40) are found to be

$$C_{\mathbf{n}} = \phi^{-1}(\mathbf{v}_0) F'_{\mathbf{n}}(\mathbf{v}_0).$$

The series (2.40) can be summed with the aid of Equation (2.36) and the result is

$$\hat{P}(\mathbf{k},\mathbf{v},t) = \exp\{-Dk^2 t + \xi^{-1}[i\mathbf{k}\cdot(\mathbf{v}_0 - \mathbf{v})]\}\, P(\mathbf{v} - 2Di\mathbf{k}, t|\mathbf{v}_0 - 2Di\mathbf{k})$$

where the last factor is the function given by Equation (2.37). Inversion of the Fourier transform yields

$$P(\mathbf{q},\mathbf{v},t|\mathbf{v}_0) = (\alpha/\pi)^{3/2}\, P(\mathbf{v},t|\mathbf{v}_0) \exp\{-\alpha[\mathbf{q} - \beta(\mathbf{v} + \mathbf{v}_0)]^2\} \qquad (2.41)$$

where

$$\alpha = [4D(t - 2\beta)]^{-1},\ \beta = \xi^{-1}(1 - e^{-\xi t})(1 + e^{-\xi t})^{-1}.$$

2.6. APPROACH TO EQUILIBRIUM

The results of the preceding section show in detail how equilibrium is approached with the passage of time. Consider first the function $P(\mathbf{v},t|\mathbf{v}_0)$. If ξt is large, we have

$$\alpha \to (4Dt)^{-1},\ \beta \to \xi^{-1},\ \gamma \to m/2\kappa T \tag{2.42}$$

and the velocity distribution becomes a Maxwellian:

$$P(\mathbf{v},t|\mathbf{v}_0) \to \phi(\mathbf{v}).$$

For small times $P(\mathbf{v},t|\mathbf{v}_0)$ has the Maxwellian form, but is centered about an average velocity $\langle \mathbf{v} \rangle$ of

$$\langle \mathbf{v} \rangle = \int d^3v\,\mathbf{v}\,P(\mathbf{v},t|\mathbf{v}_0) = e^{-\xi t}\,\mathbf{v}_0$$

and in addition T is replaced by an "effective temperature" T_{eff} given by

$$T_{\text{eff}} = T[1 - e^{-2\xi t}].$$

Thus, thermalization takes place through a process in which the average velocity goes to zero (that is, to the velocity of the surrounding fluid) while the temperature approaches that of the surroundings.

With the asymptotic values (2.42) the function (2.41) takes the form

$$P(\mathbf{q},\mathbf{v},t|\mathbf{v}_0) \to \phi(\mathbf{v})(4\pi Dt)^{-3/2}\exp\{-(4Dt)^{-1}[\mathbf{q} - \xi^{-1}(\mathbf{v} + \mathbf{v}_0)]^2\}.$$

Except for the shift of $\mathbf{q}$ by $\xi^{-1}(\mathbf{v} + \mathbf{v}_0)$ this is the product of the Maxwellian by the distribution of displacements (2.2) obtained from the theory of diffusion, and it satisfies the diffusion equation.

In the approach to equilibrium there is first a rapid thermalization of the velocity which takes place in a time of the order of ξ^{-1}. This is followed by a slower reduction of spatial homogeneities which is governed by the diffusion equation. Thermalization is governed by collisions which are fast; diffusion is slowed up by the necessity of the particles to move from one place to another. Thus, Brownian motion provides a concrete example of the ageing to hydrodynamics discussed in Chapter 1.

There is an approach to equilibrium only in the statistical sense that distribution functions approach their equilibrium forms. In any single observation one will not find a precisely uniform spatial distribution of the B-particles, and their velocities (if instantaneous velocities could be measured) will not be distributed precisely according to the Maxwellian.

That solutions to the Fokker-Planck equation approach equilibrium can be shown directly from the equation in the following way. The substitution $P = \sqrt{\phi}\,W$ gives $\Omega P = \sqrt{\phi}\,\omega W$ where the operator ω is defined by

$$\omega W = \xi\left\{\left[\frac{3}{2} - \frac{mv^2}{4\kappa T}\right]W + \frac{\kappa T}{m}\frac{\partial^2 W}{\partial v_i^2}\right\}.$$

Let brackets (,) denote the scalar product

$$(U, W) = \int d^3v \bar{U} W$$

where the bar denotes the complex conjugate. We will limit attention to functions such that (W, W) is finite, that is, to the Hilbert space with the scalar product (,). An integration by parts with neglect of the boundary term yields

$$(U, \omega W) = -(\xi\kappa T/m) \int d^3v \left(\nabla_v \bar{U} + \frac{m}{2\kappa T} \mathbf{v} \bar{U} \right) \cdot \left(\nabla_v W + \frac{m}{2\kappa T} \mathbf{v} W \right).$$

This shows that ω is symmetric, $(U, \omega W) = (\omega U, W)$, and nonpositive, $(W, \omega W) \leq 0$. If $(W, \omega W) = 0$, it is necessary that $\nabla_v W = -(m/2\kappa T)\mathbf{v}W$ which implies that except for normalization $W = \sqrt{\phi}$. To restate these results in terms of the original operator Ω, let [,] denote the scalar product

$$[P, Q] = \int d^3v \phi^{-1} \bar{P} Q.$$

Then we have

$$[P, \Omega Q] = [\Omega P, Q], \ [P, \Omega P] \leq 0.$$

Furthermore, equality in the last expression holds if and only if $P = C\phi$ where C is a constant.

Let $\mathcal{L}$ denote the functional

$$\mathcal{L}(P) = \int d^3x \ [P, P].$$

Then it follows from the Fokker-Planck equation that

$$\frac{d\mathcal{L}}{dt} = 2 \int d^3x \ [P, \Omega P]$$

and so $\mathcal{L}$ satisfies

$$\mathcal{L}(P) \geq 0, \ \frac{d}{dt} \mathcal{L}(P) \leq 0. \tag{2.43}$$

A bounded monotonic function must approach a limit, and in the limit $d\mathcal{L}/dt = 0$ which implies $P = C\phi$. Thus, we have shown that P approaches the equilibrium form as $t \to 0$.

The inequalities (2.43) show that $\mathcal{L}(P)$ is a Liapunov function for the Fokker-Planck equation. The demonstration just given of the approach to equilibrium is a modification of the argument first provided by Boltzmann in the kinetic theory of gases.

2.7. GAUSSIAN DISTRIBUTIONS

A probability density with the form

$$P(y) = (\sigma\sqrt{2\pi})^{-1} \exp\left[-(y - \bar{y})^2/2\sigma^2\right] \tag{2.44}$$

is said to be Gaussian or normal. The average of y is $\bar{y}$ and σ is the standard deviation.

The probability densities of the previous sections are in fact Gaussian. The reason for this can be traced back to the truncation of the expansion (2.24) in the derivation of the Fokker-Planck equation. The Fokker-Planck equation can be bypassed if it is assumed from the beginning that Brownian motion is a Gaussian random process. If certain symmetries are introduced and the velocity autocorrelation is given the exponential form (2.14), then the basic function (2.41) can be derived directly as will now be shown.

It can be shown by computation that the function (2.41) satisfies

$$P(\mathbf{q},\mathbf{v},t|\mathbf{v}_0)\,\phi(\mathbf{v}_0) = P(-\mathbf{q},-\mathbf{v}_0,t|-\mathbf{v})\phi(\mathbf{v}). \tag{2.45}$$

This is a consequence of microscopic reversibility and is referred to as the principle of reciprocity. An equivalent statement of reciprocity is that the joint probability

$$W(\mathbf{x},\mathbf{v},t;\mathbf{x}_0,\mathbf{v}_0,t_0) = P(\mathbf{q},\mathbf{v},t - t_0|\mathbf{v}_0)\phi(\mathbf{v}_0)$$

has the symmetry

$$W(\mathbf{x},\mathbf{v},t;\mathbf{x}_0,\mathbf{v}_0,t_0) = W(\mathbf{x}_0,-\mathbf{v}_0,t;\mathbf{x},-\mathbf{v},t_0).$$

Let $P(\mathbf{q},\mathbf{v},t|\mathbf{v}_0)$ be written in the form $P(\mathbf{q},\mathbf{v},t|\mathbf{v}_0) = A\exp(-\psi)$ where A is independent of $\mathbf{q}$, $\mathbf{v}$, and $\mathbf{v}_0$. The Gaussian assumption means that ψ is quadratic in $\mathbf{q}$, $\mathbf{v}$ and reciprocity shows that it is quadratic in $\mathbf{v}_0$ as well. Spherical symmetry implies that ψ is a linear combination of the six quadratic invariants which can be formed from $\mathbf{q}$, $\mathbf{v}$, and $\mathbf{v}_0$, namely q^2, v^2, v_0^2, $\mathbf{q}\cdot\mathbf{v}$, $\mathbf{q}\cdot\mathbf{v}_0$, and $\mathbf{v}\cdot\mathbf{v}_0$. Taking (2.45) into account, one finds the necessary form to be

$$P(\mathbf{q},\mathbf{v},t|\mathbf{v}_0) = A\exp\{-\alpha[\mathbf{q} - \beta(\mathbf{v} + \mathbf{v}_0)]^2 - \gamma[\mathbf{v} - \mu\mathbf{v}_0]^2\} \tag{2.46}$$

where $\gamma - (m/2\kappa T)[1 - \mu^2]^{-1}$. Thc normalization condition

$$\int d^3q\,d^3v\,P(\mathbf{q},\mathbf{v},t|\mathbf{v}_0) = 1$$

yields $A = (\alpha\gamma)^{3/2}\pi^{-3}$. The result (2.46) has the same general form as (2.41) but α, β, and μ are still to be determined.

The velocity autocorrelation can be calculated from

$$\langle\mathbf{v}(t)\cdot\mathbf{v}(t_0)\rangle = \int d^3v_0\,d^3v\,d^3q\,\mathbf{v}\cdot\mathbf{v}_0\phi(\mathbf{v}_0)P(\mathbf{v},\mathbf{q},t - t_0|\mathbf{v}_0).$$

A straightforward calculation gives

$$\langle \mathbf{v}(t) \cdot \mathbf{v}(0) \rangle = \langle \mathbf{v}_0^2 \rangle \, \mu$$

which identifies μ as the normalized velocity autocorrelation, $\mu = \chi(t)$. Next we calculate the correlation between $\mathbf{q}$ and $\mathbf{v}_0$ which on one hand can be obtained from the velocity autocorrelation,

$$\langle \mathbf{q}(t) \cdot \mathbf{v}(0) \rangle = \int_0^t ds \langle \mathbf{v}(s) \cdot \mathbf{v}(0) \rangle = \langle v^2 \rangle \int_0^t ds \, \chi(s). \tag{2.47}$$

On the other hand, it is also given by

$$\langle \mathbf{q}(t) \cdot \mathbf{v}(0) \rangle = \int d^3v_0 d^3v d^3q \mathbf{q} \cdot \mathbf{v} \phi(\mathbf{v}_0) P(\mathbf{q}, \mathbf{v}, t | \mathbf{v}_0)$$

for which calculation gives

$$\langle \mathbf{q}(t) \cdot \mathbf{v}(0) \rangle = \langle \mathbf{v}^2 \rangle \, \beta(1 + \mu).$$

Combining this with (2.47) we get

$$\beta = (1 + \chi)^{-1} \int_0^t ds \, \chi(s).$$

Finally we calculate the mean-square displacement,

$$\begin{aligned} \langle q^2(t) \rangle &= \int d^3v_0 \, d^3v \, d^3q \, q^2 \phi(\mathbf{v}_0) P(\mathbf{q}, \mathbf{v}, t | \mathbf{v}_0) \\ &= (3/2\alpha) + 2\langle v^2 \rangle (1 + \chi)^{-1} \left[\int_0^t ds \, \chi(s) \right]^2. \end{aligned}$$

Using the relation (2.16) between the mean-square displacement and the velocity autocorrelation, we find for α

$$\alpha = \frac{m}{4\kappa T} \left\{ \int_0^t ds (t - s) \chi(s) - (1 + \chi)^{-1} \left[\int_0^t ds \, \chi(s) \right]^2 \right\}^{-1}.$$

Thus α, β, γ, and μ are all determined by χ. It is straightforward to check that the exponential form (2.14) for χ leads to the values obtained previously for α, β, and γ.

From Equation (2.46) it can be confirmed that the Chapman-Kolmogorov equation is satisfied if and only if χ has the exponential form (2.14). If Brownian motion is assumed to be Gaussian, then it is a Markov process if and only if the velocity autocorrelation is exponential.

The distribution of displacements $P(\mathbf{q}, t) = P(\mathbf{q}, t | 0)$ can be obtained from (2.46) by integrating over the final velocity and averaging over the initial,

$$P(\mathbf{q}, t) = \int d^3v \, d^3v_0 \, \phi(\mathbf{v}_0) P(\mathbf{q}, \mathbf{v}, t | \mathbf{v}_0).$$

Alternatively, one can simply assume $\mathbf{q}$ to have a Gaussian distribution. Either way the result is

$$P(\mathbf{q},t) = (3/2\pi\langle q^2\rangle)^{3/2} \exp(-3q^2/2\langle q^2\rangle)$$

where $\langle q^2\rangle$ is determined by the velocity autocorrelation according to Equation (2.16).

For large t the function $P(\mathbf{q},t)$ approaches the solution (2.2) to the diffusion equation. For all times it satisfies the generalized diffusion equation

$$\frac{\partial P}{\partial t} = D(t)\nabla^2 P \tag{2.48}$$

where

$$D(t) = \frac{1}{6}\frac{d\langle q^2\rangle}{dt} = (\kappa T/m)\int_0^t ds\, \chi(s).$$

Thus $D(t)$ is a time-dependent diffusion constant which approaches D for large t but takes the ageing period into account for small t. The velocity autocorrelation is an even function, $\chi(t) = \chi(-t)$, from which it follows that Equation (2.48) is invariant under time reversal. For t large and positive it approaches the diffusion equation, but for t large and negative it becomes the "anti-diffusion" equation $\partial P/\partial t = -D\nabla^2 P$.

2.8. EFFECT OF GRAVITY

An external field can be included in the Fokker-Planck equation by a straightforward modification of the argument given in Section 4. There is a contribution of $\tau\mathbf{F}/m$ to the change in velocity over the (small) time τ, where $\mathbf{F}$ is the external force, and consequently the a-coefficient of Equation (2.27) gains an additional term $\mathbf{F}/m$. The Fokker-Planck equation becomes

$$\frac{\partial P}{\partial t} + \mathbf{v}\cdot\nabla P + (1/m)\nabla_{\mathbf{v}}\cdot\mathbf{F}P = \xi\nabla_{\mathbf{v}}\cdot[\mathbf{v}P + (\kappa T/m)\nabla_{\mathbf{v}}P].$$

In the general case this equation cannot be solved in closed form. If the force is derivable from a potential, $\mathbf{F} = -\nabla U$, the equilibrium solution is given by $P = \phi\exp(-U/\kappa T)$ where ϕ is the Maxwellian function of velocity.

We consider the case of gravity in more detail. The Fokker-Planck equation is

$$\frac{\partial P}{\partial t} + \mathbf{v}\cdot\nabla P + \mathbf{g}\cdot\nabla_{\mathbf{v}}P = \xi\nabla_{\mathbf{v}}\cdot[\mathbf{v}P + (\kappa T/m)\nabla_{\mathbf{v}}P].$$

Because of buoyancy, $\mathbf{g}$ is $1 - (\rho_0/\rho)$ times the acceleration of gravity, where ρ, ρ_0 are the densities of the particle and of the surrounding fluid. For clarity we will assume the particle to be more dense than the fluid so the terminal velocity is downward instead of upward.

The external force can be eliminated from the preceding equation by the change of variables

$$\mathbf{x} \rightarrow \mathbf{x}' = \mathbf{x} - \mathbf{v}_T t, \ \mathbf{v} \rightarrow \mathbf{v}' = \mathbf{v} - \mathbf{v}_T$$

where $\mathbf{v}_T = \mathbf{g}/\xi$ is the terminal velocity. It is readily confirmed that the function P' defined by $P(\mathbf{x},\mathbf{v},t) = P'(\mathbf{x}',\mathbf{v}',t)$ satisfies the field-free Fokker-Planck equation. Consequently, if the Brownian motion is viewed from a frame of reference falling at the terminal velocity, it appears the same as in a stationary frame in the field-free case. For example, Einstein's formula for the mean-square displacement becomes

$$\langle(\Delta\mathbf{x} - \mathbf{v}_T t)^2\rangle \rightarrow 6Dt.$$

Of course a B-particle confined in a container does not fall indefinitely. The description of Brownian motion in this situation is important for the theory of sedimentation. Let the bottom of the container be described as an infinite horizontal plane at $z = 0$. The Fokker-Planck equation should be solved subject to a boundary condition at the plane. The solution to boundary-value problems for the Fokker-Planck equation is quite difficult, and so we consider the simpler description provided by the theory of diffusion. Diffusion in an external field is described by the Smoluchowski equation (2.5) which for the case at hand becomes

$$\frac{\partial P}{\partial t} = D\nabla^2 P + (g/\xi)\frac{\partial P}{\partial z}$$

where the z-direction has been taken upward. To obtain the boundary condition on P, we require that the normal component of the flux (2.4) vanish at the boundary, which gives

$$\frac{\partial P}{\partial z} = -(mg/\kappa T)P, \quad \text{at} \quad z = 0.$$

We consider the function $P(\mathbf{x},t|\mathbf{x}_0)$ for which the initial condition is $P(\mathbf{x},0|\mathbf{x}_0) = \delta(\mathbf{x} - \mathbf{x}_0)$. The solution to the differential equation with the given boundary and initial condition has been obtained by Chandrasekhar, and we will only quote his result. The solution is a product $P = RZ$ where R is a function of x,y and Z depends on z. Diffusion in a plane perpendicular to the z-axis is unaffected by gravity so the first factor has the form

$$R = (4\pi Dt)^{-1}\exp(-r^2/4Dt), \quad r^2 = (x - x_0)^2 + (y - y_0)^2.$$

The remaining factor is given by

$$Z = (4\pi Dt)^{-1/2}[1 + \exp(-zz_0/Dt)] \exp[-(z - z_0 + v_T t)^2/4Dt]$$

$$+ (v_T/D\sqrt{\pi}) \exp(-v_T z/D) \int_s^\infty du \exp(-u^2)$$

where $s = (z + z_0 - v_T t)/\sqrt{4Dt}$. According to this solution, the initial pulse at z_0 begins to spread out because of diffusion, but the peak of the distribution drifts downward at the terminal velocity. The density at the bottom begins to build up and for a period of time the distribution has two maxima. Then for $t \to \infty$ the solution approaches the atmospheric law.

2.9. DIFFUSION OVER A BARRIER

Diffusion over a barrier was described by Kramers on the basis of the Fokker-Planck equation. Consider a one-dimensional situation, with the particle moving in a potential U. We suppose the diffusion to be nearly stationary so the time derivative can be neglected, in which case the equation to be solved is

$$v\frac{\partial P}{\partial x} + \frac{1}{m}\frac{\partial}{\partial v}FP = \xi\frac{\partial}{\partial v}\left[vP + \frac{\kappa T}{m}\frac{\partial P}{\partial v}\right] \tag{2.49}$$

where v is velocity in the x-direction. The potential is assumed to have the shape of a well, with bottom at $x = 0$ where we take $U(0) = 0$, and a barrier at $x = B$. In the situation envisaged, B-particles are concentrated in the well while their density elsewhere is negligible. The problem is to determine the rate of diffusion over the barrier. The height $Q = U(B)$ of the barrier will be assumed large in comparison to κT so the rate of diffusion is low.

Of course Equation (2.49) has the equilibrium solution

$$P_0 = C \exp\left[-\left(\frac{1}{2}mv^2 + U\right)/\kappa T\right] \tag{2.50}$$

where C is a normalizing constant, but this does not meet the condition of large concentration within the well. However since the flow is assumed to be nearly steady, the particles within the well should have a near-equilibrium distribution, so this form is a good approximation in the well. There we approximate U by a parabola,

$$U = \frac{1}{2}m\,\omega^2 x^2$$

where ω is the frequency of oscillation associated with the well. The total number N of B-particles (within the well) is

$$N = \int dv\, dx\, P_0.$$

Since the height of the barrier is large, little error is involved in extending the integration over x to infinity on both sides, which yields

$$N = (2\pi\kappa T/m\omega)C.$$

This determines C in terms of N.

Particles diffusing over the barrier do not have an equilibrium distribution, so Equation (2.50) is not a valid approximation near the barrier. In that region we write

$$P = P_0 F$$

where F is a function still to be determined. Near the barrier we approximate U by a parabola, now pointing downwards,

$$U = Q - \frac{1}{2} m\,\omega_B^2 (x - B)^2.$$

(Although ω_B has the dimension of frequency, of course no oscillations are possible near the barrier.) Equation (2.49) then yields the equation for F:

$$v\frac{\partial F}{\partial x} + \omega_B^2(x - B)\frac{\partial F}{\partial v} = \xi\left[\frac{\kappa T}{m}\frac{\partial^2 F}{\partial v^2} - v\frac{\partial F}{\partial v}\right].$$

Let a variable V be defined by

$$V = v - \lambda\xi(x - B)$$

where

$$\lambda = \frac{1}{2}\left[\sqrt{1 + (2\omega_B/\xi)^2} - 1\right].$$

Then F satisfies

$$VF' = -(\kappa T\lambda\xi^2/m\,\omega_B^2)F''$$

where the prime denotes differentiation with respect to V. This equation is solved by

$$F = \sqrt{m/2\pi\kappa T\lambda}(\omega_B/\xi)\int_{-\infty}^{V} du\, \exp[-mu^2\omega_B^2/2\kappa T\lambda\xi^2].$$

The constants of integration have been chosen so that $F \rightarrow 1$ for x large and negative, so we regain the equilibrium distribution on the left of the barrier, and $F \to 0$ for x large and positive, corresponding to the small density beyond the barrier.

The flux of B-particles over the barrier is given by

$$j = \int_{-\infty}^{\infty} dv\, vP$$

where P is evaluated at $x = B$. Carrying out the integral, we get

$$j = N(\omega\omega_B/2\pi\lambda\xi)\exp(-Q/\kappa T).$$

The ratio j/N is the rate constant for diffusion of a particle out of the well. This has the characteristic temperature dependence of a thermally activated process, with activation energy Q. In the limit of low viscosity, or small ξ, we have $\lambda \to \omega_B/\xi$ and the pre-exponential factor is just the frequency of oscillation in the well. In the limit of large damping, $\lambda \to 1$ and the pre-exponential factor is changed to $\omega\omega_B/2\pi\xi$.

2.10. NON-GAUSSIAN EFFECTS

In the theory as described to this point, it has been assumed that Brownian motion is both Gaussian and Markovian. Careful observations have shown that there are small corrections due to a breakdown of both of these assumptions. The description of non-Gaussian effects wil be considered in this section, and of non-Markovian effects in the next section.

Corrections to Gaussian behavior can be described by a standard technique of probability theory. For a random variable y the characteristic function is defined by

$$f(k) = \langle e^{iky} \rangle. \tag{2.51}$$

This is the Fourier transform of the probability density,

$$f(k) = \int dy\, e^{iky} P(y).$$

The logarithm of $f(k)$ is called the generating function,

$$g(k) = \ln f(k).$$

If $g(k)$ is expanded in a power series in k,

$$g(k) = \sum_{n=1}^{\infty} \frac{(ik)^n}{n!} C_n$$

the coefficients C_n are called cumulants. (In general there is no guarantee that the series converge or that all of the cumulants exist.) The first cumulants are

$$C_1 = \langle y \rangle,\ C_2 = \langle (\Delta y)^2 \rangle,\ C_3 = \langle (\Delta y)^3 \rangle,\ C_4 = \langle (\Delta y)^4 \rangle - 3\langle (\Delta y)^2 \rangle^2$$

where $\Delta y = y - \langle y \rangle$. Thus C_1 and C_2 are respectively the mean and the variance of y. As can be seen from the preceding examples and shown in higher orders, the cumulants except for C_1 are unchanged if a constant is added to y.

If $P(y)$ is Gaussian, the cumulants of order higher than 2 vanish. A Gaussian distribution is determined by C_1 and C_2 but in general all of the cumulants are needed to determine the distribution. The cumulant expansion provides a natural way of making corrections to the Gaussian approximation.

In the case of several random variables, say y_α, the characteristic function is defined by

$$f = \langle \exp\left(i \sum_\alpha k_\alpha y_\alpha\right) \rangle.$$

We expand in powers of each of the k_α

$$g = \sum (n_1! n_2! \ldots)^{-1} (ik_1)^{n_1} (ik_2)^{n_2} \ldots C_{n_1 n_2 \ldots}$$

In the sum the n's take all values 0,1,2... except that the term with all n's vanishing simultaneously is omitted. A different way of writing the sum is

$$g = \sum_\alpha (ik_\alpha) C^{(1)}_\alpha + \frac{1}{2} \sum_{\alpha\beta} (ik_\alpha)(ik_\beta) C^{(2)}_{\alpha\beta} + \ldots$$

The coefficients $C^{(2)}$, $C^{(3)}$, and so on, are called correlations. The first few are given by

$$C^{(1)}_\alpha = \langle y_\alpha \rangle,\ C^{(2)}_{\alpha\beta} = \langle \Delta y_\alpha \Delta y_\beta \rangle,\ C^{(3)}_{\alpha\beta\gamma} = \langle \Delta y_\alpha\ \Delta y_\beta\ \Delta y_\gamma \rangle,$$

$$C^{(4)}_{\alpha\beta\gamma\delta} = \langle \Delta y_\alpha\ \Delta y_\beta\ \Delta y_\lambda\ \Delta y_\delta \rangle - \langle \Delta y_\alpha\ \Delta y_\beta \rangle \langle \Delta y_\gamma\ \Delta y_\delta \rangle$$
$$- \langle \Delta y_\alpha\ \Delta y_\gamma \rangle \langle \Delta y_\beta\ \Delta y_\delta \rangle - \langle \Delta y_\alpha\ \Delta y_\delta \rangle \langle \Delta y_\beta\ \Delta y_\gamma \rangle.$$

For Brownian motion y is the vector displacement $\mathbf{q}$ in time t. The probability density and characteristic function are functions of the time,

$$f(\mathbf{k},t) = \int d^3 q e^{i\mathbf{k}\cdot\mathbf{q}} P(\mathbf{q},t).$$

We will assume P to satisfy a diffusion equation but with higher-order terms such as occur in the expansion (2.24)

$$\partial P/\partial t = D\, \nabla^2 P + D'\, \nabla^4 P + \ldots \tag{2.52}$$

Cubic and other terms of odd order do not occur because of spherical symmetry. The corresponding equation for f is

$$\partial f/\partial t = (-Dk^2 + D'k^4 + \ldots)f.$$

With the initial condition $P(\mathbf{q},0) = \delta(\mathbf{q})$, $f(\mathbf{k},0) = 1$, the solution is

$$f(\mathbf{k},t) = \exp[(-Dk^2 + D'k^4 + \ldots)t]$$

and the generating function is

$$g(\mathbf{k},t) = (-Dk^2 + D'k^4 + \ldots)t. \tag{2.53}$$

To avoid complications due to the vector nature of $\mathbf{q}$, we consider only the component of motion along a single axis, say the x-axis. Denoting this component by q, we obtain for its cumulants

$$C_2 = 2Dt,\ C_4 = 4!D't, \text{ etc.}$$

Actually the diffusion equation holds only after an ageing period and these formulas are only asymptotically valid for long times. Hence, we should write instead

$$\begin{aligned} D &= \lim_{t\to\infty} \frac{1}{2t} C_2(t) = \lim_{t\to\infty} \frac{1}{2t}\langle q^2(t)\rangle, \\ D' &= \lim_{t\to\infty} \frac{1}{4!t}[\langle q^4(t)\rangle - 3\langle q^2(t)\rangle^2]. \end{aligned} \tag{2.54}$$

Thus, the second-order diffusion constant D' is related to deviations from Gaussian behavior. However, there are complications which are unexpected on the basis of this brief discussion. Persistent correlations, or long-time tails, which will be discussed briefly in Section 12, have the effect that the limit for D' does not exist. Hence the expansion (2.52) is not legitimate; or in other words, the generating function is not analytic in $\mathbf{k}$. The changes in the theory which are necessary to accommodate this nonanalyticity are rather complex and will not be given here.

2.11. VELOCITY AUTOCORRELATION

The velocity autocorrelation was calculated in Section 3 on the basis of the Langevin equation, with a particular assumption for the macroscopic drag. In the following section a different drag law will be used to determine the velocity autocorrelation. For that purpose we now show that in general the time dependence of the average velocity follows that of the velocity autocorrelation.

The case when the B-particle has a nonvanishing average velocity $\mathbf{u}$ but is otherwise in equilibrium is described by a Maxwellian with the velocity $\mathbf{v}$ replaced by $\mathbf{v} - \mathbf{u}$:

$$\phi_{\mathbf{u}}(\mathbf{v}) = (m/2\pi\kappa T)^{-3/2} \exp[-(m/2\kappa T)(\mathbf{v} - \mathbf{u})^2].$$

We assume the average velocity to be small in comparison to $\sqrt{\kappa T/m}$ and expand:

$$\phi_{\mathbf{u}} = [1 + (m/\kappa T)\mathbf{u} \cdot \mathbf{v}]\phi.$$

Let the motion of the B-particle be governed by microscopic equations of motion which include its interaction with the atoms of the surrounding fluid. Let F denote the distribution function for the system consisting of the B-particle and the surrounding fluid; F is a function of the coordinates and momenta of the B-particle and the atoms in the fluid. Suppose the system is in equilibrium, with distribution function denoted by F_{eq}, but the B-particle is given a nonvanishing average velocity. According to the preceding equation, this initial condition is described by

$$F = [1 + (m/\kappa T)\mathbf{u} \cdot \mathbf{v}]F_{\text{eq}}.$$

The average velocity of the B-particle at time t (which vanishes in equilibrium) is given by

$$u_i(t) = (m/\kappa T)u_j\langle v_i(t)v_j\rangle$$

where the brackets denote an average over F_{eq}. In terms of the normalized velocity autocorrelation, this can be written

$$\mathbf{u}(t) = \chi(t)\mathbf{u}(0) \tag{2.55}$$

which is the desired equation for the average velocity. The drag law $d\mathbf{u}(t)/dt = -\xi\mathbf{u}(t)$ is recovered if $\chi(t)$ is given the exponential form of Section 3.

The velocity autocorrelation also determines the coefficient of diffusion in the general case. Einstein's formula (2.3) is a direct consequence of the equation of diffusion, without reference to the particular mechanism underlying the diffusion process. From it and (2.16) we obtain

$$D = \lim_{t\to\infty} (1/6t)\langle q^2(t)\rangle = \frac{1}{3}\int_0^\infty d\tau\langle\mathbf{v}(\tau) \cdot \mathbf{v}(0)\rangle. \tag{2.56}$$

This reduces to Einstein's relation for D when the velocity autocorrelation is that obtained from the Langevin equation, but is valid in general. The same equation holds for molecular diffusion when the velocity autocorrelation is calculated by the microscopic equations of motion. This is one example of a class of formulas, called Green-Kubo formulas, which relate coefficients for transport processes to the time dependence of spontaneous fluctuations.

2.12. LONG-TIME TAILS

Stokes' law is derived from the equations of fluid dynamics for the case of steady flow. For B-particles the flow is in fact not steady and a better description is obtained by using an equation of motion which takes this into account. The drag force for nonsteady flow, which was derived by Basset and Boussinesq, is given by

$$\frac{d\mathbf{u}}{dt} = -\lambda\xi\mathbf{u} - \nu\sqrt{\xi/\pi}\int_{-\infty}^{t} d\tau(t-\tau)^{-1/2}\frac{d\mathbf{u}(\tau)}{d\tau}. \tag{2.57}$$

As in the previous section, we have written the macroscopic equation of motion in terms of the average velocity $\mathbf{u}$. The parameters λ and ν are given by

$$\lambda = [1 + (m_f/2m)]^{-1},\ \nu = 3\sqrt{\lambda(1-\lambda)}$$

where m_f is the mass of fluid displaced by the particle. Stokes' law is regained if $\lambda = 1$, or when the density of the fluid is negligible in comparison to that of the particle.

The reason for Stokes' law is that the moving particle sets up a shear flow, or vortex, around itself. Viscous dissipation in this vortex removes energy from the particle, which causes the drag. Equation (2.57) contains two additional effects. One, described by λ, is an effective mass of the particle due to the mass of fluid which is pulled along with it. The last term occurs because the vortex is not fixed but is related dynamically to the motion of the particle.

Since Equation (2.57) depends on the entire past history of the particle, it cannot be solved subject to an initial condition on the particle at a fixed time. (The initial value problem requires specification of the state of the fluid as well as the particle.) To obtain a determined problem, we will suppose the particle (and fluid) to be at rest until $t = 0$ when the particle is suddenly given a velocity $\mathbf{u}_0$. For this case Equation (2.57) becomes

$$\frac{d\mathbf{u}}{dt} = -\lambda\xi\mathbf{u} - \nu\sqrt{\xi/\pi t}\,\mathbf{u}_0 - \nu\sqrt{\xi/\pi}\int_{0}^{t} d\tau(t-\tau)^{-1/2}\frac{d\mathbf{u}(\tau)}{d\tau}.$$

The second term occurs because of the discontinuity at $t - 0$.

It follows from Equation (2.55) that the velocity autocorrelation satisfies the same equation, but with initial value $\chi(0) = 1$:

$$\frac{d\chi}{dt} = -\lambda\xi\chi - \nu\sqrt{\xi/\pi t} - \nu\sqrt{\xi/\pi}\int_{0}^{t} d\tau(t-\tau)^{-1/2}\frac{d\chi(\tau)}{d\tau}. \tag{2.58}$$

To solve this equation let $\tilde{\chi}$ denote the Laplace transform

$$\tilde{\chi} = \int_{0}^{\infty} dt\, e^{-st}\,\chi(t).$$

The transform of $1/\sqrt{t}$ is $\sqrt{\pi/\mathrm{s}}$ and after transforming Equation (2.58) we find

$$\tilde{\chi} = (\mathrm{s} + \lambda\xi + \nu\sqrt{\xi \mathrm{s}})^{-1}.$$

Let a, b be the roots to the equation $x^2 - \nu\sqrt{\xi}x + \lambda\xi = 0$ (so a, b have positive real parts). Inversion of the transform yields

$$\chi(t) = [\sqrt{\pi}(a-b)]^{-1}\int_0^\infty d\tau(\tau + t)^{-1/2}[a^2\exp(-a^2\tau) - b^2\exp(-b^2\tau)]. \quad (2.59)$$

Equation (2.59) shows significant departures from exponential form, particularly for small and for large times. For small t one finds

$$\chi(t) = 1 - 2\nu\sqrt{\xi t/\pi} + \ldots$$

which shows that $\chi(t)$ has an infinite slope at $t = 0$, as can be seen directly from Equation (2.58). (This anomaly occurs because of the unphysical nature of the initial condition we have used.) For large t the expansion is

$$\chi(t) = (\pi\nu/2\lambda^2)(\pi\xi t)^{-3/2} + \ldots$$

Thus $\chi(t)$ decays more slowly than an exponential; the particle remembers its initial velocity longer than is the case when Stokes' law applies. This slow decay is usually referred to as the persistent correlation, or the long-time tail.

Because of the nonexponential form of (2.59), the Chapman-Kolmogorov equation is not satisfied; the particle's "memory" of its initial velocity prevents the process from being Markovian. This is true independently of the Gaussian assumption. The random force cannot be characterized as white noise; indeed the memory effect produces correlations between the random force and the velocity.

The coefficient of diffusion is obtained from Equation (2.56) as $D = \kappa T/(m\lambda\xi)$. Except for the effective-mass correction, this is the same as Einstein's formula. The mean-square displacement as a function of time can be obtained by using Equation (2.16), but the resulting formula is rather long and will not be written out in full. Its asymptotic form for long times is found to be

$$\langle q^2(t)\rangle = 6Dt[1 - (2\nu/\lambda)(\pi\xi t)^{-1/2} + \ldots].$$

Einstein's result is approached more slowly than in the Ornstein-Furth formula (2.10), so much so that there is no longer a sharp distinction between an ageing period and a hydrodynamic stage.

Higher moments of the displacement, such as that occurring in Equation (2.54) for D', also have long-time tails. As already mentioned, the net effect is that the limit given for D' does not exist, so a different kind of expansion is necessary for the equation governing diffusion.

Notes on Chapter 2

The early history of Brownian motion is described in Perrin [1914] and Chapter VII of Millikan [1935]. Einstein's papers on Brownian motion are collected in Furth [1956]. For an account of Smoluchowski's contributions, see his biography by Teske [1977]. Wax [1954] contains reprints of a number of important papers. Some recent general references are Fox [1978], Wyllie [1980], van Kampen [1975, 1981], and Risken [1984]. For applications, see Hidy and Brock [1970].

If the particle density is high, there is a hydrodynamic interaction because the flow field around one particle is perturbed by the presence of another. For recent references on Brownian motion with interactions (hydrodynamic or otherwise), see Bedeaux, Kapral, and Mazur [1977], Felderhof [1978], Herczynski and Pienkowska [1980], van den Broeck, Lostak, and Lekkerkerker [1981], and Mazur and van Saarloos [1982]. Fluctuations are modified by the presence of a hydrodynamic flow; for a recent reference see Ronis [1984a].

A subject not discussed here is rotational Brownian motion. For recent work on this subject, see Hubbard [1981], Reichl [1981, 1982], and Masters and Keyes [1985].

2.3 The statistical properties of the random force are discussed in Uhlenbeck and Ornstein [1930]. In current work the ad hoc introduction of the drag law and Langevin's equation is avoided by using methods based on the microscopic equations of motion. For example, see Mori [1965a], Hynes, Kapral, and Weinberg [1975, 1979], van Beijeren and Dorfman [1980a, b], and Cukier, Kapral, Lebenhaft, and Mehaffey [1980].

2.4 See Wang and Uhlenbeck [1945].

2.5 The eigenvalue problem for Ω is solved in Wang Chang and Uhlenbeck [1970]. The fundamental solution (2.41) to the Fokker-Planck equation is obtained in Chandrasekhar [1943]. See also McLennan [1966], Titulaer [1978], and Kuscer and Illner [1980]. For solutions to the Fokker-Planck equation subject to boundary conditions, see Harris [1980] and Burschka and Titulaer [1981a, 1981b].

2.8 See Chandrasekhar [1943].

2.9 See Kramers [1940], Visscher [1976], Matkowsky, Schuss, and Tier [1984], and Dekker [1986]. The Brownian motion of coupled oscillators is worked out in Wang and Uhlenbeck [1945].

2.10 For the expansion (2.52) or (2.53), see McLennan [1973], Keyes and Oppenheim [1973], Alder and Alley [1978, 1981], and Alley and Alder [1979]. For the failure of the expansion, see Ernst and Dorfman [1972, 1975], Dufty [1972], de Schepper, van Beijeren, and Ernst [1974], Dufty and McLennan [1974], de Schepper and Ernst [1979], Ernst and van Beijeren [1981b], and the reviews by Dorfman [1981] and Keyes and Masters [1985].

2.11 See Kubo [1966]. The diffusion constant in the form (2.56) was first obtained by Taylor [1920] for turbulent diffusion, when the particle velocity changes because of the fluctuating fluid velocity. Kirkwood [1946] obtained a formula for the friction constant in Brownian motion as a time integral of the correlation function for the random force.

2.12 Long-time tails were discovered by Alder and Wainwright [1969, 1970], Alder, Gass, and Wainwright [1971], in computer experiments on systems of identical particles. The case of Brownian motion was treated by Zwanzig and Bixon [1970, 1975], Widom [1971], Chow and Hermans [1972], and Michaels and Oppenheim [1975a]. For a derivation of the Basset-Boussinesq equation, see Lamb [1932] or Landau and Lifshitz [1959]. There is an extensive literature on long-time tails in more general situations than Brownian motion, some of which is cited in Chapter 12. For a recent review, see Keyes and Masters [1985].

For experimental results on long-time tails, see Kim and Matta [1973], Boon and Bouiller [1976], Fedele and Kim [1980], and Paul and Pusey [1981].

3

THE BOLTZMANN EQUATION

3.1. DERIVATION OF THE BOLTZMANN EQUATION

Kinetic theory for gases at low density is based on an integro-differential equation usually called the Boltzmann equation; the contributions of Maxwell to its development were so substantial that sometimes it is also, with justification, called the Maxwell-Boltzmann equation. In this chapter the Boltzmann equation will be derived and some of its general properties discussed. The construction of a solution to describe hydrodynamics will be carried out in the following chapter.

Consider a low-density gas of identical particles. Internal structure of the particles will not be taken into account until a later chapter, so here the discussion applies primarily to a monatomic gas. The particles will be assumed to interact by central forces with a finite range; the long-ranged Coulomb interaction between charged particles cannot be handled by the Boltzmann equation. (Technically the condition of finite range requires the force to decrease more rapidly than r^{-2} when the separation r between particles becomes large.)

Let $f(\mathbf{x}, \mathbf{v}, t)$ be the distribution function, defined so that $f(\mathbf{x}, \mathbf{v}, t)d^3x\, d^3v$ is the number of particles with positions in d^3x and velocities in d^3v at time t. The Boltzmann equation determines the rate of change of f with time, which has two sources: streaming in and out of d^3x because the particles are moving, and "streaming" in and out of d^3v because interactions change the velocities.

If interactions are neglected, a differential equation for f can be obtained as follows. Consider those atoms which at time t are located at $\mathbf{x},\mathbf{v}$. Then τ seconds later they will be at $\mathbf{x}',\mathbf{v}'$ where

$$\mathbf{x}' = \mathbf{x} + \mathbf{v}\tau, \ \mathbf{v}' = \mathbf{v}. \tag{3.1}$$

The number of atoms at the two times is the same, so

$$f(\mathbf{x}',\mathbf{v}',t + \tau)\, d^3x'\, d^3v' = f(\mathbf{x},\mathbf{v},t)d^3x\, d^3v.$$

It is readily confirmed that the transformation (3.1) has unit Jacobian so the volume elements on both sides can be canceled. This gives

$$f(\mathbf{x} + \mathbf{v}\tau,\mathbf{v},t + \tau) = f(\mathbf{x},\mathbf{v},t) \tag{3.2}$$

from which follows

$$\frac{\partial f}{\partial t} + \mathbf{v}\cdot\nabla f = 0. \tag{3.3}$$

This is called the collisionless Boltzmann equation, or the Knudsen equation. It applies when conditions are such that an atom makes more collisions with the walls of the container, or other physical body, than with other atoms.

Interactions give an additional contribution to $\partial f/\partial t$ which will be denoted by J. Thus when interactions are included, Equation (3.3) becomes

$$\frac{\partial f}{\partial t} + \mathbf{v}\cdot\nabla f = J.$$

In the next section, J will be worked out as an integral operator acting on f, and the resulting equation is Boltzmann's equation.

3.2. THE COLLISION OPERATOR

At low density any particle spends most of its time in free motion, occasionally coming close enough to another particle for the two to interact. The interaction can be described as an elastic collision whose duration is short in comparison to the time between successive collisions. Only binary collisions need be considered, since it is unlikely that three or more particles will simultaneously be close enough to interact.

Suppose two particles are initially far apart so the interaction between them can be neglected but are moving towards one another. Let their initial velocities be $\mathbf{v},\mathbf{v}_1$, and let $\boldsymbol{g} = \mathbf{v} - \mathbf{v}_1$ denote their relative velocity. The particle with velocity $\mathbf{v}_1$ will be referred to as the target particle, and the other as the incident particle. As the particles move within their range of interaction, they will be deflected and will then move apart again. Let $\mathbf{v}',\mathbf{v}_1'$ denote the final velocities after they have moved beyond the range of interaction, and let $\mathbf{g}' = \mathbf{v}' - \mathbf{v}_1'$ denote the final relative velocity. It is convenient to view the collision in a frame of reference fixed on the target particle; in this coordinate system the incident particle approaches with velocity $\mathbf{g}$ and then moves away with velocity $\mathbf{g}'$. As will be shown, the magnitudes of $\mathbf{g}'$ and $\mathbf{g}$ are equal as a consequence of conservation of energy and momentum. To determine the direction of $\mathbf{g}'$, it is necessary to specify two parameters. These can be taken to be the polar angles θ, φ of $\mathbf{g}'$ in a coordinate system with z-axis along $\mathbf{g}$, which are called the scattering angles.

The scattering angles are defined with reference to the initial relative velocity $\mathbf{g}$. It is also possible to specify the collision by parameters which are referred to a coordinate system fixed in space. Since the magnitudes of $\mathbf{g}'$ and $\mathbf{g}$ are the same, we may write

$$\mathbf{g}' = \mathbf{g} - 2\mathbf{e}\mathbf{e}\cdot\mathbf{g} \tag{3.4}$$

where $\mathbf{e}$ is a unit vector. It follows from this equation that

$$\mathbf{e}\cdot\mathbf{g}' = -\mathbf{e}\cdot\mathbf{g},\ \mathbf{g} = \mathbf{g}' - 2\mathbf{e}\mathbf{e}\cdot\mathbf{g}'.$$

The angles θ, φ range over the sphere, $0 \leq \theta \leq \pi$, $0 \leq \varphi \leq 2\pi$, but $\mathbf{e}$ is restricted to a hemisphere since $\mathbf{e}$ and $-\mathbf{e}$ determine the same collision. The scattering angles and $\mathbf{e}$ are related as follows. Let χ denote the angle between $\mathbf{e}$ and $\mathbf{g}$. Then we have

$$\mathbf{g}'\cdot\mathbf{g} = g^2\cos\theta = g^2(1 - 2\cos^2\chi)$$

so $\theta = \pi - 2\chi$. If $d\Omega = \sin\theta\, d\theta\, d\varphi$ is an element of solid angle about $\mathbf{g}'$, then

$$g\, d\Omega = 4|\mathbf{e}\cdot\mathbf{g}|d^2e$$

where $d^2e = \sin\chi d\chi d\varphi$ is an element of solid angle about $\mathbf{e}$.

To define the scattering cross section, let a uniform beam of particles be incident on a target. The incident flux is ng where n is the density of incident particles and g their speed relative to the target. The number of particles scattered per second into the element of solid angle $d\Omega$ about $\mathbf{g}'$ is proportional to the incident flux and to $d\Omega$, with the constant of proportionality being the cross section σ:

$$\text{Number scattered into } d\Omega \text{ per second} = n\,g\,\sigma\, d\Omega.$$

If there are many target particles scattering independently, and if multiple scattering can be ignored, the total scattering rate is obtained by multiplying by the number of target particles.

The cross section and element of solid angle depend on the frame of reference but the product $\sigma\, d\Omega$ is unchanged by a transformation to a different frame. In the following σ and $d\Omega$ will be defined with respect to the center-of-mass frame, as this choice is most convenient. For particles without internal structure, the center-of-mass cross section is determined by the initial and final relative velocities $\mathbf{g}$, $\mathbf{g}'$ and this functional dependence will be indicated by writing $\sigma(\mathbf{g} \to \mathbf{g}')$. If the particles interact by a central force, σ is actually a function only of g and the scattering angle θ and so may also be written as $\sigma(g, \theta)$. The total cross section is $\sigma_T = \int \sigma \, d\Omega$ and when the distinction is necessary σ is called the differential cross section. (In some cases the total cross section does not exist because of a singularity in the differential cross section at $\theta = 0$.)

Now consider particles with velocity in the range d^3v, incident on target particles with velocity in the range d^3v_1. The flux of incident particles is $g f(\mathbf{v}) d^3v$, and the number of target particles in a volume element d^3x is $f(\mathbf{v}_1) d^3v_1 d^3x$. Hence the rate at which particles are scattered from $\mathbf{v}$ into $d\Omega$ is

$$f(\mathbf{v}) f(\mathbf{v}_1) g \, \sigma \, d\Omega \, d^3v \, d^3v_1 \, d^3x = f(\mathbf{v}) f(\mathbf{v}_1) 4|\mathbf{e} \cdot \mathbf{g}| \sigma \, d^2e \, d^3v \, d^3v_1 d^3x.$$

This expression can also be viewed as the rate at which particles with velocity $\mathbf{v}_1$ are scattered by particles with velocity $\mathbf{v}$. On integrating over d^2e and d^3v_1 we obtain the net rate at which particles in $d^3v \, d^3x$ undergo collisions. Since they will generally finish with a velocity different from $\mathbf{v}$, this is a rate of loss from $d^3v \, d^3x$:

$$\text{rate of loss} = \left[\int f(\mathbf{v}) f(\mathbf{v}_1) 4|\mathbf{g} \cdot \mathbf{e}| \sigma d^2e \, d^3v_1 \right] d^3v \, d^3x. \tag{3.5}$$

The rate of gain can be calculated in a similar way. Consider an inverse collision such that the velocities after the collision are $\mathbf{v}, \mathbf{v}_1$. The initial velocities which result in these final velocities (for a given $\mathbf{e}$) will now be denoted by $\mathbf{v}', \mathbf{v}_1'$. The collision rate is

$$f(\mathbf{v}') f(\mathbf{v}_1') 4|\mathbf{g}' \cdot \mathbf{e}| \sigma \, d^2e \, d^3v_1' d^3v' d^3x.$$

In this expression we change variables from $\mathbf{v}', \mathbf{v}_1'$ to $\mathbf{v}, \mathbf{v}_1$. As will be shown, the Jacobian determinant is equal to 1 so $d^3v' d^3v_1' = d^3v \, d^3v_1$. Furthermore, we have $|\mathbf{g} \cdot \mathbf{e}| = |\mathbf{g}' \cdot \mathbf{e}|$ so the rate of gain in a form corresponding to (3.5) is

$$\text{rate of gain} = \left[\int f(\mathbf{v}') f(\mathbf{v}_1') 4|\mathbf{g} \cdot \mathbf{e}| \sigma \, d^2e \, d^3v_1 \right] d^3v \, d^3x. \tag{3.6}$$

Here $\mathbf{v}', \mathbf{v}_1'$ are to be considered as functions of $\mathbf{v}, \mathbf{v}_1$, and $\mathbf{e}$.

The difference between (3.6) and (3.5) is the net rate of change of the number of particles in $d^3x \, d^3v$ due to collisions. Dividing by $d^3x \, d^3v$ we obtain the collision rate

$$J[f] = \int d^3v_1 d^2e \, 4|\mathbf{g} \cdot \mathbf{e}| \sigma [f' f_1' - f f_1] \tag{3.7}$$

where we have used the abbreviations

$$f' = f(\mathbf{v}'),\ f_1' = f(\mathbf{v}_1'),\ f_1 = f(\mathbf{v}_1).$$

The dependence on position and time has not been indicated, but all f's are evaluated at the same $\mathbf{x},t$. We recall that $\mathbf{e}$ is restricted to a hemisphere, which can be taken to be the one with $\mathbf{e}\cdot g \geq 0$. The collision operator has the equivalent form

$$J[f] = \int d^3v_1\, d\Omega\, g\,\sigma\,[f'f_1' - ff_1]. \tag{3.8}$$

The form containing $\mathbf{e}$ is preferable for the derivation, since the scattering angles would be referred to different axes for the inverse and original collisions.

A collision can be specified by the impact parameter b rather than the scattering angle θ; the impact parameter is the distance from the target to a line tangent to the initial trajectory of the incident particle. The rate at which incident particles cross an element of area $bdb\,d\varphi$ is the same as the rate at which they scatter into the corresponding element of solid angle $d\Omega$ so $g\,\sigma\, d\Omega = g\, bdb\, d\varphi$ which gives $\sigma \sin\theta d\theta = bdb$. For a given force law, θ can be worked out as a function of g and b and the cross section determined by

$$\sigma(g,\theta) = \frac{1}{\sin\theta}\, b\, \frac{db}{d\theta}. \tag{3.9}$$

In terms of the impact parameter, the collision operator is

$$J[f] = \int d^3v_1\, g\, bdb\, d\varphi[f'f_1' - ff_1].$$

An external force can be included in the following way. During a short time interval τ the velocity is changed by a force $\mathbf{F}$ to $\mathbf{v}' = \mathbf{v} + m^{-1}\mathbf{F}\tau$ so Equation (3.2) is changed to

$$f(\mathbf{x} + \mathbf{v}\tau,\ \mathbf{v} + m^{-1}\mathbf{F}\tau,\ t + \tau) = f(\mathbf{x},\mathbf{v},t).$$

Proceeding in the same way as before, we obtain

$$\frac{\partial f}{\partial t} + \mathbf{v}\cdot\nabla f + \nabla_{\mathbf{v}}\cdot m^{-1}\mathbf{F}f = J[f]. \tag{3.10}$$

As previously noted, the Boltzmann equation applies only when the density is low; the dimensionless parameter $\nu = nD^3$, where n is the density and D the force range, must be small. Under this condition, deviations from the ideal gas law are negligible. Generally speaking, the Boltzmann equation applies to a monatomic gas at pressures up to several atmospheres, and density-dependent corrections begin to become significant when the pressure exceeds about ten atmospheres.

In order for the force range, and so the parameter ν, to be well defined it is necessary that the force decrease sufficiently rapidly with increasing separation. A detailed argument shows that the decrease must be more rapid than r^{-2} where r is the distance between the particles, which is the case for neutral atoms.

The two distribution functions in the collision integral are evaluated at the same position and time. This gives an adequate description if f does not vary appreciably over distances of the order of the force range or over times of the order of the duration of a collision; the spatial extent of a collision and its duration in time must be small. In addition the two colliding particles have been represented by a product of one-particle distribution functions, which requires that statistical correlations between the two particles be negligible. These limitations are also met when ν is small.

3.3. MECHANICS OF A COLLISION

Let the velocities before and after a collision be denoted by $\mathbf{v}'$, $\mathbf{v}_1'$ and $\mathbf{v},\mathbf{v}_1$. These are limiting values in configurations such that the two particles are well separated so the potential energy is negligible. The energy and momentum are

$$E = \frac{1}{2}m(v^2 + v_1^2),\ \mathbf{P} = m(\mathbf{v} + \mathbf{v}_1)$$

and conservation of energy and momentum give

$$v'^2 + v_1'^2 = v^2 + v_1^2,\ \mathbf{v}' + \mathbf{v}_1' = \mathbf{v} + \mathbf{v}_1.$$

Let $\boldsymbol{g}$, $\boldsymbol{G}$ denote the relative and mass-center velocities:

$$\boldsymbol{g} = \mathbf{v} - \mathbf{v}_1,\ \boldsymbol{G} = \frac{1}{2}(\mathbf{v} + \mathbf{v}_1)$$

with similar definitions for $\boldsymbol{g}'$, $\boldsymbol{G}'$. The particle velocities are given in terms of $\boldsymbol{g}$, $\boldsymbol{G}$ by

$$\mathbf{v} = \boldsymbol{G} + \frac{1}{2}\boldsymbol{g},\ \mathbf{v}_1 = \boldsymbol{G} - \frac{1}{2}\boldsymbol{g}$$

and the energy is

$$E = m\left(G^2 + \frac{1}{4}g^2\right).$$

Conservation of momentum requires $\boldsymbol{G}' = \boldsymbol{G}$ and conservation of energy then gives $g' = g$.

Using Equation (3.4) we get

$$\mathbf{v}' = \mathbf{v} - \mathbf{ee}\cdot(\mathbf{v} - \mathbf{v}_1),\ \mathbf{v}_1' = \mathbf{v}_1 + \mathbf{ee}\cdot(\mathbf{v} - \mathbf{v}_1). \tag{3.11}$$

These are the functions of $\mathbf{v}$, $\mathbf{v}_1$, and $\mathbf{e}$ which occur as the arguments of f', f_1' in the collision operator.

The transformation from $\mathbf{v}$, $\mathbf{v}_1$ to $\boldsymbol{g}$, $\boldsymbol{G}$ is readily shown to have unit Jacobian,

$$\frac{\partial(\mathbf{v},\mathbf{v}_1)}{\partial(\boldsymbol{g},\boldsymbol{G})} = 1. \tag{3.12}$$

The same result also holds for the primed velocities. In addition it follows from Equation (3.4) that

$$\frac{\partial(\boldsymbol{G}',\boldsymbol{g}')}{\partial(\boldsymbol{G},\boldsymbol{g})} = \frac{\partial(\boldsymbol{g}')}{\partial(\boldsymbol{g})} = 1 \tag{3.13}$$

where the differentiations are taken with $\mathbf{e}$ held constant. Consequently, the transformation from $\mathbf{v}',\mathbf{v}_1'$ to $\mathbf{v},\mathbf{v}_1$ also has unit Jacobian, which fact was used in the derivation of the collision operator.

For some purposes it is convenient to use a form for the collision operator in which the velocities before and after a collision are treated as independent variables. This form is

$$J = \frac{1}{2}m\int d^3v_1\, d^3v'\, d^3v_1'[f'f_1' - ff_1]\sigma\,\delta(\boldsymbol{G}' - \boldsymbol{G})\delta(E - E'). \tag{3.14}$$

Conservation of energy and momentum is imposed by the delta functions, and the cross section is considered to be a function of the four initial and final velocities. To show the equivalence with the original form, we first change variables from $\mathbf{v}',\mathbf{v}_1'$ to $\boldsymbol{g}'$, $\boldsymbol{G}'$ and then carry out the integration over $\boldsymbol{G}'$. This yields

$$J = \frac{1}{2}m\int d^3v_1\, d^3\boldsymbol{g}'\ [f'f_1' - ff_1]\sigma\,\delta(E - E') \tag{3.15}$$

where $\mathbf{v}',\mathbf{v}_1'$ are restricted by conservation of momentum. The energy difference is $E - E' = \frac{1}{4}m(g^2 - g'^2)$ and so

$$\delta(E - E') = (2/mg)\delta(g - g').$$

For the integration over $\boldsymbol{g}'$ we use polar coordinates with z-axis along $\boldsymbol{g}$. The polar angles θ, φ are then the same as the scattering angles previously introduced. The volume element is $d^3g' = g'^2dg'\, d\Omega$ and the remaining delta function can be used to integrate over g'. The result is the form (3.8) for J.

3.4. A LEMMA ON THE COLLISION OPERATOR

We use the form (3.14) for J to prove a fundamental lemma which will be needed in the following. For two functions h, f of velocity, let $K[h, f]$ denote

$$K[h, f] = \int d^3v\, h J[f]. \tag{3.16}$$

Using Equation (3.14) for J we get

$$K[h, f] = \frac{1}{2}m\int d^3v\, d^3v_1\, d^3v'\, d^3v_1'\sigma\delta(\mathbf{G} - \mathbf{G}')\delta(E - E')h[f'f_1' - ff_1].$$

If the interparticle potential is invariant to the inversion $\mathbf{x} \to -\mathbf{x}$ (as it is for central forces) then the cross section satisfies

$$\sigma(\boldsymbol{g} \to \boldsymbol{g}') = \sigma(-\boldsymbol{g} \to -\boldsymbol{g}'). \tag{3.17}$$

Thus, the cross section is unchanged by an interchange of the two particles; interchanging them we obtain

$$K[h, f] = \frac{1}{2}m\int d^3v\, d^3v_1\, d^3v'\, d^3v_1'\sigma\delta(\mathbf{G} - \mathbf{G}')\delta(E - E')h_1[f'f_1' - ff_1].$$

It can be shown to follow from microscopic reversibility that (for particles without internal angular momentum) the cross section is invariant to an interchange of the initial and final velocities:

$$\sigma(\boldsymbol{g} \to \boldsymbol{g}') = \sigma(\boldsymbol{g}' \to \boldsymbol{g}).$$

This is known as the principle of detailed balancing. The change of variables between primed and unprimed variables gives two more forms for K:

$$\begin{aligned} K[h, f] &= -\frac{1}{2}m\int d^3v\, d^3v_1\, d^3v'\, d^3v_1'\sigma\,\delta(\mathbf{G} - \mathbf{G}')\delta(E - E')h'[f'f_1' - ff_1] \\ &= -\frac{1}{2}m\int d^3v\, d^3v_1\, d^3v'\, d^3v_1'\sigma\,\delta(\mathbf{G} - \mathbf{G}')\delta(E - E')h_1'[f'f_1' - ff_1] \end{aligned}$$

If the four forms are averaged, still another is obtained:

$$\begin{aligned} K[h, f] = \frac{1}{8}m\int d^3v\, d^3v_1\, d^3v'\, d^3v_1'\sigma\,\delta(\mathbf{G} - \mathbf{G}')\delta(E - E') \\ \times [h + h_1 - h' - h_1'][f'f_1' - ff_1]. \end{aligned}$$

In the original way of writing the collision operator, this is

$$\int d^3v\, h J[f] = \frac{1}{4}\int d^3v\, d^3v_1\, d\Omega\, g\sigma[h + h_1 - h' - h_1'][f'f_1' - ff_1] \tag{3.18}$$

which is the desired lemma.

3.5. CONSERVATION LAWS

Let ψ denote a function of velocity with the property that the sum of its values for the two particles is the same before and after a collision:

$$\psi(\mathbf{v}) + \psi(\mathbf{v}_1) = \psi(\mathbf{v}') + \psi(\mathbf{v}_1').$$

Such a quantity is referred to as a summational invariant. There are only five independent quantities having this property, namely the energy, the components of momentum, and any constant. Denote these by $\psi_\alpha, \alpha = 1,..,5$:

$$\psi_\alpha = 1, \frac{1}{2}mv^2, m\mathbf{v}. \tag{3.19}$$

It is easily seen that there can be no summational invariants which are independent of these. Conservation of energy and momentum imposes four conditions relating $\mathbf{v}', \mathbf{v}_1'$ to $\mathbf{v}, \mathbf{v}_1$, and specification of the scattering angles imposes two more. There is no freedom remaining for additional constraints.

It follows from the lemma (3.18) that if ψ is a summational invariant, then

$$\int d^3v\, \psi J = 0.$$

Hence, the Boltzmann equation gives

$$\frac{\partial}{\partial t}\int d^3v\, \psi f + \nabla \cdot \int d^3v\, \psi\, \mathbf{v} f = 0.$$

The conservation laws for particle number, energy, and momentum are obtained by setting ψ equal to the five functions (3.19) respectively.

Taking $\psi = 1$ we get the equation of continuity

$$\frac{\partial n}{\partial t} + \nabla \cdot \mathbf{j} = 0$$

where

$$n = \int d^3v f, \; \mathbf{j} = \int d^3v\, \mathbf{v} f.$$

This can also be written in terms of the mass density ρ and momentum density $\mathbf{g}$ as

$$\frac{\partial \rho}{\partial t} + \nabla \cdot \mathbf{g} = 0$$

where

$$\rho = mn, \mathbf{g} = m\mathbf{j} = \int d^3v\, m\, \mathbf{v} f.$$

Next we take $\psi = m\mathbf{v}$ to obtain the local conservation law for momentum

$$\frac{\partial g_i}{\partial t} + \frac{\partial t_{ij}}{\partial x_j} = 0 \tag{3.20}$$

where

$$t_{ij} = \int d^3v\, m\, v_i v_j f.$$

Finally we use $\psi = \frac{1}{2}mv^2$ which gives conservation of energy:

$$\frac{\partial \epsilon}{\partial t} + \nabla \cdot \mathbf{s} = 0. \tag{3.21}$$

Here the energy density and its flux are given by

$$\epsilon = \int d^3v \frac{1}{2} m v^2 f, \quad \mathbf{s} = \int d^3v \frac{1}{2} m v^2 \mathbf{v} f.$$

The same procedure can be followed for the Boltzmann equation (3.10) with an external force. The continuity equation has the same form but Equations (3.20) and (3.21) now contain source terms

$$\frac{\partial \mathfrak{g}}{\partial t} + \nabla \cdot \mathbf{t} = \mathfrak{F}, \quad \frac{\partial \epsilon}{\partial t} + \nabla \cdot \mathbf{s} = w.$$

Here $\mathfrak{F}$ is a body force given by

$$\mathfrak{F} = \int d^3v\, \mathbf{F} f$$

and w is the rate at which work is done,

$$w = \int d^3v\, \mathbf{v} \cdot \mathbf{F} f.$$

3.6. EQUILIBRIUM

In equilibrium f is independent of time and, in the absence of external forces, of position. Boltzmann's equation reduces to $J = 0$. To solve, multiply by $\ln f$ and integrate. Using the lemma (3.18) with $h = \ln f$ we get

$$\int d^3v \ln f J[f] = \frac{1}{4} \int d^3v\, d^3v_1\, d\Omega\, \mathfrak{g}\, \sigma [f'f_1' - ff_1] \ln (ff_1/f'f_1') = 0.$$

For positive x, y the function $(y - x)\ln(x/y)$ is never positive, so the integral can vanish only if its integrand vanishes. This gives the condition

$$f'f_1' = ff_1.$$

The quantity $\psi = \ln f$ satisfies

$$\psi' + \psi_1' = \psi + \psi_1$$

and so is a summational invariant. Hence, f must have the form

$$f = \exp[A + B\frac{1}{2}mv^2 + \mathbf{C}\cdot m\mathbf{v}]$$

where $A, B, \mathbf{C}$ are constants to be determined. The average density n, flow velocity $\mathbf{u}$, and average energy density ϵ are related to f by

$$n = \int d^3v f,\ n\mathbf{u} = \int d^3v\,\mathbf{v}f,\ \epsilon = \int d^3v \frac{1}{2}mv^2 f. \tag{3.22}$$

According to thermodynamics the average energy density of a gas is

$$\epsilon = \frac{3}{2}n\kappa T + \frac{1}{2}nmu^2.$$

The integrals (3.22) are readily evaluated and the constants $A, B, \mathbf{C}$ can then be determined in terms of n, T, and $\mathbf{u}$. The resulting form for the distribution function, which we denote by f_0, is

$$f_0 = n(m/2\pi\kappa T)^{3/2}\exp[-(m/2\kappa T)(\mathbf{v} - \mathbf{u})^2]. \tag{3.23}$$

The equilibrium distribution function was first obtained by Maxwell by a different method; the argument just given is due to Boltzmann. Generally we denote the Maxwellian without the density factor by ϕ, so $f_0 = n\phi$.

3.7. H-THEOREM

Consider a spatially uniform gas, with f depending on t but not on $\mathbf{x}$. The Boltzmann equation reduces to

$$\frac{\partial f}{\partial t} = J.$$

Let $\mathcal{H}$ denote the quantity

$$\mathcal{H} = \int d^3v f \ln f.$$

Its time derivative is

$$\frac{d\mathcal{H}}{dt} = \int d^3v(1 + \ln f)\frac{\partial f}{\partial t} = \int d^3v\,(1 + \ln f)J.$$

Using Equation (3.18) with $h = \ln f$ we get

$$\frac{d\mathcal{H}}{dt} = \frac{1}{4}\int d^3v\,d^3v_1\,d\Omega\,g\,\sigma\,(f'f_1' - ff_1)\ln(ff_1/f'f_1').$$

According to the argument of Section 6, the integrand is never positive and we obtain the inequality known as Boltzmann's H-theorem:

$$\frac{d\mathcal{H}}{dt} \leq 0. \tag{3.24}$$

Equality holds if and only if f has the equilibrium form (3.23).

It can be shown that $\mathcal{H}$ has a lower bound as follows. Consider a Maxwellian f_0 in which the parameters n, T and $\mathbf{u}$ are chosen so that both f_0 and the actual distribution function f give the same average density, momentum density, and energy density. This means that

$$\int d^3v\, \psi(f - f_0) = 0 \tag{3.25}$$

where ψ is any summational invariant. Let $\mathcal{H}_0$ be the value of the functional $\mathcal{H}$ at f_0:

$$\mathcal{H}_0 = \int d^3v\, f_0 \ln f_0.$$

It follows from the conservation laws that $n, \mathbf{u}$, and ϵ are constant in the spatially uniform case and so $\mathcal{H}_0$ is also a constant. Because of Equations (3.25) and the form of f_0 we have

$$\int d^3v\, f \ln f_0 = \int d^3v\, f_0 \ln f_0$$

and it follows that

$$\mathcal{H} - \mathcal{H}_0 = \int d^3v\, f \ln (f/f_0) = \int d^3v\, f_0\{1 + (f/f_0)[\ln (f/f_0) - 1]\}.$$

For non-negative x the quantity $1 - x + x \ln x$ is never negative and so $\mathcal{H}$ is bounded from below by $\mathcal{H}_0$:

$$\mathcal{H} \geq \mathcal{H}_0. \tag{3.26}$$

A bounded monotonic function must approach a limit. In the limit $d\mathcal{H}/dt = 0$ which implies that f has the Maxwellian form (3.23) and since n, T and $\mathbf{u}$ are constant the particular Maxwellian must be the one fixed by Equation (3.25). Thus, f approaches f_0 with the passage of time.

The H-theorem can be extended to include spatial variations as follows. Suppose the gas to be enclosed in a rigid, stationary container. Under the most general conditions, for example, if the walls have an uneven or time-dependent temperature, an approach to equilibrium is not expected. We therefore assume the interaction with the walls to be such that the gas is thermodynamically isolated. This is the case if the atoms are specularly reflected, which means that a component of velocity parallel to the wall is unchanged while the perpendicular component is reversed. Thus if $\mathbf{v}$ is the velocity just before colliding with a wall, then the velocity after the collision is $\mathbf{v}' = \mathbf{v} - 2\mathbf{n}\mathbf{n}\cdot\mathbf{v}$ where $\mathbf{n}$ is a unit vector perpendicular to the wall. The interaction with the wall can be described by the boundary condition $f(\mathbf{x},\mathbf{v}) = f(\mathbf{x},\mathbf{v}')$ where $\mathbf{x}$ is on the wall. We now show that this boundary condition leads to an H-theorem.

Define H by

$$H = \int d^3v\, d^3x\, f \ln f$$

where the integration on $\mathbf{x}$ extends over the interior of the container. The Boltzmann equation gives

$$\frac{dH}{dt} = \int d^3v\, d^3x (1 + \ln f)(J - \mathbf{v}\cdot\nabla f).$$

The last term can be converted to a surface integral,

$$\int d^3v\, d^3x (1 + \ln f)\mathbf{v}\cdot\nabla f = \int d^3v \int d\mathbf{A}\cdot\mathbf{v}(1 + \ln f) f.$$

Because of the boundary condition on f, $(1 + \ln f)f$ is an even function of the normal component of $\mathbf{v}$. The factor $d\mathbf{A}\cdot\mathbf{v}$ makes the integrand odd in the normal component so the integral vanishes. We are left with

$$\frac{dH}{dt} = \int d^3v\, d^3x (1 + \ln f) J$$

and it follows in the same way as for the spatially homogeneous case that

$$\frac{dH}{dt} \leq 0. \tag{3.27}$$

Because of the boundary condition, the total particle number, energy, and momentum are time independent, and their values determine an equilibrium state with Maxwellian f_0. We now obtain

$$\int d^3v\, d^3x\, f \ln f_0 = \int d^3v\, d^3x\, f_0 \ln f_0$$

and

$$H - H_0 = \int d^3v\, d^3x\, f \ln(f/f_0) = \int d^3v\, d^3x\, f_0\{1 + (f/f_0)[\ln(f/f_0) - 1]\}.$$

Thus H is bounded from below by H_0:

$$H \geq H_0. \tag{3.28}$$

Again we can conclude that H must approach a limit, and in the limit the distribution function must have the Maxwellian form (3.23). However, we have not shown that this Maxwellian must be the same as f_0 introduced in the preceding paragraph or that the limiting value for H is H_0; the condition $dH/dt = 0$ is satisfied by a local Maxwellian in which n, T and $\mathbf{u}$ are functions of position and time. It will be shown in Section 10 that in a container with fixed walls, the only locally Maxwellian solution to the Boltzmann equation is either a strict Maxwellian (with constant $n, T, \mathbf{u}$) or a uniform rotation if such is allowed by the shape of the container. Hence, the gas approaches either a uniform equilibrium state (for which the distribution function must be the same as f_0 defined in the preceding paragraph) or, if possible, a state of uniform rotation. In the latter case the angular velocity is determined by conservation of angular momentum.

The inequalities (3.27), (3.28) show that H is a Liapunov function for the Boltzmann equation; indeed it was historically the first Liapunov function.

3.8. H AND ENTROPY

Computation shows that the value of H in equilibrium is

$$H_0 = V[n \ln n - \frac{3}{2} n \ln (2\pi\kappa Te/m)]$$

where V is the volume. Except for an additive constant, this is related to the thermodynamic entropy of an ideal monatomic gas by

$$S_0 = -\kappa H_0.$$

For a nonequilibrium state, the thermodynamic entropy as defined in Chapter 1 can be related to the distribution function in the following way. Let f_L denote a distribution function with the form (3.23) where n, T and $\mathbf{u}$ are determined from f. Although f_L has the Maxwellian dependence on $\mathbf{v}$, it need not correspond to equilibrium since n, T and $\mathbf{u}$ may vary with position and time. To distinguish it from strict equilibrium, f_L is referred to as the local-equilibrium distribution function. Let H_L be the H-function corresponding to local equilibrium:

$$H_L = \int d^3x\, d^3v\, f_L \ln f_L.$$

Then the entropy is related to H_L by

$$S = -\kappa H_L.$$

An argument like that leading to the inequality (3.26) shows that $H \geq H_L$ where H is calculated with the actual distribution function f. This is equivalent to

$$S \geq -\kappa H.$$

Thus, the entropy has a monotonically increasing lower bound (which does not imply that the entropy must be monotonic).

3.9. LOCALLY-MAXWELLIAN SOLUTIONS

If f has the Maxwellian form as a function of velocity, but n, T and $\mathbf{u}$ vary with position and time, then Boltzmann's equation reduces to the collisionless equation (3.3). In this case Euler's equations are satisfied so there is no dissipation. The general form of a Maxwellian which satisfies the collisionless equation will be determined systematically, but first we consider three simple examples.

The first example is a gas in uniform rotation, with $\mathbf{u} = \boldsymbol{\omega} \times \mathbf{x}$ where $\boldsymbol{\omega}$ is the angular velocity. There is no time dependence, so the equation for f is $\mathbf{v} \cdot \nabla f = 0$. It is readily checked that $\mathbf{v} \cdot \nabla \mathbf{v} \cdot \mathbf{u} = 0$. Consequently,

$$f = A \exp[-(m/2\kappa T)(v^2 - 2\mathbf{v} \cdot \mathbf{u})],$$

where A is a constant, is a solution with the Maxwellian dependence on velocity. This can be put in the standard form (3.23) with

$$n(\mathbf{x}) = n(0) \exp(mu^2/2\kappa T).$$

The temperature is constant, but the density and pressure vary with position. If the gas is in a fixed container, uniform rotation is possible if the container has an axis of symmetry and the rotation axis coincides with the symmetry axis.

Another example is obtained from the observation that any function $f(\mathbf{x} - \mathbf{v}t)$ is a solution to (3.3). Suppose the flow velocity is $\mathbf{u} = \mathbf{x}/t$ which describes a uniform dilation. If $T = A/t^2, n = B/t^3$ where A, B are constants, then the Maxwellian takes the form

$$f = B(m/2\pi\kappa A)^{3/2} \exp[-(m/2\kappa A)(\mathbf{x} - \mathbf{v}t)^2]$$

which is a solution. This solution is possible because, as will be shown in the next chapter, the gas has no bulk viscosity; a nonvanishing bulk viscosity leads to dissipation in a dilating flow.

Similarly, it is seen immediately that the function

$$f = C \exp[\mathbf{b} \cdot (\mathbf{x} - \mathbf{v}t)] \exp(-mv^2/2\kappa T)$$

where C, $\mathbf{b}$ and the temperature are constants, is a solution. Putting it in the form (3.23), we get

$$\mathbf{u} = \mathbf{a}t, \; n(\mathbf{x},t) = n_0 \exp[-(m/\kappa T)\mathbf{a} \cdot (\mathbf{x} - \frac{1}{2}\mathbf{a}t^2)]$$

where $\mathbf{a} = -(\kappa T/m)\mathbf{b}$. This describes a uniformly accelerated motion which is driven by a pressure gradient.

The general form of a locally Maxwellian solution to the collisionless equation can be determined as follows. When (3.23) (with nonuniform n, T and $\mathbf{u}$) is substituted into the left-hand side of the Boltzmann equation, the result is f_0 times a sum of terms which are linear, quadratic, and cubic in the components of **v**, plus a term which is independent of **v**. The equation must hold identically in **v** so each of these terms vanishes separately. The term which is independent of **v** can be picked out by setting $\mathbf{v} = 0$ which yields

$$\frac{\partial}{\partial t} n(m/2\pi\kappa T)^{3/2} \exp(-mu^2/2\kappa T) = 0. \tag{A}$$

The cubic term can also be picked out by inspection and its vanishing shows that the temperature is spatially uniform,

$$\nabla T = 0. \tag{B}$$

The remaining two equations are

$$\nabla \ln [n(m/2\pi\kappa T)^{3/2} \exp(-mu^2/2\kappa T)] + \frac{\partial}{\partial t}(m\mathbf{u}/\kappa T) = 0, \tag{C}$$

$$\frac{1}{2}\left(\frac{\partial u_i}{\partial x_j} + \frac{\partial u_j}{\partial x_i}\right) - \delta_{ij}\frac{\partial \ln T}{\partial t} = 0. \tag{D}$$

To solve these equations let χ denote the quantity following the time derivative in (A), so χ is time independent. It then follows from (C) that

$$\mathbf{u} = (\kappa T/m)(-t\nabla \ln \chi + \mathbf{A})$$

where **A** is a time-independent constant of integration. Using this form for **u** and (B) in (D) we find first of all that $T^{-1}\partial \ln T/\partial t$ is a linear function of t which means that T has the form

$$T = [\alpha + \beta t + \gamma t^2]^{-1}$$

where α, β, γ are constants; in addition we get

$$\frac{\partial^2 \ln \chi}{\partial x_i \partial x_j} = (2m\gamma/\kappa)\delta_{ij}, \quad \frac{1}{2}\left(\frac{\partial A_i}{\partial x_j} + \frac{\partial A_i}{\partial x_j}\right) = -(m\beta/\kappa)\delta_{ij}.$$

These equations can be solved to give

$$\ln \chi = a + \mathbf{b}\cdot\mathbf{x} + (m\gamma/\kappa)x^2, \quad \mathbf{A} = \mathbf{B} - (m\beta/\kappa)\mathbf{x} + \omega \times \mathbf{x}$$

where $a, \mathbf{b}, \mathbf{B}$ and ω are constants. The final form for **u** is

$$\mathbf{u} = (\kappa T/m)[\mathbf{B} - (m\beta/\kappa)\mathbf{x} + \omega \times \mathbf{x} - t[\mathbf{b} + (2m\gamma/\kappa)\mathbf{x}]].$$

The distribution function is now determined in terms of the constants of integration. The general form is a combination of motion with constant velocity, uniform acceleration, uniform rotation, and uniform dilation. If the gas is enclosed in a container with fixed walls, only strict equilibrium and uniform rotation are possible, and the latter only if the container has an axis of symmetry.

In the presence of an external force $\mathbf{F}$, the collisionless Boltzmann equation is

$$\partial f/\partial t + \mathbf{v}\cdot\nabla f + \frac{1}{m}\nabla_{\mathbf{v}}\cdot\mathbf{F}f = 0.$$

If $\mathbf{F}$ is derivable from a potential, $\mathbf{F} = -\nabla U$, then a stationary solution is

$$f = n(m/2\pi\kappa T)^{-3/2}\exp[-(\tfrac{1}{2}mv^2 + U)/\kappa T].$$

In a uniform gravitational field, $\mathbf{F} = m\mathbf{g} =$ constant, the spatially homogeneous equation

$$\partial f/\partial t + \mathbf{g}\cdot\nabla_{\mathbf{v}}f = 0$$

is solved by any function $f(\mathbf{v} - \mathbf{g}t)$; in particular the Maxwellian

$$f = n(m/2\pi\kappa T)^{3/2}\exp[-(m/2\kappa T)(\mathbf{v} - \mathbf{g}t)^2]$$

describes a gas which is in equilibrium but falling like a rock.

3.10. COLLISION FREQUENCY

The rate at which an atom makes collisions can be obtained from the argument which led to the Boltzmann collision operator. If the expression (3.5) for the rate of loss is divided by $f(\mathbf{v})d^3v\,d^3x$ the result is the number of collisions made per second by a single atom. This collision rate for a gas in equilibrium will be denoted by $\nu(v)$:

$$\nu(v) = \int d^3v_1\,d\Omega\,g\,\sigma f_0(v_1) = \int d^3v_1\,g\,\sigma_T f_0(v_1).$$

To estimate the collision frequency, consider a gas of hard spheres. These undergo elastic collisions when their surfaces come in contact, or when their centers are separated by their diameter D. The total cross section is the area blocked out by a sphere of radius D, or

$$\sigma_T = \pi D^2.$$

In this case $\nu(v)$ is given by

$$\nu(v) = \sigma_T\int d^3v_1\,f_0(v_1)|\mathbf{v} - \mathbf{v}_1|.$$

Evaluation of the integral is straightforward and the result is

$$\nu(v) = \frac{1}{2}\nu_0\left\{e^{-\xi^2} + (2\xi + \xi^{-1})\int_0^{\xi}dt\,e^{-t^2}\right\}. \qquad (3.29)$$

Here ξ is a dimensionless velocity given by

$$\xi = \sqrt{m/2\kappa T}\,v$$

and

$$\nu_0 = 2n\sigma_T\sqrt{2\kappa T/\pi m}. \tag{3.30}$$

We note that $\nu_0 = \nu(0)$ so ν_0 is the collision frequency for atoms at rest. As v increases, $\nu(v)$ increases monotonically, eventually becoming linear in v:

$$\nu(v) \to \frac{1}{2}\sqrt{\pi}\,\nu_0\xi = n\,\sigma_T v, \text{ as } v \to \infty.$$

The inverse of $\nu(v)$ is the time between collisions for an atom with speed v. If this is multiplied by v, the product is the distance covered between collisions:

$$l(v) = v/\nu(v).$$

This ranges from 0 for the slowest atoms to a limiting value of

$$l = \lim_{v \to \infty} l(v) = 1/n\,\sigma_T.$$

This limiting value is essentially the same as the mean free path.

Atomic diameters are about 10^{-10} m. At standard conditions one finds a value for ν_0 of about 5×10^9 sec^{-1}. Thus, the slow atoms spend about 2×10^{-10} sec between collisions, and faster atoms correspondingly less. The mean free path is about 2×10^{-7} m or 20 microns.

3.11. MIXTURES

Suppose the gas to be a mixture of atoms of different kinds, and let f_s denote the distribution function for atoms of the s'th kind. The Boltzmann equation for f_s is

$$\frac{\partial f_s}{\partial t} + \mathbf{v}_s \cdot \nabla f_s = J_s$$

where J_s describes collisions of an s-type atom with other atoms of all kinds. Let the contribution due to collisions of an atom of kind s with an atom of kind r be denoted by J_{sr}. An argument similar to that of Section 2 shows that

$$J_{sr} = \int d^3v_r\, d\Omega\, g\, \sigma_{sr}[f_s' f_r' - f_s f_r] \tag{3.31}$$

where σ_{sr} is the cross section for a collision between atoms of types s and r. Collisions of an s-type atom with atoms of all kinds are taken into account by summing over r,

$$J_s = \sum_r J_{sr}.$$

The energy and momentum of a pair of particles are

$$E = \frac{1}{2}(m_r v_r^2 + m_s v_s^2),\ \mathbf{P} = m_r \mathbf{v}_r + m_s \mathbf{v}_s$$

where m_s is the mass of a particle of the s'th kind. The relative and mass-center velocities are defined by

$$\boldsymbol{g} = \mathbf{v}_r - \mathbf{v}_s,\ \boldsymbol{G} = \frac{1}{M_{rs}}(m_r \mathbf{v}_r + m_s \mathbf{v}_s)$$

where $M_{rs} = m_r + m_s$ is the total mass. In terms of $\boldsymbol{g}, \boldsymbol{G}$ the energy is given by

$$E = \frac{1}{2}(M_{rs} G^2 + \mu_{rs} g^2)$$

where $\mu_{rs} = m_r m_s / M_{rs}$. Again it follows from conservation of energy and momentum that the magnitude of the relative velocity is the same before and after a collision, so a collision can be specified by the unit vector $\mathbf{e}$ as in Equation (3.4). Equation (3.11) is replaced by

$$\mathbf{v}_r' = \mathbf{v}_r - 2\frac{m_s}{M_{rs}}\mathbf{ee}\cdot\boldsymbol{g},\ \mathbf{v}_s' = \mathbf{v}_s + 2\frac{m_r}{M_{rs}}\mathbf{ee}\cdot\boldsymbol{g}. \tag{3.32}$$

The Jacobians (3.12) and (3.13) are again equal to 1, as is the Jacobian for the transformation between initial and final velocities.

In analogy to Equation (3.16) we define the quantity

$$K = \sum_s \int d^3v_s h_s J_s.$$

Detailed balance takes the form

$$\sigma_{sr}(\boldsymbol{g}\rightarrow\boldsymbol{g}') = \sigma_{sr}(\boldsymbol{g}'\rightarrow\boldsymbol{g})$$

while Equation (3.17) becomes

$$\sigma_{sr}(\boldsymbol{g}\rightarrow\boldsymbol{g}') = \sigma_{sr}(-\boldsymbol{g}\rightarrow\boldsymbol{g}').$$

Manipulations similar to those used in Section 3 show that

$$K = \frac{1}{4}\sum_{r,s}\int d^3v_r d^3v_s d\Omega\, g\, \sigma[h_s + h_r - h_s' - h_r'][f_s' f_r' - f_s f_r].$$

The conservation laws can be obtained by a procedure similar to that used previously. The continuity equation for particles of type s is

$$\partial n_s/\partial t + \nabla\cdot\mathbf{j}_s = 0$$

where

$$n_s = \int d^3v_s f_s, \; \mathbf{j}_s = \int d^3v_s \mathbf{v}_s f_s.$$

The conservation laws for energy and momentum have the same form as before, with the densities and fluxes given by

$$\epsilon = \sum_s \int d^3v_s \frac{1}{2} m_s v_s^2 f_s, \; \mathbf{s} = \sum_s \int d^3v_s \frac{1}{2} m_s v_s^2 \mathbf{v}_s f_s,$$

$$\mathbf{g} = \sum_s \int d^3v_s m_s \mathbf{v}_s f_s, \; t_{ij} = \sum_s \int d^3v_s m_s v_{si} v_{sj} f_s.$$

In equilibrium each species has a Maxwellian distribution

$$f_{s0} = n_s (m_s/2\pi\kappa T)^{3/2} \exp[-m_s(\mathbf{v}_s - \mathbf{u})^2/2\kappa T].$$

The discussion of the H-theorem is similar to that given previously. For example, in the spatially homogeneous case the H-function is defined by

$$\mathcal{H} = \sum_s \int d^3v_s f_s \ln f_s.$$

This satisfies the inequalities (3.24) and (3.26) so the approach to equilibrium is established as before.

3.12. LORENTZ GAS

The Lorentz gas is a binary mixture with one kind of particle being much heavier than the other, and with the density of light particles being so small that collisions between them can be neglected. Let f_l, f_h denote the distribution functions for the light and heavy particles. Since the light particles collide only with the heavy ones, their Boltzmann equation is

$$\frac{\partial f_l}{\partial t} + \mathbf{v}_l \cdot \nabla f_l = J_{lh}$$

where J_{lh} is given by a formula like (3.31). This simplifies if the recoil of the heavy particles is neglected; then $\mathbf{v}_h' = \mathbf{v}_h$ which gives

$$J_{lh} = \int d^3v_h \, d\Omega \, g \, \sigma_{lh} [f_l' - f_l] f_h.$$

The distribution function for heavy particles is not significantly affected by collisions with the light particles so J_{lh} may be neglected in comparison to J_{hh}. Hence, we get

$$\frac{\partial f_h}{\partial t} + \mathbf{v}_h \cdot \nabla f_h = J_{hh}. \tag{3.33}$$

There is further simplification if the heavy particles have a fixed equilibrium distribution which we denote by f_0. There is then no equation corresponding to (3.33) but only one for the light particles. Denoting their distribution function now by f, we get

$$\frac{\partial f}{\partial t} + \mathbf{v} \cdot \nabla f = \int d^3 v_h \, d\Omega \, g \, \sigma [f' - f] f_0. \tag{3.34}$$

Equations similar to this are used for the "gas" of conduction electrons in a metal, and for neutrons diffusing through a moderator of heavy particles.

A different extreme occurs when the heavy particles are so dilute that collisions between them can be neglected. The motion of a heavy particle through the gas of light particles resembles Brownian motion and, as shown in the next section, the collision operator has the Fokker-Planck form.

Following a slightly different line of argument, we consider a mixture of two kinds of particles with one being so dilute that collisions between them can be neglected while the other has a fixed equilibrium distribution f_0. The same conditions were previously introduced, but now we do not require that one particle be much heavier than the other. Thus, we have a dilute gas interacting with scattering centers. The collisions between the scattering centers maintain their equilibrium distribution f_0, while the latter is not affected by collisions between the gas particles and the scattering centers. The Boltzmann equation for this case is similar to (3.34) except that the recoil cannot be neglected for either species:

$$\frac{\partial f}{\partial t} + \mathbf{v} \cdot \nabla f = \int d^3 v_1 \, d\Omega \, g \, \sigma [f' f'_{01} - f f_{01}].$$

In the case of self-diffusion, the "gas" particles and the "scattering centers" are mechanically identical but the former are distinguished in some manner.

3.13. FOKKER-PLANCK EQUATION

A particle undergoing Brownian motion has only small deflections when colliding with molecules of the surrounding fluid. If the scattering is small except near the forward direction, then the Boltzmann equation describes a similar situation and should take on a form like that of the Fokker-Planck equation. This will be shown for the case of a binary mixture with one kind of particle being much heavier than the other and with the density of heavy particles being so low that collisions between them can be neglected.

Let $M, \mathbf{V}, F$ denote the mass, velocity, and distribution function for the heavy particles, with those for light particles being $m, \mathbf{v}, f$. The Boltzmann equation for the heavy particles is

$$\frac{\partial F}{\partial t} + \mathbf{V} \cdot \nabla F = J$$

where the collision operator is given by

$$J = \int d^3v \, d\Omega \, g \, \sigma [F(\mathbf{V}')f(\mathbf{v}') - F(\mathbf{V})f(\mathbf{v})].$$

According to Equation (3.32), the change in velocity of a heavy particle on collision is

$$\delta\mathbf{V} = \mathbf{V} - \mathbf{V}' = 2(m/M_T)\mathbf{ee} \cdot \mathbf{g}$$

where $\mathbf{g} = \mathbf{V} - \mathbf{v}$ and $M_T = M + m$. We write the corresponding equation for the light particles as

$$\mathbf{v}' = \mathbf{v}_0 - \delta\mathbf{V}$$

where $\mathbf{v}_0 = \mathbf{v} + 2\mathbf{ee} \cdot \mathbf{g}$ is the limiting value of $\mathbf{v}'$ for the case that the mass ratio m/M vanishes.

We wish to expand the collision operator in powers of the small quantity $\delta\mathbf{V}$. The leading term is

$$J\Big|_{\delta V = 0} = F\int d^3v \, d\Omega \, g \, \sigma [f(\mathbf{v}_0) - f(\mathbf{v})].$$

This expression is shown to vanish by changing variables in the first term from $\mathbf{v}$ to $\mathbf{v}_0$. (This is the same as the earlier transformation from velocities after a collision to those before, but now it is somewhat simpler since $\mathbf{V}' = \mathbf{V}$ for the heavy particles.) The term linear in $\delta\mathbf{V}$ is

$$-\int d^3v \, d\Omega \, g \, \sigma \, \delta\mathbf{V} \cdot \{F(\mathbf{V})\nabla_{\mathbf{v}_0} f(\mathbf{v}_0) + f(\mathbf{v}_0)\nabla_{\mathbf{V}} F(\mathbf{V})\}. \qquad (3.35)$$

Again we change variables from $\mathbf{v}$ to $\mathbf{v}_0$, which changes the sign of $\delta\mathbf{V}$. The result is

$$\int d^3v \, d\Omega \, g \, \sigma \, \delta\mathbf{V} \cdot \{F(\mathbf{V})\nabla_{\mathbf{v}} f(\mathbf{v}) + f(\mathbf{v})\nabla_{\mathbf{V}} F(\mathbf{V})\}.$$

We next integrate by parts on $\mathbf{v}$ in the first term. Since $\delta\mathbf{V}$, g and σ_T depend on $\mathbf{v}$ only in the combination $\mathbf{v} - \mathbf{V}$, the derivative $\nabla_{\mathbf{v}}$ can be replaced by $-\nabla_{\mathbf{V}}$. The linear contribution (3.35) is found to reduce to

$$-\nabla_{\mathbf{V}} F\int d^3v \, d\Omega \, g \, \sigma \, \delta\mathbf{V} \, f(\mathbf{v}).$$

The quadratic contribution to J is

$$\frac{1}{2}\int d^3v \, d\Omega \, g\sigma \, \delta V_i \delta V_j \left\{ f(\mathbf{v}_0) \frac{\partial^2 F(V)}{\partial V_i \partial V_j} + F(\mathbf{V}) \frac{\partial^2 f(\mathbf{v}_0)}{\partial v_{0i} \partial v_{0j}} + 2\frac{\partial F(\mathbf{V})}{\partial V_i} \frac{\partial f(\mathbf{v}_0)}{\partial v_{0j}} \right\}.$$

We follow the same procedure: change variables from $\mathbf{v}$ to $\mathbf{v}_0$, integrate by parts, then convert derivatives with respect to $\mathbf{v}$ to those with respect to $\mathbf{V}$

(with a change in sign). The foregoing expression then reduces to

$$\frac{\partial^2}{\partial V_i \partial V_j} F(\mathbf{V}) \int d^3v \, d\Omega \, g\sigma \, \delta V_i \delta V_j f(\mathbf{v}).$$

Collecting terms together, we find the collision operator through second order in $\mathbf{V}$ to have the Fokker-Planck form (2.26):

$$J = \frac{\partial}{\partial V_i} \left\{ -a_i F + \frac{1}{2} \frac{\partial}{\partial V_j} b_{ij} F \right\}.$$

Here a_i and b_{ij} are given by

$$a_i = \int d^3v \, d\Omega \, g \sigma \, \delta V_i f(\mathbf{v}), \quad b_{ij} = \int d^3v \, d\Omega \, g \sigma \, \delta V_i \delta V_j f(\mathbf{v}). \tag{3.36}$$

This can be reduced further if the cross section and the distribution function for light particles are known.

Physically there is a significant difference between Brownian motion and a gas described by the Boltzmann equation. The latter case is characterized by impulsive collisions which are widely separated in time, while a B-particle experiences frequent soft collisions. Hence, the coefficients (3.36) should not be compared directly with the friction constant of the theory of Brownian motion.

Notes on Chapter 3

General references on the Boltzmann equation are Chapman and Cowling [1970], Hirschfelder, Curtiss, and Bird [1954], Grad [1958], Waldmann [1958], Cercignani [1969], Kogan [1969], and Ferziger and Kaper [1972]. A useful overview is provided by Uhlenbeck and Ford [1963]. Brush [1972] gives an interesting historical account with reprints of basic papers by Hilbert, Enskog, and Chapman.

For mathematical discussions of existence and uniqueness of solutions to the Boltzmann equation, and properties of the solution, see Carleman [1933, 1957], Wild [1951], Morgenstern [1954, 1955], Grad [1963a, b], Filippov and Il'inskaya [1966], Arkeryd [1972a, 1972b, 1982], Glikson [1972], Ukai [1976], Simons [1978], Kaniel and Shinbrot [1978], Cercignani, Greenberg, and Zweifel [1979], Spohn [1979], Bobylev [1981], Cercignani [1983], and Greenberg, Polewczak, and Zweifel [1983].

The collision operator in the theory of neutron transport has some analytical similarities with the gas-kinetic operator. For example, see Duderstadt and Martin [1979], and Kaper, Lekkerkerker, and Hejtmanek [1982].

3.3 The expression (3.14) for the collision operator is due to Waldmann [1958].

3.7 As a general measure of disorder, Boltzmann's H occurs in a variety of contexts. For example, see Jaynes [1963, 1979], and Mead [1982].

3.8 Attempts to identify $-\kappa H$ with the thermodynamic entropy away from equilibrium date back to Boltzmann. However, it was shown by Prigogine [1949] and Grad [1949] that while the two quantities are equal in the approximation corresponding to the Navier-Stokes equations, they differ in higher approximations.

3.9 The general solution to the collisionless Boltzmann equation is worked out in Chapman and Cowling [1970] and Grad [1949].

3.13 See Green [1951a], Wang Chang and Uhlenbeck [1970], and Chapman and Cowling [1970].

4

NORMAL SOLUTION TO THE BOLTZMANN EQUATION

4.1. INTRODUCTION

In the previous chapter the local conservation laws were derived from the Boltzmann equation. To complete the hydrodynamic equations, it is necessary to relate the fluxes to spatial derivatives of the thermodynamic variables, that is, to derive the constitutive equations. This is accomplished by the construction of a particular solution, called the normal solution, to the Boltzmann equation. The normal solution is a functional of the thermodynamic variables with the basic property that the functional dependence is independent of position and time; the distribution function depends on position and time only through its dependence on the thermodynamic variables. For a simple gas the thermodynamic variables are n, T and $\mathbf{u}$, and this property of the normal solution can be stated as $f(\mathbf{x}, \mathbf{v}, t) = f[n(\mathbf{x}, t), T(\mathbf{x}, t), \mathbf{u}(\mathbf{x}, t), \mathbf{v}]$. Clearly if f has this form, then not only the fluxes but the average of any function of the velocity is determined by the thermodynamic variables.

Because of its functional nature, the normal solution is actually a class of solutions (but not the most general class). A specific solution is known only when the thermodynamic variables are determined as functions of position and time. The hydrodynamic equations are obtained in the process of working out the normal solution, and the thermodynamic variables can then be determined by solving the hydrodynamic equations subject to given initial and boundary values.

As a particular solution (or class of solutions), the normal solution cannot be forced to fit arbitrary initial or boundary conditions on the distribution function. Near a physical boundary the distribution function will not in general have the normal form. If the values of the hydrodynamic variables in the interior of the gas are extrapolated to a boundary on the basis of the hydrodynamic equations, the extrapolated values will differ from the actual values. This difference is referred to as slip, and the region over which slip takes place is called the Knudsen layer. We will be concerned only with the normal solution itself and will not investigate the structure of the Knudsen layer or of the ageing period following an assigned initial condition.

In practice the construction of the normal solution is based on an expansion developed by Chapman and Enskog which is ordered according to the degree of the spatial derivatives of the thermodynamic variables. The expansion of the distribution function leads to a corresponding expansion of the hydrodynamic equations, which in lowest order are the Euler equations, in next order the Navier-Stokes equations, then the Burnett equations, and so forth.

The locally Maxwellian solutions obtained in Chapter 3 are normal solutions which have a particularly simple form because of the special nature of the solutions to the hydrodynamic equations. They nevertheless illustrate the point that the dependence on position and time occurs only through the thermodynamic variables.

4.2. CHAPMAN-ENSKOG EXPANSION

Let f_L denote the local-equilibrium distribution function introduced in Chapter 3:

$$f_L = n(m/2\pi\kappa T)^{3/2}\exp[-(m/2\kappa T)(\mathbf{v} - \mathbf{u})^2].$$

Here $n, \mathbf{u}$ and T are the density, flow velocity, and temperature determined from the actual distribution function f,

$$n = \int d^3v f,\ n\mathbf{u} = \int d^3v\mathbf{v}f,\ \frac{3}{2}n\kappa T + \frac{1}{2}nmu^2 = \int d^3v\frac{1}{2}mv^2 f.$$

It follows from the form of f_L that these equations still hold if f is replaced by f_L so we have

$$\int d^3v\psi(f - f_L) = 0 \tag{4.1}$$

where ψ is any summational invariant.

Let f be written in the form

$$f = f_L(1 + \Phi) \tag{4.2}$$

where Φ will be determined by the requirement that f be a solution to the Boltzmann equation. If n, T and $\mathbf{u}$ are constants, then f_L reduces to the equilibrium distribution function, and f is a solution with $\Phi = 0$. For the nonequilibrium case, Φ is expanded as

$$\Phi = \Phi^{(1)} + \Phi^{(2)} + \ldots \tag{4.3}$$

where $\Phi^{(1)}$ is linear in the first spatial derivatives of the thermodynamic variables, $\Phi^{(2)}$ is quadratic in the first derivatives and linear in second derivatives, and so on. The derivatives are multiplied by coefficients which are functions of n, T and $\mathbf{u}$. Because of Equation (4.1), the Φ's satisfy

$$\int d^3v \psi \Phi f_L = \int d^3v \psi \Phi^{(1)} f_L = \int d^3v \psi \Phi^{(2)} f_L = \ldots = 0. \tag{4.4}$$

In working out the Chapman-Enskog expansion one encounters time derivatives of the thermodynamic variables because of the time derivative in the Boltzmann equation. To obtain an expansion which involves only spatial derivatives, the time derivatives must be eliminated by using the hydrodynamic equations. These involve the fluxes $\mathbf{s}$ and $\mathbf{t}$ which in turn depend on the unknown distribution function f. One proceeds as follows: the lowest approximations for $\mathbf{s}$ and $\mathbf{t}$ are obtained from f_L and are used to determine $\Phi^{(1)}$. Once $\Phi^{(1)}$ is known, the next approximation for $\mathbf{s}$ and $\mathbf{t}$ can be calculated and then used in the evaluation of $\Phi^{(2)}$. Continuing, $\mathbf{s}$ and $\mathbf{t}$ at any stage are available to proceed to the next stage.

Let $\mathbf{s}_L, \mathbf{t}_L$ denotes the fluxes obtained from f_L. A straightforward integration gives

$$\begin{aligned} \mathbf{s}_L &= \int d^3v \frac{1}{2} mv^2 \mathbf{v} f_L = \left(h + \frac{1}{2}\rho u^2\right)\mathbf{u}, \\ t_{Lij} &= \int d^3v m v_i v_j f_L = p\delta_{ij} + \rho u_i u_j. \end{aligned} \tag{4.5}$$

Here $\rho = nm$ is the mass density, $h = \frac{5}{2} n\kappa T$ is the enthalpy density, and $p = n\kappa T$ is the pressure. These are just the fluxes for an ideal fluid (specialized to the case of an ideal gas) and if substituted into the conservation laws, they yield the Euler equations.

The irreversible fluxes denoted by $\mathbf{t}^*$ and $\mathbf{s}^*$ in Chapter 1 are given by

$$\begin{aligned} t^*_{ij} &= t_{ij} - t_{Lij} = \int d^3 v m v_i v_j \Phi f_L, \\ s^*_i &= s_i - s_{Li} - t^*_{ij} u_j = \int d^3v \frac{1}{2} mv^2 v_i \Phi f_L - t^*_{ij} u_j. \end{aligned} \tag{4.6}$$

A more convenient form is obtained by changing variables from $\mathbf{v}$ to $\mathbf{c} = \mathbf{v} - \mathbf{u}$. When the condition (4.4) is taken into account, the result is found to be

$$t^*_{ij} = \int d^3c m c_i c_j \Phi f_L, \quad s^*_i = \int d^3c \frac{1}{2} mc^2 c_i \Phi f_L. \tag{4.7}$$

(The change of variables $\mathbf{v} \to \mathbf{c}$ in a summational invariant yields a summational invariant, so (4.4) can be expressed in terms of either $\mathbf{v}$ or $\mathbf{c}$.) As is seen from its form, f_L depends on $\mathbf{v}$ only through its dependence on $\mathbf{c}$, and it turns out that the same is true of Φ. Hence, the integration removes any dependence on $\mathbf{u}$ so the irreversible fluxes are independent of $\mathbf{u}$. This is expected from the discussion of Chapter 1 where the irreversible fluxes were defined in the local rest frame.

It follows from Equations (4.4) and (4.7) that $t_{ii}^* = 0$. This implies, among other things, that the bulk viscosity vanishes. This result is specific to a low-density monatomic gas. The bulk viscosity is actually negligible in monatomic gases except at high densities but if the gas is polyatomic, rotational and vibrational relaxation processes produce a bulk viscosity at low density.

4.3. FIRST APPROXIMATION

To determine $\Phi^{(1)}$, the approximation

$$f = f_L[1 + \Phi^{(1)}] \tag{4.8}$$

is substituted into the Boltzmann equation. In the collision operator terms which are quadratic in $\Phi^{(1)}$ are neglected at this point but must be retained in higher approximations. We obtain for J

$$J = \int d^3v_1\, d\Omega\, g\sigma \{f'_L f'_{L1}[1 + \Phi^{(1)\prime} + \Phi^{(1)\prime}_1] - f_L f_{L1}[1 + \Phi^{(1)} + \Phi^{(1)}{}_1]\}$$

The local Maxwellian satisfies the equation for equilibrium, $f'_L f'_{L1} = f_L f_{L1}$, so this expression reduces to

$$J = -n\, f_L\, I\Phi^{(1)}$$

where I is a linear operator defined by

$$nI\Phi = \int d^3v_1\, d\Omega\, g\sigma f_{L1}[\Phi + \Phi_1 - \Phi' - \Phi'_1].$$

The factor n has been included to make I independent of density.

When the approximation (4.8) is substituted into the left-hand side of the Boltzmann equation, contributions are obtained from f_L and from the product $f_L \Phi^{(1)}$. The contributions from the latter are either quadratic in first derivatives or linear in second derivatives and so are to be neglected at the present order. Thus, the Boltzmann equation reduces to

$$\left(\frac{\partial}{\partial t} + \mathbf{v}\cdot\nabla\right)\ln f_L = -nI\Phi^{(1)}.$$

Since f_L is known, this equation determines $\Phi^{(1)}$.

Using the explicit form for f_L we obtain for the left-hand side

$$\left(\frac{\partial}{\partial t} + \mathbf{v}\cdot\nabla\right)\ln f_L =$$
$$\frac{1}{n}\left(\frac{\partial}{\partial t} + \mathbf{v}\cdot\nabla\right)n + \left(\frac{m}{2\kappa T^2}(\mathbf{v} - \mathbf{u})^2 - \frac{3}{2\kappa T}\right)\left(\frac{\partial}{\partial t} + \mathbf{v}\cdot\nabla\right)T$$
$$- \frac{m}{\kappa T}(u_i - v_i)\left(\frac{\partial}{\partial t} + \mathbf{v}\cdot\nabla\right)u_i. \tag{4.9}$$

The Euler equations are

$$\frac{\partial n}{\partial t} = -\nabla\cdot n\mathbf{u}, \quad \frac{\partial T}{\partial t} = -\mathbf{u}\cdot\nabla T - \frac{2}{3}T\nabla\cdot\mathbf{u}, \quad \frac{\partial \mathbf{u}}{\partial t} = -\rho^{-1}\nabla p - \mathbf{u}\cdot\nabla\mathbf{u}$$

and when these are used in Equation (4.9) the result reduces to

$$\left(\frac{\partial}{\partial t} + \mathbf{v}\cdot\nabla\right)\ln f_L = \frac{1}{\kappa T^2}\left(\frac{1}{2}mc^2 - \frac{5}{2}\kappa T\right)\mathbf{c}\cdot\nabla T$$
$$+ \frac{\mathrm{m}}{\kappa \mathrm{T}}\left(\mathrm{c}_i\mathrm{c}_j - \frac{1}{3}\delta_{ij}c^2\right)\frac{\partial u_i}{\partial x_j}.$$

Thus, the equation satisfied by $\Phi^{(1)}$ is

$$I\Phi^{(1)} = -\frac{1}{n\kappa T^2}\left(\frac{1}{2}mc^2 - \frac{5}{2}\kappa T\right)\mathbf{c}\cdot\nabla T - \frac{m}{n\kappa T}\left(c_ic_j - \frac{1}{3}\delta_{ij}c^2\right)\frac{\partial u_i}{\partial x_j}. \tag{4.10}$$

The velocity $\mathbf{v}$ occurs on the right-hand side of the preceding equation, and in the operator I, only through $\mathbf{c}$, which confirms that $\Phi^{(1)}$ depends on $\mathbf{v}$ only through $\mathbf{c}$. In calculating the irreversible fluxes we may take $\mathbf{u} = 0$ and write Equation (4.7) as

$$t_{ij}^* = \int d^3v\, mv_iv_j\Phi^{(1)}f_0, \quad s_i^* = \int d^3v\,\frac{1}{2}mv^2v_i\Phi^{(1)}f_0 \tag{4.11}$$

where f_0 now denotes the Maxwellian with vanishing flow velocity. We must also set $\mathbf{u} = 0$ in Equation (4.10) so it becomes

$$I\Phi^{(1)} = -\frac{1}{n\kappa T^2}\mathbf{S}\cdot\nabla T - \frac{1}{n\kappa T}T_{ij}\frac{\partial u_i}{\partial x_j} \tag{4.12}$$

where $\mathbf{S}$ and $\mathbf{T}$ are the abbreviations

$$\mathbf{S} = \left(\frac{1}{2}mv^2 - \frac{5}{2}\kappa T\right)\mathbf{v}, \quad T_{ij} = m\left(v_iv_j - \frac{1}{3}\delta_{ij}v^2\right). \tag{4.13}$$

In addition I is now given by

$$nI\Phi = \int d^3v_1\, d\Omega\, g\sigma f_{01}[\Phi + \Phi_1 - \Phi' - \Phi_1']$$

and the condition (4.4) on $\Phi^{(1)}$ takes the form

$$\int d^3v \, \psi \Phi^{(1)} f_0 = 0. \tag{4.14}$$

The solution to Equation (4.12) can be written as

$$\Phi^{(1)} = -(1/n\kappa T^2)\mathcal{S}_i \frac{\partial T}{\partial x_i} - (1/n\kappa T)\mathcal{T}_{ij}\frac{\partial u_i}{\partial x_j} \tag{4.15}$$

where $\mathcal{S}_i$ and $\mathcal{T}_{ij}$ are solutions to

$$I\mathcal{S}_i = S_i, \; I\mathcal{T}_{ij} = T_{ij}. \tag{4.16}$$

In order to satisfy Equation (4.14), $\mathcal{S}$ and $\mathcal{T}$ will be restricted by

$$\int d^3v \, \psi \mathcal{S}_i f_0 = \int d^3v \, \psi \, \mathcal{T}_{ij} f_0 = 0. \tag{4.17}$$

The solution f as determined so far is

$$f = f_L\left\{1 - \frac{1}{n\kappa T^2}\mathcal{S}_i \frac{\partial T}{\partial x_i} - \frac{1}{n\kappa T}\mathcal{T}_{ij}\frac{\partial u_i}{\partial x_j} + \ldots\right\}. \tag{4.18}$$

(In the distribution function it is essential to retain the flow velocity, so here the argument of $\mathcal{S}$ and $\mathcal{T}$ is $\mathbf{v} - \mathbf{u}$.) Further terms in the series involve higher inverse powers of n; the expansion is one in powers of the mean free path ℓ which is inversely proportional to the density. The terms proportional to ℓ^2 will be worked out in Section 10 and an estimate can then be obtained of the error involved in the preceding approximation. Roughly speaking the relevant dimensionless parameter is the ratio $\ell/\mathcal{L}$ where $\mathcal{L}$ is a hydrodynamic length which characterizes the scale of variation of the thermodynamic variables; if this ratio is small, then the leading terms in the expansion provide a good approximation.

Paradoxically, the Boltzmann equation holds at low density but the Chapman-Enskog expansion is one in inverse powers of the density. Normally there is a wide range of density for which both the equation and the expansion are valid because the hydrodynamic length is usually so much larger than the size D of the atom. The condition for validity of the Boltzmann equation is $nD^3 << 1$ which is equivalent to $\ell >> D$, and this can be satisfied simultaneously with $\ell << \mathcal{L}$.

4.4. TRANSPORT COEFFICIENTS

It is convenient to use an abbreviated notation for certain integrals which occur repeatedly. For two functions h,k of velocity, let (h,k) denote

$$(h,k) = n^{-1}\int d^3v \, f_0 \bar{h} k. \tag{4.19}$$

At this point we are concerned only with real functions but have introduced the complex conjugate, denoted by the bar, for later use. The form (,) has the properties of a scalar product and so defines a Hilbert space of functions of velocity. The factor n^{-1} makes the scalar product independent of the density.

The conditions (4.17) can be abbreviated as

$$(\psi, \mathcal{S}_i) = (\psi, \mathcal{T}_{ij}) = 0 \tag{4.20}$$

while (4.11) takes the form

$$t_{ij}^* = n(mv_iv_j, \Phi^{(1)}), \quad s_i^* = n(\tfrac{1}{2}\, mv^2 v_i, \Phi^{(1)}).$$

In view of Equation (4.20), these equations can also be written as

$$t_{ij}^* = n(T_{ij}, \Phi^{(1)}), \quad s_i^* = n(S_i, \Phi^{(1)}).$$

Under rotations of $\mathbf{v}$, $\mathcal{S}_i$ transforms as a vector and $\mathcal{T}_{ij}$ as a second-rank tensor. (Actually this statement needs a proof, which follows.) It follows that the first term in Equation (4.15) for $\Phi^{(1)}$ does not contribute to $\mathbf{t}^*$, nor the second term to $\mathbf{s}^*$. We obtain

$$t_{ij}^* = -\frac{1}{\kappa T}(T_{ij}, \mathcal{T}_{kl})\frac{\partial u_k}{\partial x_l}, \quad s_i^* = -\frac{1}{\kappa T^2}(S_i, \mathcal{S}_j)\frac{\partial T}{\partial x_j}.$$

Using $(S_i, \mathcal{S}_j) = \frac{1}{3}\delta_{ij}(S_k, \mathcal{S}_k)$ we obtain Fourier's law, $\mathbf{s}^* = -\lambda\nabla T$, where the thermal conductivity is given by

$$\lambda = (1/3\kappa T^2)(S_i, \mathcal{S}_i). \tag{4.21}$$

This can also be written as, say,

$$\lambda = (1/\kappa T^2)(S_x, \mathcal{S}_x).$$

From its form, T_{ij} is a symmetric tensor with vanishing trace. Spherical symmetry then requires that

$$(T_{ij}, \mathcal{T}_{kl}) = A\left(\delta_{ik}\delta_{jl} + \delta_{il}\,\delta_{jk} - \frac{2}{3}\,\delta_{ij}\,\delta_{kl}\right) \tag{4.22}$$

where A is a constant whose value can be determined by setting $i = k$ and $j = l$:

$$A = \frac{1}{10}(T_{ij}, \mathcal{T}_{ij}).$$

The equation for t_{ij}^* reduces to Newton's viscosity law, $t_{ij}^* = -2\eta\tilde{D}_{ij}$, where

$$\tilde{D}_{ij} = \frac{1}{2}\left(\frac{\partial u_i}{\partial x_j} + \frac{\partial u_j}{\partial x_i} - \frac{2}{3}\,\delta_{ij}\nabla\cdot\mathbf{u}\right)$$

and the viscosity η is given by

$$\eta = (1/10\kappa T)(T_{ij}, \mathcal{T}_{ij}). \tag{4.23}$$

As noted in Section 2, $\mathbf{t}^*$ does not contain a bulk-viscosity term.

For later use we rewrite Equation (4.22) as

$$(T_{ij}, \mathcal{T}_{kl}) = \eta\kappa T\left(\delta_{ik}\delta_{jl} + \delta_{il}\delta_{jk} - \frac{2}{3}\delta_{ij}\delta_{kl}\right).$$

Equivalent equations for η can be obtained by setting, say, $i = k = x$, $j = l = y$:

$$\eta = (1/\kappa T)(T_{xy}, \mathcal{T}_{xy}).$$

It is evident from the preceding formulas that λ and η are independent of the density. This has come about because the Boltzmann equation is valid only at low density, so the formulas obtained are the limiting values of λ and η for $n \to 0$. This result may seem paradoxical, particularly if the density is actually zero. However, as the density is decreased the mean free path eventually becomes larger than the dimensions of the container. This is the Knudsen case, for which the transport of energy and momentum is dominated by the interaction with the walls.

4.5. PROPERTIES OF THE INTEGRAL EQUATIONS

The evaluation of the thermal conductivity and viscosity depends on the solution to the integral equations (4.16). These have the generic form

$$IX = \chi \tag{4.24}$$

where χ is known and X is to be determined. A method for constructing solutions will be discussed, but first we develop certain general properties of the equation.

Consider the scalar product

$$(k, Ih) = n^{-2}\int d^3v\, d^3v_1\, d\Omega\, g\sigma f_0 f_{01}\bar{k}[h + h_1 - h' - h_1']. \tag{4.25}$$

Following manipulations similar to those leading to Equation (3.18) we obtain

$$(k, Ih) = (1/4n^2)\int d^3v\, d^3v_1\, d\Omega\, g\sigma f_0 f_{01}[\bar{k} + \bar{k}_1 - \bar{k}' - \bar{k}_1'] \times [h + h_1 - h' - h_1']. \tag{4.26}$$

This shows that I is symmetric,

$$(k, Ih) = (Ik, h) \tag{4.27}$$

and non-negative,

$$(h, Ih) \geq 0. \tag{4.28}$$

Here equality holds if and only if h is a summational invariant. In addition, any summational invariant is an eigenvector of I with eigenvalue 0: $I\psi = 0$. Conversely, any eigenvector with zero eigenvalue must be a summational invariant since from $I\psi = 0$ we can conclude $(\psi, I\psi) = 0$.

Taking the scalar product of Equation (4.24) with a summational invariant, we obtain on the left-hand side $(\psi, IX) = (I\psi, X) = 0$. Hence existence of a solution requires that the right-hand side satisfy the solvability conditions $(\psi, \chi) = 0$. This is an example of the familiar situation in linear algebra that existence of a solution to an inhomogeneous equation requires the inhomogeneous term to be orthogonal to the solutions to the homogeneous equation. It is straightforward to verify that in the case at hand the conditions are in fact satisfied,

$$(\psi, S_i) = 0, \ (\psi, T_{ij}) = 0.$$

The sufficiency of these conditions for the existence of a solution will be discussed in Chapter 6.

The solution to Equation (4.24) is not unique. If X and X' are both solutions, it can be concluded that $I(X - X') = 0$ which implies that $X - X'$ is a summational invariant, not necessarily 0. The solution can be fixed uniquely by requiring that it be orthogonal to the summational invariants:

$$(\psi, X) = 0.$$

Thus, the conditions (4.20) determine $\mathcal{S}_i$ and $\mathcal{T}_{ij}$ uniquely.

It was assumed in working out the formulas (4.21) and (4.23) for the thermal conductivity and viscosity that $\mathcal{S}_i$ transforms as a vector and $\mathcal{T}_{ij}$ as a second-order tensor. Actually this is not required by the integral equations (4.16) alone since the solutions could contain summational invariants with the "wrong" transformation properties. However, once the solutions are fixed by the conditions (4.20) it is necessary that they transform in the manner assumed.

From the integral equations the thermal conductivity and viscosity can be written as

$$\lambda = (1/3\kappa T^2)(\mathcal{S}_i, I\mathcal{S}_i), \ \eta = (1/10\kappa T)(\mathcal{T}_{ij}, I\mathcal{T}_{ij}).$$

The inequality (4.28) shows that λ and η cannot be negative, in accordance with the second law of thermodynamics.

4.6. VARIATIONAL PRINCIPLE

Equation (4.24) has a variational principle which provides a useful method for constructing approximations to the solution. Let ξ denote a real function which is orthogonal to the summational invariants but is otherwise arbitrary. The inequality (4.28) shows that

$$(X - \xi, I(X - \xi)) \geq 0.$$

Expanding the scalar product, we find

$$(X, IX) \geq (\xi, I[2X - \xi]),$$

or

$$(X, \chi) \geq (\xi, 2\chi - I\xi). \tag{4.29}$$

Let $\Lambda[\xi]$ denote the functional

$$\Lambda[\xi] = (\xi, 2\chi - I\xi).$$

Equation (4.24) shows that $\Lambda[X] = (X, \chi)$ so the inequality (4.29) can be stated

$$\Lambda[X] \geq \Lambda[\xi]. \tag{4.30}$$

Furthermore, by retracing the argument we see that $\Lambda[\xi] = \Lambda[X]$ implies $\xi = X$.

The inequality (4.30) is utilized by choosing a trial function ξ which depends on one or more adjustable parameters and then varying the parameters to maximize $\Lambda[\xi]$. A choice which suggests itself for the trial function is the inhomogeneous term χ multiplied by a constant a. We obtain

$$\Lambda[a\chi] = 2a(\chi, \chi) - a^2(\chi, I\chi)$$

which is a maximum at $a = (\chi, \chi)/(\chi, I\chi)$. The value of the functional at this maximum is

$$\Lambda[a\chi] = (\chi, \chi)^2/(\chi, I\chi).$$

In applying the variational principle to evaluate the thermal conductivity, we take $\chi = S_i$, $X = \mathcal{S}_i$ and denote the trial function by ξ_i. Instead of the functional $\Lambda[\xi]$ defined previously, we use one which includes the numerical factor in Equation (4.21):

$$\lambda[\xi_i] = (1/3\kappa T^2)(\xi_i, 2S_i - I\xi_i).$$

If ξ_i is equal to the solution $\mathcal{S}_i$ then this functional is the same as the thermal conductivity λ while otherwise it is a lower bound:

$$\lambda \geq \lambda[\xi_i].$$

Proceeding as before we take $\xi_i = aS_i$ for the trial function. The functional is maximized at

$$a = (S_i, S_i)/(S_j, IS_j)$$

and its maximum is

$$\lambda_0 = (1/3\kappa T^2)(S_i, S_i)^2/(S_j, IS_j). \tag{4.31}$$

The evaluation of λ_0 requires only the calculation of the integrals involved, which follows.

The viscosity can be treated in a similar way. Let ξ_{ij} denote the trial function and $\eta[\xi_{ij}]$ the functional

$$\eta[\xi_{ij}] = (1/10\kappa T)(\xi_{ij}, 2T_{ij} - I\xi_{ij})$$

which is a lower bound for the viscosity, $\eta \geq \eta[\xi_{ij}]$. We use the trial function $\xi_{ij} = bT_{ij}$ and maximizing the lower bound we get

$$b = (1/10\kappa T)(T_{ij}, T_{ij})/(T_{kl}, IT_{kl}).$$

The corresponding approximation for the viscosity is

$$\eta_0 = (1/10\kappa T)(T_{ij}, T_{ij})^2/(T_{kl}, IT_{kl}). \tag{4.32}$$

The approximations λ_0 and η_0 are actually accurate to within a few percent and can be improved by using more elaborate trial functions.

The variational principle can be formulated as a principle of "maximum entropy production." Using the integral equations (4.16) and Equations (4.21) and (4.23) for the thermal conductivity and viscosity, we get

$$n^2\kappa(\Phi^{(1)}, I\Phi^{(1)}) = -T^{-2}\,\mathbf{s}^* \cdot \nabla T - T^{-1}\, t_{ij}^* \tilde{D}_{ij}.$$

The right-hand side is the entropy production σ as defined in Chapter 1:

$$\sigma = n^2\kappa(\Phi^{(1)}, I\Phi^{(1)}).$$

The functional

$$\sigma[\xi] = -n(\xi, 2[T^{-2}\mathbf{S} \cdot \nabla T + T^{-1}T_{ij}\tilde{D}_{ij}] + I\xi)$$

is a lower bound to the entropy production, $\sigma \geq \sigma[\xi]$, with equality holding if and only if $\xi = \Phi^{(1)}$.

4.7. EVALUATION OF THERMAL CONDUCTIVITY AND VISCOSITY

The numerators in Equations (4.31) and (4.32) can be evaluated by straightforward integration. We have

$$(S_i, S_i) = n^{-1}\int d^3v \left[\frac{1}{2}mv^2 - \frac{5}{2}\kappa T\right]^2 v^2 f_0, \quad (T_{ij}, T_{ij}) = n^{-1}\int d^3v \frac{2}{3}m^2v^4 f_0.$$

These integrals are cases of the general formula

$$n^{-1}\int d^3v \left(\frac{1}{2}mv^2\right)^p f_0 = \frac{3}{2}\cdot\frac{5}{2}\cdots\frac{2p+1}{2}(\kappa T)^p$$

which yields

$$(S_i, S_i) = 15\,(\kappa T)^3/2m, \quad (T_{ij}, T_{ij}) = 10\,(\kappa T)^2.$$

Using Equation (4.26) we write

$$(\chi, I\chi) = \frac{1}{4} n^{-2} \int d^3v \, d^3v_1 \, d\Omega \, g \sigma f_0 \, f_{01} [\chi]^2 \tag{4.33}$$

where $[\chi]$ is the abbreviation

$$[\chi] = \chi + \chi_1 - \chi' - \chi_1'.$$

This is most easily dealt with by transforming to the relative and mass-center velocities,

$$\boldsymbol{g} = \mathbf{v} - \mathbf{v}_1, \; \boldsymbol{G} = \frac{1}{2}(\mathbf{v} + \mathbf{v}_1).$$

Using conservation of energy and momentum, we find

$$[\mathbf{S}] = \frac{1}{2} m(\boldsymbol{G} \cdot \boldsymbol{g} \, \boldsymbol{g} - \boldsymbol{G} \cdot \boldsymbol{g}' \, \boldsymbol{g}'), \; [T_{ij}] = \frac{1}{2} m \, (g_i g_j - g_i' g_j'). \tag{4.34}$$

In the integral (4.33) we change variables from $\mathbf{v}, \mathbf{v}_1$ to $\boldsymbol{g}, \boldsymbol{G}$ and then integrate over $\boldsymbol{G}$. The product $f_0 \, f_{01}$ in terms of $\boldsymbol{g}, \boldsymbol{G}$ is

$$f_0 \, f_{01} = n^2 \phi_g \phi_G$$

where ϕ_g, ϕ_G are Maxwellians for the relative and mass-center velocities:

$$\phi_g = (m/4\pi\kappa T)^{3/2} \exp(-mg^2/4\kappa T), \; \phi_G = (m/\pi\kappa T)^{3/2} \exp(-mG^2/\kappa T).$$

The integrals over $\boldsymbol{G}$ needed are

$$\int d^3G \, \phi_G = 1, \; \int d^3G \, \phi_G G_i G_j = (\kappa T/2m)\delta_{ij}.$$

Equation (4.34) yields

$$[S_i][S_i] = (m^2/4)[(\boldsymbol{G} \cdot \boldsymbol{g})^2 g^2 + (\boldsymbol{G} \cdot \boldsymbol{g}')^2 \, g'^2 - 2\boldsymbol{G} \cdot \boldsymbol{g} \, \boldsymbol{G} \cdot \boldsymbol{g}' \boldsymbol{g} \cdot \boldsymbol{g}']$$

and averaging over $\boldsymbol{G}$ we get

$$\int d^3G \phi_G [S_i][S_i] = \frac{1}{4} m\kappa T [g^4 - (\boldsymbol{g} \cdot \boldsymbol{g}')^2] = \frac{1}{4} m\kappa T g^4 \sin^2\theta.$$

Similarly we find

$$\int d^3G \, \phi_G [T_{ij}][T_{ij}] = \frac{1}{2} m^2 \, g^4 \, \sin^2\theta.$$

On the basis of the preceding results we have at this point

$$(T_{ij}, IT_{ij}) = \frac{1}{8} m^2 \int d^3g \, d\Omega \, \phi_g g^5 \, \sigma \, \sin^2\theta$$

and the corresponding expression involving S_i can be obtained from

$$(S_i, IS_i) = (\kappa T/2m)(T_{ij}, IT_{ij}). \tag{4.35}$$

Integration over the directions of $\mathcal{g}$ gives a factor 4π. Then the angle φ in $d\Omega$ can be integrated which gives an additional factor 2π. The result is

$$\begin{aligned}(T_{ij}, IT_{ij}) = (\pi m)^2(m/4\pi\kappa T)^{3/2} \int_0^\infty d\mathcal{g}\,\mathcal{g}^7 \exp(-m\mathcal{g}^2/4\kappa T) \\ \times \int_0^\pi \sigma(\mathcal{g},\theta) \sin^3\theta\, d\theta.\end{aligned} \tag{4.36}$$

Further reduction of this expression will be considered in the next section.

Because of Equation (4.35) the thermal conductivity and viscosity in the present approximation are related by

$$\lambda_0 = \frac{15\kappa\eta_0}{4m} = \frac{5}{2}\, C_v\eta_0 \tag{4.37}$$

where $C_v = 3\kappa/2m$ is the specific heat. The ratio $\lambda/C_v\eta$ is called the Eucken number, so for a monatomic gas the Eucken number is approximately $\frac{5}{2}$. A related dimensionless number is the Prandtl number, defined by $Pr = \eta C_p/\lambda$ where C_p is the specific heat at constant pressure; the Prandtl number is approximately $\frac{2}{3}$.

4.8. CONNECTION WITH FORCE LAW

The preceding formulas for the thermal conductivity and viscosity can be reduced further if the cross section is known; the cross section can be calculated by methods of scattering theory if the interaction between particles is known. Since internal structure has not been taken into account, the interaction is treated by an effective force instead of the actual electromagnetic forces between the electrons and nuclei composing the particles. In practice the effective force is determined by fitting to experimental results, and once known it can be used to extrapolate through regions where experimental data for the transport coefficients are unavailable. There is an extensive literature on the comparison with experiment, including the derivation of better approximations than introduced in Section 6. Here only a brief discussion will be given for particles obeying classical mechanics; quantum effects however are significant for the lighter gases, hydrogen and helium.

We first consider hard spheres for which $\sigma = \frac{1}{4}D^2$ where D is the diameter. Evaluating the integral (4.36) we get

$$(T_{ij}, IT_{ij}) = 32D^2(\kappa T)^3 \sqrt{\pi/m\kappa T}$$

which gives for the viscosity

$$\eta_0 = \frac{5}{16} D^{-2} \sqrt{m\kappa T/\pi}.$$

The thermal conductivity is obtained from Equation (4.37),

$$\lambda_0 = \frac{75}{64} (\kappa/mD^2) \sqrt{m\kappa T/\pi}.$$

The parameters a, b in the trial functions used to evaluate the thermal conductivity and viscosity have the values

$$a = \frac{15}{32} D^{-2} \sqrt{m/\pi\kappa T} = \frac{15}{16} \sqrt{2}\, (n/\nu_0), \; b = \frac{2}{3} a$$

where ν_0 is the collision frequency (3.30).

For other interatomic forces the cross section can be obtained as follows. Using relative coordinates, let r, φ be two-dimensional polar coordinates of a point on the trajectory. Conservation of energy is expressed by

$$\frac{1}{4} m (\dot{r}^2 + r^2\dot{\varphi}^2) + U(r) = \frac{1}{4} m g^2$$

where $U(r)$ is the potential energy and the dot denotes the time derivative. Conservation of angular momentum gives

$$mr^2 \dot{\varphi} = mgb$$

where b is the impact parameter. Combining these equations, we obtain the differential equation for the trajectory,

$$\frac{dr}{d\varphi} = \dot{r}/\dot{\varphi} = \frac{r}{b}\left\{ r^2[1 - (4U/mg^2)] - b^2 \right\}^{1/2}. \tag{4.38}$$

Let r_0, φ_0 be coordinates of the apsis where $dr/d\varphi = 0$. Then r_0 is a root to the equation

$$r_0^2\{1 - [4U(r_0)/mg^2]\} - b^2 = 0.$$

For simplicity we assume the force to be repulsive so this equation has only a single root. To determine φ_0, integrate Equation (4.38):

$$\varphi_0 = \int_{r_0}^{\infty} \frac{d\varphi}{dr} dr = b \int_{r_0}^{\infty} dr\, r^{-1}\{r^2[1 - (4U/mg^2)] - b^2\}^{-1/2}.$$

The scattering angle is $\theta = \pi - 2\varphi_0$ so we obtain the relation between θ and b:

$$\theta = \pi - 2b \int_{r_0}^{\infty} dr\, r^{-1} \{r^2[1 - (4U/mg^2)] - b^2\}^{-1/2}. \tag{4.39}$$

The cross section can be obtained from Equation (3.9). If the impact parameter is used as the integration variable, the integral occurring in Equation (4.36) is written as

$$\int_0^\pi \sin^3\theta\, \sigma(g,\theta) d\theta = \int_0^D \sin^2\theta\, b db$$

where θ is given as a function of b by Equation (4.39) and D is the force-range. In principal the quantity (4.36) is now determined, but in most cases a numerical evaluation is necessary. Generally the transport coefficients are found to vary with temperature more rapidly than the $\sqrt{T}$-dependence for hard spheres.

Maxwellian molecules, for which $U(r)$ varies as r^{-4}, is a case of special interest. The collision operator simplifies because the product $g\,\sigma(g,\theta)$ is independent of g. It then turns out that the trial functions used in Section 6 are exact solutions to the integral equations, and there are other respects in which the analysis can be carried further. Maxwellian molecules will be treated separately in Chapter 5.

4.9 NORMAL SOLUTION FOR A MIXTURE

The normal solution for a mixture is constructed by a procedure similar to that used previously. The local-equilibrium distribution function for the s'th species is

$$f_{sL} = n_s(m_s/2\pi\kappa T)^{3/2} \exp[-(m_s/2\kappa T)(\mathbf{v}_s - \mathbf{u})^2]. \tag{4.40}$$

The contribution of local equilibrium to the energy and momentum fluxes has the same form as in Equation (4.5) while for the particle fluxes

$$\mathbf{j}_{sL} = n_s \mathbf{u}.$$

We write f_s in the form

$$f_s = f_{sL}(1 + \Phi_s)$$

and substitute into the Boltzmann equation, retaining only linear terms in Φ_s. The resulting equation, like (4.10), contains $\mathbf{v}_s$ only in the combination $\mathbf{c}_s = \mathbf{v}_s - \mathbf{u}$. Hence we may put $\mathbf{u} = 0$, later replacing $\mathbf{v}_s$ by $\mathbf{c}_s$ where necessary. The equation for Φ_s is found to be

$$-\sum_r \int d^3v_r\, d\Omega\, g\, \sigma_{sr} f_{r0}[\Phi_s + \Phi_r - \Phi_s' - \Phi_r']$$

$$= \frac{1}{\kappa T^2}\left[\frac{1}{2}m_s v_s^2 - \left(\frac{3}{2} + \frac{nm_s}{\rho}\right)\kappa T\right] \mathbf{v}_s \cdot \nabla T \tag{4.41}$$

$$+ \frac{1}{\kappa T} m_s \left[v_{si}v_{sj} - \frac{1}{3}\delta_{ij}\, v_s^2\right]\frac{\partial u_i}{\partial x_j} + \mathbf{v}_s \cdot \left[\frac{1}{n_s}\nabla n_s - \sum_r \frac{m_s}{\rho}\nabla n_r\right].$$

Here $n = \Sigma n_s$ is the total number density, $\rho = \Sigma m_s n_s$ is the mass density, and f_{s0} denotes the Maxwellian (4.40) with $\mathbf{u} = 0$. We are concerned only with the first approximation $\Phi_s^{(1)}$ so the superscript has been dropped to simplify the notation.

The equations can be formally simplified by taking Φ_s to be the components of a vector Φ in a space of S-dimensions where S is the number of species. If χ is another such vector with components χ_s then we define the scalar product (χ, Φ) by

$$(\chi, \Phi) = n^{-1} \sum_s \int d^3v_s \, f_{s0} \overline{\chi}_s \, \Phi_s. \tag{4.42}$$

The factor n^{-1} makes the scalar product independent of density but it still depends on the relative concentrations n_s/n. If the linearized collision operator is defined by its components according to

$$(I\Phi)_s = n^{-1} \sum_r \int d^3v_r \, d\Omega \, g \, \sigma_{sr} \, f_{r0}[\Phi_s + \Phi_r - \Phi'_s - \Phi'_r]$$

then the left-hand side of Equation (4.41) is $-n(I\Phi)_s$.

As in Chapter 1, the quantities $\nu_s = \mu_s/T$ will be used as the thermodynamic variables in place of the particle densities. For an ideal gas

$$d \ln n_s = d(\mu_s/\kappa T) + \frac{3}{2} d \ln \kappa T.$$

and Equation (4.41) can be reduced to

$$n \, I\Phi = -\frac{1}{\kappa T^2} S_i \frac{\partial T}{\partial x_i} - \frac{1}{\kappa T} T_{ij} \frac{\partial u_i}{\partial x_j} - \kappa^{-1} \sum_r J_{ri} \frac{\partial \nu_r}{\partial x_i}. \tag{4.43}$$

Here S_i, T_{ij} and J_{ri} are vectors in the S-component space with components given by

$$(S_i)_s = m_s \left[\frac{1}{2} v_s^2 - \frac{5}{2} \frac{n \kappa T}{\rho} \right] v_{si},$$

$$(T_{ij})_s = m_s (v_{si} \, v_{sj} - \frac{1}{3} \delta_{ij} v_s^2), \tag{4.44}$$

$$(J_{ri})_s = \left[\delta_{rs} - \frac{n_r m_s}{\rho} \right] v_{si}.$$

The solution to Equation (4.43) can be written as

$$\Phi = -\frac{1}{n\kappa T^2} \mathcal{S}_i \frac{\partial T}{\partial x_i} - \frac{1}{n\kappa T} \mathcal{T}_{ij} \frac{\partial u_i}{\partial x_j} - \frac{1}{n\kappa} \sum_r \mathcal{J}_{ri} \frac{\partial \nu_r}{\partial x_i} \tag{4.45}$$

where $\mathcal{S}_i$, $\mathcal{T}_{ij}$, and $\mathcal{J}_{ri}$ satisfy

$$I\mathcal{S}_i = S_i, \; I\mathcal{T}_{ij} = T_{ij}, \; I\mathcal{J}_{ri} = J_{ri}. \tag{4.46}$$

It can be verified that the conditions of solvability for the Equations (4.45) are satisfied, that is, the right-hand sides are orthogonal [with the scalar product (4.42)] to the summational invariants. To illustrate the calculations, we show that J_{ri} is orthogonal to the summational invariant for momentum. This requires that

$$\sum_s \int d^3v_s\, f_{s0}\, m_s v_{si}\, (J_{rj})_s = 0.$$

On introducing the formula (4.44) for $(J_{rj})_s$ we obtain for the left-hand side

$$\sum_s \int d^3v_s\, f_{s0}\, m_s v_{si} v_{sj} [\delta_{rs} - (n_r m_s/\rho)] = \delta_{ij} \kappa T \sum_s n_s [\delta_{rs} - (n_r m_s/\rho)] = 0.$$

The remaining conditions are verified in a similar way.

As before the solutions to the integral equations (4.46) are determined only up to the addition of a summational invariant but become unique when it is required that they be orthogonal to the summational invariants.

Having determined Φ, we can now obtain the constitutive equations by a straightforward calculation. The contributions of Φ to the fluxes are given by

$$t_{ij}^* = -2\eta \tilde{D}_{ij}, \quad s^* = -\lambda \nabla T - \sum_r L_r \nabla \nu_r, \quad \mathbf{j}_s^* = -L_s' \nabla T - \sum_r L_{sr} \nabla \nu_r.$$

The transport coefficients are given by

$$\eta = (1/10\kappa T)(T_{ij}, \mathcal{T}_{ij}), \quad \lambda = (1/3\kappa T^2)(S_i, \mathcal{S}_i),$$

$$L_r = (1/3\kappa)(S_i, \mathcal{J}_{ri}), \quad L_r' = (1/3\kappa T^2)(J_{ri}, \mathcal{S}_i), \quad L_{sr} = (1/3\kappa)(J_{si}, \mathcal{J}_{ri}).$$

Equations (4.27) and (4.28) can be shown to hold for the mixture, and these lead to restrictions on the transport coefficients. For example Equation (4.27) yields

$$(S_i, \mathcal{J}_{ri}) = (I\mathcal{S}_i, \mathcal{J}_{ri}) = (\mathcal{S}_i, I\mathcal{J}_{ri}) = (\mathcal{S}_i, J_{ri}).$$

This shows that $L_r = T^2 L_r'$ and in a similar way it can be shown that $L_{rs} = L_{sr}$. This demonstrates Onsager's theorem for the case at hand.

It follows from Equation (4.28) that the thermal conductivity and viscosity are non-negative. A more general relation can be obtained as follows. For arbitrary real numbers a_s, b we have

$$\left(\sum_s a_s J_{si} + bS_i, \sum_r a_r \mathcal{J}_{ri} + b\mathcal{S}_i\right) = \left(\sum_s a_s \mathcal{J}_{si} + b\mathcal{S}_i, I\left[\sum_r a_r \mathcal{J}_{ri} + b\mathcal{S}_i\right]\right) \geq 0.$$

This is equivalent to

$$\sum_{r,s} a_r a_s L_{rs} + 2b \sum_r a_r L_r + T^2 b^2 \lambda \geq 0$$

which implies that the entropy production is non-negative. More specific inequalities can be obtained from the one preceding such as

$$L_{ss} \geq 0, \quad |L_s| \leq T\sqrt{\lambda L_{ss}}, \quad |L_{rs}| \leq \sqrt{L_{ss}L_{rr}}.$$

These results simplify in the case of self-diffusion. As pointed out in Chapter 1, the coefficients L and L' vanish and this can be confirmed from the formulas as given. The coefficient of self-diffusion is given by

$$D_S = \frac{1}{3n}(v_i, \mathcal{J}_i)$$

where the scalar product is now the same as for a simple gas and $\mathcal{J}_i$ satisfies

$$I_S \mathcal{J}_i = v_i, \quad (\mathcal{J}_i, 1) = 0.$$

Here I_S is the linearized collision operator for self-diffusion which is given by

$$I_S \Phi = n^{-1} \int d^3v_1\, d\Omega\, g\, \sigma f_0(\mathbf{v}_1)[\Phi(\mathbf{v}) - \Phi(\mathbf{v}')]. \tag{4.47}$$

The use of a variational principle to obtain solutions to the integral equations (4.46) follows similar lines to those for the simple gas.

4.10. BURNETT APPROXIMATION

The Burnett approximation is next in the Chapman-Enskog expansion beyond the order already developed. In this order terms in the distribution function which are linear in second spatial derivatives, or quadratic in first derivatives, of the thermodynamic variables are retained. Time derivatives are then to be evaluated on the basis of the Navier-Stokes (rather than the Euler) equations, which give

$$\begin{aligned} \frac{\partial T}{\partial t} &= -\mathbf{u}\cdot\nabla T - \frac{2}{3}T\nabla\cdot\mathbf{u} + \frac{2}{3n}[\nabla\cdot(\lambda\nabla T) + 2\eta\, \tilde{D}_{ij}\tilde{D}_{ij}], \\ \frac{\partial u_i}{\partial t} &= -\mathbf{u}\cdot\nabla u_i - \frac{1}{\rho}\frac{\partial p}{\partial x_i} + \frac{2}{\rho}\frac{\partial}{\partial x_j}\eta\tilde{D}_{ij}. \end{aligned} \tag{4.48}$$

Of course $\partial n/\partial t = -\nabla\cdot n\mathbf{u}$ in all orders. Since λ and η are functions only of T, it follows that for example $\partial\lambda/\partial t = (d\lambda/dT)\partial T/\partial t$.

In the left-hand side of the Boltzmann equation we substitute $f = f_L(1 + \Phi^{(1)})$ where $\Phi^{(1)}$ is given by Equation (4.15), and then use (4.48) for the time derivatives. In the collision operator f is replaced by $f_L[1 + \Phi^{(1)} + \Phi^{(2)}]$ but the quadratic terms in $\Phi^{(2)}$ are neglected at this order. The result is a linear integral equation for $\Phi^{(2)}$ which contains terms depending on the now-known quantity $\Phi^{(1)}$. The calculation is tedious but straightforward and the result is

$$\begin{aligned}
n^2 T^2 I\Phi^{(2)} = & -(\kappa T/\rho)\mathcal{T}_{ij}(p_{;ij} - p^{-1}p_{;i}p_{;j}) + \left(\mathcal{S}_i v_j - \frac{2}{3}\lambda\epsilon'\delta_{ij}\right)T_{;ij} \\
& - T^{-2}\left[L(\mathcal{S}_s, \mathcal{S}_j) + \frac{2}{3}T^2\epsilon'\frac{d\lambda}{dT}\delta_{ij} - S_i\mathcal{S}_j + Tv_i\left(\mathcal{S}_j - T\frac{\partial\mathcal{S}_j}{\partial T}\right)\right]T_{;i}T_{;j} \\
& -p^{-1}\left[\mathcal{S}_i v_j + \frac{\kappa T}{m}\left(\mathcal{T}_{ij} - \frac{\partial\mathcal{S}_i}{\partial v_j}\right)\right]T_{;i}p_{;j} - n^{-1}\left[v_k\mathcal{T}_{ij} - \frac{\kappa T}{m}\frac{\partial\mathcal{T}_{ij}}{\partial x_k}\right]u_{i;j}p_{;k} \\
& + \left\{-T\frac{d\eta}{DT}\Delta_{ijkl}v_l + T^{-1}\mathcal{S}_k\mathcal{T}_{ij} + T^{-1}T_{ij}\mathcal{S}_k + \frac{5}{3}\delta_{ij}\mathcal{S}_k\right. \\
& \left. - \delta_{ik}\mathcal{S}_j + Tv_k\frac{\partial\mathcal{T}_{ij}}{\partial T} - \frac{2}{3}T\delta_{ij}\frac{\partial\mathcal{S}_k}{\partial T} - v_j\frac{\partial\mathcal{S}_k}{\partial v_i} - \frac{2}{T}L(\mathcal{T}_{ij}, \mathcal{S}_k)\right\}u_{i;j}T_{;k} \\
& - T\left[\eta\left(v_i\delta_{jk} + \frac{1}{3}v_k\delta_{ij}\right) + \frac{2}{3}\mathcal{S}_j\delta_{ik} - \mathcal{T}_{ij}v_k\right]u_{i;jk} \\
& + \left\{-\frac{2}{3}\eta\epsilon'\Delta_{ijkl} + T_{ij}\mathcal{T}_{kl} - \frac{2}{3}T^2\delta_{kl}\frac{\partial\mathcal{T}_{ij}}{\partial T} - Tv_j\frac{\partial\mathcal{T}_{kl}}{\partial v_i}\right. \\
& \left. + \frac{5}{3}T\mathcal{T}_{ij}\delta_{kl} - T\delta_{jk}\mathcal{T}_{il} - L(\mathcal{T}_{ij}, \mathcal{T}_{kl})\right\}u_{i;j}u_{k;l}.
\end{aligned} \tag{4.49}$$

Here subscripts have been used to indicate derivatives with respect to the coordinates, for example $T_{;ij} = \partial^2 T/\partial x_i \partial x_j$ and $u_{i;j} = \partial u_i/\partial x_j$. The quantities S_i, T_{ij}, $\mathcal{S}_i$ and $\mathcal{T}_{ij}$ are those defined previously in Equations (4.13) and (4.16). In addition we have introduced the abbreviations

$$\epsilon' = \frac{1}{2}mv^2 - \frac{3}{2}\kappa T, \quad \Delta_{ijkl} = \delta_{ik}\,\delta_{jl} + \delta_{il}\,\delta_{jk} - \frac{2}{3}\delta_{ij}\,\delta_{kl}$$

and L is a bilinear form derived from the collision operator,

$$L(A, B) = \frac{1}{2n}\int d^3v_1\, d\Omega\, g\, \sigma\, f_{L1}[AB_1 + A_1B - A'B_1' - A_1'B']. \tag{4.50}$$

Equation (4.49) has been written for $\mathbf{u} = 0$ but the general case can be recovered by replacing $\mathbf{v}$ by $\mathbf{v} - \mathbf{u}$. It is straightforward to verify that the right-hand side of (4.49) is orthogonal to the summational invariants and so the conditions for solvability of the integral equation are satisfied.

The contribution of $\Phi^{(2)}$ to the fluxes can now be determined. Actually it is not necessary to solve the integral equation since, for example, $(S_i, \Phi^{(2)}) = (I\mathcal{S}_i, \Phi^{(2)}) = (\mathcal{S}_i, I\Phi^{(2)})$ and $I\Phi^{(2)}$ is given directly by the integral equation.

We state the result of the calculation, including the Navier-Stokes contributions for completeness. The heat flux through the Burnett order is

$$s_i^* = -\lambda T_{;i} + \gamma_1 \nabla^2 u_i + \gamma_2 D_{;i} + \gamma_3 \tilde{D}_{ij} T_{;j} + \gamma_4 A_{ij} T_{;j} + \gamma_5 D T_{;i} + \gamma_6 \tilde{D}_{ij} p_{;j}. \tag{4.51}$$

Here $D = \nabla \cdot \mathbf{u}$, A_{ij} denotes the vorticity

$$A_{ij} = \frac{1}{2}\left(\frac{\partial u_i}{\partial x_j} - \frac{\partial u_j}{\partial x_i}\right)$$

and the γ's are higher-order transport coefficients which will be defined in the following. The corresponding equation for the momentum flux is

$$\begin{aligned} t_{ij}^* &= 2\eta \tilde{D}_{ij} + \gamma_7\left[T_{;ij} - \frac{1}{3}\delta_{ij}\nabla^2 T\right] \\ &+ \gamma_8\left\{p_{;ij} - \frac{1}{3}\delta_{ij}\nabla^2 p - p^{-1}\left[p_{;i}p_{;j} - \frac{1}{3}\delta_{ij}(\nabla p)^2\right]\right\} \\ &+ \gamma_9\left[T_{;i}T_{;j} - \frac{1}{3}\delta_{ij}(\nabla T)^2\right] + \gamma_{10}\left[T_{;i}p_{;j} + T_{;j}p_{;i} - \frac{2}{3}\delta_{ij}\nabla T\cdot\nabla p\right] \\ &+ \gamma_{11} D\tilde{D}_{ij} + \gamma_{12}\left[D_{ik}D_{jk} - \frac{1}{3}\delta_{ij}D_{kl}D_{kl}\right] \\ &+ \gamma_{13}\left[D_{ik}A_{jk} + D_{jk}A_{ik} + A_{ik}A_{jk} - \frac{1}{3}\delta_{ij}A_{kl}A_{kl}\right]. \end{aligned} \tag{4.52}$$

The γ-coefficients can be expressed in terms of β-coefficients which are defined by the following list of integrals:

$$(\mathcal{S}_i, \mathcal{S}_j) = \beta_1\delta_{ij}, \qquad (\mathcal{S}_i, \mathcal{S}_j T_{kl}) = \beta_2\Delta_{ijkl}, \qquad (\mathcal{S}_i, S_j\mathcal{T}_{kl}) = \beta_3\Delta_{ijkl},$$

$$(\mathcal{S}_i, \partial\mathcal{S}_j/\partial T) = \beta_4\delta_{ij}, \qquad (\mathcal{S}_i, v_j\mathcal{T}_{kl}) = \beta_5\Delta_{ijkl}, \qquad (\mathcal{S}_i, v_j\partial\mathcal{T}_{kl}/\partial T) = \beta_6\Delta_{ijkl},$$

$$(\mathcal{S}_i, \partial\mathcal{T}_{kl}/\partial v_j) = \beta_7\,\Delta_{ijkl}, \qquad (\mathcal{S}_i, v_j\partial\mathcal{S}_k/\partial v_l) = \beta_8(\delta_{jl}\delta_{ik} + \delta_{il}\delta_{jk}) + \beta_9\delta_{ij}\delta_{kl},$$

$$(\mathcal{S}_i, L(\mathcal{S}_j, \mathcal{T}_{kl})) = \beta_{10}\Delta_{ijkl}, \qquad (\mathcal{T}_{ij}, \mathcal{T}_{kl}) = \beta_{11}\Delta_{ijkl},$$

$$(\mathcal{T}_{ij}, L(\mathcal{S}_k, \mathcal{S}_l)) = \beta_{12}\,\Delta_{ijkl}, \qquad (\mathcal{T}_{ij}, v_k\partial\mathcal{S}_l/\partial T) = \beta_{13}\,\Delta_{ijkl},$$

$$(\mathcal{T}_{ij}, \partial\mathcal{S}_k/\partial v_l) = \beta_{14}\,\Delta_{ijkl}, \qquad (\mathcal{T}_{ij}, \mathcal{T}_{kl}\,T_{mn}) = \beta_{15}\,\Delta_{ijklmn},$$

$$(\mathcal{T}_{ij}, L(\mathcal{T}_{kl}, \mathcal{T}_{mn})) = \beta_{16}\,\Delta_{ijklmn}, \qquad (\mathcal{T}_{ij}, \partial\mathcal{T}_{kl}/\partial T) = \beta_{17}\,\Delta_{ijkl},$$

$$\begin{aligned} (\mathcal{T}_{ij}, v_k\partial\mathcal{T}_{mn}/\partial v_l) &= \beta_{18}\,\delta_{kl}\,\Delta_{ijmn} + \beta_{19}\left[\delta_{lm}\,\Delta_{knij} + \delta_{ln}\,\Delta_{kmij} - \frac{2}{3}\delta_{mn}\,\Delta_{klij}\right] \\ &+ \beta_{20}\left[\delta_{km}\,\Delta_{lnij} + \delta_{kn}\,\Delta_{lmij} - \frac{2}{3}\,\delta_{mn}\,\Delta_{klij}\right]. \end{aligned}$$

The forms of the right-hand sides follow from spherical symmetry plus the condition $\mathcal{T}_{it} = 0$. In terms of the β's, the γ's are given by

$$\gamma_1 = \beta_5/p, \qquad \gamma_2 = (\beta_5 - 2\beta_1)/3p,$$
$$\gamma_3 = (pT^2)^{-1}(2\beta_2 + 2\beta_3 - T\beta_1 + 2T^2\beta_6 - T\beta_8 - T\beta_9 - 4\beta_{10}),$$
$$\gamma_4 = (\beta_1 - \beta_8 + \beta_9)/pT, \qquad \gamma_5 = (4\beta_1 - 2T\beta_4 - 4\beta_8 - \beta_9)/3pT,$$
$$\gamma_6 = (2/mp^2)(T\beta_7 - m\beta_5), \qquad \gamma_7 = (2/T)\gamma_1,$$
$$\gamma_8 = -(2/\rho p)\beta_{11}, \qquad \gamma_9 = -(2/pT^3)(\beta_{12} - \beta_3 + T\beta_5 - T^2\beta_{13}),$$
$$\gamma_{10} = -(1/Tp^2)[\beta_5 + (T/m)(\beta_{11} - \beta_{14})],$$
$$\gamma_{11} = (2/3pT)(5T\beta_{11} - 8\beta_{15} + 8\beta_{16} - 2T^2\beta_{17} - 3T\beta_{18} + 2T\beta_{19} + 2T\beta_{20}),$$
$$\gamma_{12} = (2/pT)[4(\beta_{15} - \beta_{16} - T\beta_{19}) + T\beta_{11}], \qquad \gamma_{13} = -\rho\gamma_8.$$

The β's are not all independent as certain relations among them can be obtained as follows. It follows from spherical symmetry and the condition $\mathcal{T}_{ii} = 0$ that $\mathcal{S}_i$ and $\mathcal{T}_{ij}$ have the forms

$$\mathcal{S}_i = \mathcal{S}(v^2)v_i, \qquad \mathcal{T}_{ij} = \mathcal{T}(v^2)(v_iv_j - \frac{1}{3}\delta_{ij}v^2).$$

In addition for any A, B

$$\frac{\partial}{\partial T}(A, B) = (\partial A/\partial T, B) + (A, \partial B/\partial T) + (\kappa T)^{-2}(A, B\epsilon').$$

One can then show that

$$\begin{aligned}
&2\beta_4 = \frac{\partial\beta_1}{\partial T} + \frac{3\beta_1}{2T} - \frac{5\beta_2}{2T^2}, \quad 2\beta_8 = \frac{\beta_2}{T} - \beta_1, \quad 2\beta_9 = T^{-1}\beta_2 + \beta_1,\\
&\beta_{13} = \frac{\partial\beta_5}{\partial T} - \frac{\beta_5}{T} - \beta_6 - \frac{\beta_3}{T^2}, \quad \beta_{14} = \frac{m}{T}\beta_5 - \beta_7,\\
&2\beta_{17} = \frac{\partial\beta_{11}}{\partial T} + \frac{3\beta_{11}}{2T} - \frac{7\beta_{15}}{2T^2}, \quad 2\beta_{18} = T^{-1}\beta_{15} - \beta_{11},\\
&2\beta_{19} = T^{-1}\beta_{15} + \beta_{11}, \quad \beta_{20} = \beta_{18}.
\end{aligned} \tag{4.53}$$

Thus, there are no more than eleven independent β's. There follow certain relations among the γ's:

$$\gamma_4 = T^{-1}(\gamma_1 - 3\gamma_2), \qquad \gamma_5 = \frac{1}{3}\left(\gamma_4 - \frac{1}{2}T\frac{\partial\gamma_4}{\partial T}\right),$$
$$\gamma_7 = (2/T)\gamma_1, \qquad \gamma_{10} = -(1/pT)\left[\gamma_1 + \frac{1}{2}p(\gamma_6 - \gamma_8)\right],$$
$$\gamma_{12} = -\frac{3}{2}\gamma_{11} - \rho\left(\gamma_8 - \frac{1}{2}T\frac{\partial\gamma_8}{\partial T}\right), \qquad \gamma_{13} = -\rho\gamma_8.$$

It is customary to introduce dimensionless parameters defined by

$$\theta_1 = (\rho T/\eta^2)\left(\frac{2}{3}\gamma_4 + \gamma_5\right), \qquad \theta_2 = (\rho T/2\eta^2)\gamma_4, \qquad \theta_3 = (\rho p/\eta^2)\gamma_6,$$

$$\theta_4 = (2\rho/\eta^2)\gamma_1, \qquad \theta_5 = (\rho T/\eta^2)(\gamma_3 + \gamma_4), \qquad \omega_2 = -(\rho p/\eta^2)\gamma_8,$$

$$\omega_1 = (p/\eta^2)\left(\gamma_{11} + \frac{2}{3}\gamma_{12} - \frac{4}{3}\rho\gamma_8\right), \qquad \omega_3 = (2\rho/\eta^2)\gamma_1,$$

$$\omega_4 = -(\rho p/\eta^2)\left(\gamma_6 + \frac{2}{p}\gamma_1\right), \qquad \omega_5 = (\rho T^2/\eta^2)\gamma_9,$$

$$\omega_6 = (p/\eta^2)(\gamma_{12} - 3\rho\gamma_8).$$

The θ's and ω's are not all independent, since from Equations (4.53) we obtain

$$\omega_3 = \theta_4, \qquad \omega_4 = -(\theta_3 + \theta_4),$$

$$\frac{3}{2}\theta_1 = \theta_2\left(\frac{7}{2} - \frac{T}{\eta}\frac{\partial\eta}{\partial T}\right) - \frac{1}{2}T\frac{\partial\theta_2}{\partial T}, \qquad \frac{3}{2}\omega_1 = \omega_2\left(\frac{7}{2} - \frac{T}{\eta}\frac{\partial\eta}{\partial T}\right) - \frac{1}{2}T\frac{\partial\omega_2}{\partial T}.$$

The Burnett-order terms fall into two classes, those which are linear in second derivatives of n, T and $\mathbf{u}$ and those which are quadratic in the first derivatives. The former are important for small amplitude but rapidly varying processes. The existence of a stress due to temperature variations (the term with coefficient γ_7) was deduced by Maxwell and Reynolds. This term describes flow through a porous block with equal pressure but different temperature on opposite faces; such a flow was observed by Reynolds and given the name thermal transpiration by him. Maxwell and Reynolds used this effect to explain the Crookes radiometer. The linear terms are also important for sound at wavelengths which are comparable to the mean free path. An analysis based on the linearized Boltzmann equation but going well beyond the second-derivative terms was made by Foch. Comparing the Burnett and higher-order (or super-Burnett) equations with experiment, he found them to be clearly superior to the Navier-Stokes equations.

The nonlinear terms are usually too small to be important. Consider for example the ratio of the sixth to the first term in $\mathbf{s}^*$:

$$\gamma_5 D/\lambda = (\eta^2/\rho T\lambda)\left(\theta_1 - \frac{2}{3}\theta_2\right)D.$$

At standard conditions the value of the coefficient $\eta^2/\rho\lambda T$ is some 10^{-10} sec and so D would have to be huge for this term to be significant. However, the nonlinear terms are qualitatively different from the linear ones; terms similar to these are known to be important in some materials other than gases.

Notes on Chapter 4

4.1 Boundary-value problems and the Knudsen layer are discussed in Mott-Smith [1954], Welander [1954], Loyalka and Ferziger [1967], Cercignani [1969], Makashev [1979], and Pasternak, Senkevich, Iushkanov, and Ialomov [1980].

4.2 The Chapman-Enskog method was preceded by an expansion due to Hilbert. Roughly the difference is that Chapman and Enskog expand the hydrodynamic equations, whereas Hilbert expands the hydrodynamic variables themselves. For discussions of Hilbert's approach, see Grad [1958, 1963a], Struminski [1965], and Delale [1982].

It was shown by Grad [1963a] that the Chapman-Enskog expansion is asymptotically (in the mean free path) equivalent to a solution to the Boltzmann equation. For other work related to the validity of the Chapman-Enskog expansion and the hydrodynamic equations, see Grad [1965], Kaniel [1977], Nishida [1978], Kawashima, Matsumura, and Nishida [1979], Caflisch [1980, 1983], and Bobylev [1982].

4.6 The variational principle is discussed in Kohler [1948], Ziman [1956], Snider [1964a,b], and Benin [1970]. Ziman explains the relation between the variational principle and the principle of minimum entropy production which applies to the macroscopic equations.

Chapman and Cowling [1970] solve the integral equations by expanding the solution in Sonine polynomials and truncating the expansion after a finite number of terms; this can be formulated in terms of the variational principle by using a trial function which is a linear combination of Sonine polynomials. For another method applicable to hard spheres, see Pekeris [1955] who transforms the integral equation into a differential equation which is then solved numerically; Pekeris notes that the reduction to a differential equation was discovered by Boltzmann.

4.7 The comparison with experimental values for the transport coefficients is reviewed in Kestin and Mason [1973] and Kestin [1977].

4.10 Numerical values of the θ's and ω's for Maxwellian molecules and hard spheres are given in Chapman and Cowling [1970] and Wang Chang and Uhlenbeck [1970]. The Burnett coefficients are more important in polyatomic gases; see the review by Moraal [1975]. For the application of the Burnett (and super-Burnett) equations to sound, see Foch [1967a], Foch and Uhlenbeck [1967b], and Foch and Ford [1970].

5

MAXWELLIAN MOLECULES

5.1. MAXWELL'S FORCE LAW

It was noticed by Maxwell that the description of collisions is greatly simplified if the force between particles varies inversely as the fifth power of the distance between them. In this case the integral equations (4.16) can be solved analytically; as will be shown, the functions S_i and T_{ij} are eigenfunctions of the linearized collision operator I. More generally the full set of eigenfunctions of I is known, and these functions are useful in the analysis of the Boltzmann equation for other force laws.

The potential energy for Maxwellian molecules is given by

$$U(r) = \epsilon_0(D/r)^4 \tag{5.1}$$

where ϵ_0 and D are constants with dimensions of energy and length respectively. The change of variables $r = bx$ transforms the integral (4.39) into

$$\theta = \pi - 2\int_{x_0}^{\infty} dx[x^4 - x^2 - s^2]^{-1/2}$$

where

$$s = \alpha/gb^2, \ \alpha = 2D^2\sqrt{\epsilon_0/m}$$

and x_0 is the positive real root to the equation $x^4 - x^2 - s^2 = 0$. This shows that the scattering angle θ depends on g and b only in the combination gb^2.

According to Equation (3.9) the product of relative velocity times cross section is

$$g\sigma = (1/\sin\theta)\, d(gb^2)/d\theta$$

which is a function only of θ; for fixed θ, the cross section varies inversely with g. Stated differently, an integral of the form

$$F = \int_0^\infty f(\theta)\, g\, b db$$

can be transformed into

$$F = \frac{1}{2}\alpha \int_0^\infty f(\theta)\, s^{-2}\, ds \tag{5.2}$$

which is independent of g.

One can consider a generalized Maxwell model for which $g\sigma$ is a function only of θ, even though the cross section is not derived from the potential (5.1) or indeed from any potential.

5.2. THERMAL CONDUCTIVITY AND VISCOSITY

Considering Equations (4.16), we now show that T_{ij} is an eigenfunction of the linearized collision operator I; a similar argument shows the same for S_i. Thus, the trial functions used are actually exact solutions. We have

$$IT_{ij} = n^{-1} \int d^3v_1\, g\, b db\, d\varphi\, f_{01}\, [T_{ij}]$$

where $[T_{ij}]$ is given by Equation (4.34):

$$[T_{ij}] = \frac{1}{2} m(g_i\, g_j - g_i'\, g_j').$$

Let $\mathbf{a}$ be a unit vector perpendicular to g and in the plane containing g and g' so

$$g' = g \cos\theta + g \sin\theta\, \mathbf{a}.$$

Since the integral over φ is an integral over the directions of $\mathbf{a}$, we have

$$\int_0^{2\pi} \mathbf{a}\, d\varphi = 0.$$

Since $\mathbf{a}$ is constrained to the plane perpendicular to g, it also follows that

$$\frac{1}{2\pi}\int_0^{2\pi} a_i a_j\, d\varphi = \frac{1}{2} g^{-2} (g^2\, \delta_{ij} - g_i\, g_j).$$

Using these results we get

$$\int_0^{2\pi} [T_{ij}]\, d\varphi = \frac{3}{2}\pi m \sin^2\theta\, (g_i g_j - \frac{1}{3}\delta_{ij}\, g^2).$$

This gives

$$\int_0^{2\pi} [T_{ij}]\, g\, bdb\, d\varphi = m(g_i g_j - \frac{1}{3}\delta_{ij}\, g^2) \int_0^\infty \frac{3}{2}\pi \sin^2\theta\, g\, bdb. \qquad (5.3)$$

The remaining integral is a constant for Maxwell's force law. For reasons which will appear, we denote this constant by λ_{02}:

$$\lambda_{02} = \frac{3}{2}\pi \int_0^\infty \sin^2\theta\, g\, bdb = \frac{3}{4}\pi\alpha\beta.$$

In analogy to (5.2), β can be written as

$$\beta = \int_0^\infty \sin^2\theta\, s^{-2}\, ds$$

for which a numerical evaluation yields $\beta = 1.23$.

At this point we have

$$IT_{ij} = \lambda_{02}\, n^{-1} \int d^3v_1\, f_{01}\, m(g_i g_j - \frac{1}{3}\delta_{ij}\, g^2).$$

The remaining integration is straightforward and we get

$$IT_{ij} = \lambda_{02}\, T_{ij}.$$

Thus T_{ij} is an eigenfunction of I with the eigenvalue λ_{02}.

A similar calculation shows that

$$I\, S_i = \lambda_{11}\, S_i$$

where

$$\lambda_{11} = \frac{1}{2}\pi\alpha\beta = \frac{2}{3}\lambda_{02}. \qquad (5.4)$$

Thus, for Maxwellian molecules the trial functions aS_i and bT_{ij} are exact solutions to the integral equations (4.16) provided that the constants a,b are given by

$$a = 1/\lambda_{11}, \quad b = 1/\lambda_{02}.$$

The corresponding values for the thermal conductivity and viscosity are

$$\lambda = 5\kappa^2 T/(2m\lambda_{11}) = 5\kappa^2 T/(\pi m\alpha\beta), \; \eta = \kappa T/\lambda_{02} = 4\kappa T/(3\pi\alpha\beta). \qquad (5.5)$$

The eigenvalues λ_{02} and λ_{11} are independent of the temperature so the thermal conductivity and viscosity increase linearly with temperature. For other force laws the temperature dependence generally falls between the $\sqrt{T}$ dependence found for hard spheres and the linear dependence of Maxwellian molecules.

5.3. MOMENT EQUATION FOR SHEAR FLOW

The average of a product of n velocity components is called an n'th order moment. For example, the density, momentum density and energy density are moments of order 0, 1, and 2, and their fluxes are moments of order one greater, respectively. The procedure used to derive the conservation laws can be extended to obtain equations for moments of higher orders. The transport coefficients can then be obtained from the moment equations, and in addition it can be seen in detail how the fluxes approach the normal form during the ageing period.

Moment equations can be derived for any force law but the procedure is particularly effective with Maxwellian molecules because, as will be seen, the n'th order moment of the collision operator reduces to a simple combination of moments of order n and less. The equation for the n'th order moment is coupled to higher moments only through the convective term, which involves the moment of order $n + 1$.

The use of moment equations will be illustrated by calculation of the viscosity. Let the Boltzmann equation be multiplied by mv_iv_j and then integrated over velocity. The result is

$$\frac{\partial}{\partial t} t_{ij} + \frac{\partial}{\partial x_k} M_{ijk} = J_{ij} \tag{5.6}$$

where

$$M_{ijk} = \int d^3v\, m\, v_i\, v_j\, v_k\, f(\mathbf{v}),\; J_{ij} = \int d^3v\, m\, v_i\, v_j\, J.$$

Using Equation (5.3) we obtain

$$J_{ij} = \int d^3v\, d^3v_1\, f(\mathbf{v})f(\mathbf{v}_1) \int g\, bdb\, d\varphi\, \frac{1}{2}\, [T_{ij}]$$

$$= -\lambda_{02} \int d^3v\, d^3v_1\, f(\mathbf{v})f(\mathbf{v}_1)\, \frac{1}{2}\, m\, [g_i\, g_j - \frac{1}{3}\, \delta_{ij}\, g^2].$$

On writing g as $\mathbf{v} - \mathbf{v}_1$ and expanding we get

$$J_{ij} = -\lambda_{02}\, n\, t^*_{ij}$$

where

$$t^*_{ij} = \int d^3v\, f\, m\left(c_i\, c_j - \frac{1}{3}\, \delta_{ij}\, c^2\right),$$

$\mathbf{c} = \mathbf{v} - \mathbf{u}$, $\mathbf{u}$ being the average velocity, and $\mathbf{t}^*$ is the same as in Equation (4.6).

In Equation (5.6) we use

$$t_{ij} = p\,\delta_{ij} + \rho\,u_i\,u_j + t^*_{ij}$$

and write

$$M_{ijk} = M^0_{ijk} + u_i t^*_{jk} + u_j t^*_{ik} + u_k t^*_{ij} + \rho\,u_i u_j u_k + p(u_i\,\delta_{jk} + u_j\,\delta_{ik} + u_k\,\delta_{ij}).$$

where M^0 is calculated in the local rest frame:

$$M^0_{ijk} = \int d^3v\, f\, m\, c_i\, c_j\, c_k.$$

To simplify the resulting equation, we consider a uniform shear flow with $u_x = K\,y$, $u_y = u_z = 0$. We suppose the density to be independent of position and the equation of continuity then shows it to be time independent as well; we also assume the temperature and M^0 to be independent of position. If K is small the product $K\mathbf{t}^*$ can be neglected and taking $i = x, j = y$ in Equation (5.6) we get

$$\frac{\partial}{\partial t}\, t^*_{xy} + Kp = -n\lambda_{02}\, t^*_{xy}. \tag{5.7}$$

In the small-K approximation the pressure is time independent. The preceding equation can be solved to give

$$t^*_{xy}\,(t) = e^{-t/\tau}\, t^*_{xy}\,(0) - \eta\, K(1 - e^{-t/\tau}).$$

where the viscosity η is the same as given by (5.5) and the relaxation time is

$$\tau = 1/n\lambda_{02} = \eta/p.$$

For large t a steady state is approached with $t^*_{xy} = -\eta\,K$. This approach to the normal form illustrates ageing to hydrodynamics.

This calculation is due to Maxwell, who also obtained the thermal conductivity and coefficient of diffusion with a similar approach. The method has the remarkable feature that it is not necessary to determine the non-equilibrium distribution function. Maxwell's argument showed the thermal conductivity and viscosity to be independent of the pressure, contrary to the then current belief that they decreased linearly with pressure. By comparing his results with experimental data, Maxwell made the first realistic determination of atomic dimensions and Avogadro's number.

Actually viscous heating causes the temperature, and so the pressure, to increase with time. The solution to the moment equations with this taken into account has been obtained by Ikenberry and Truesdell. Under the same conditions as before but with the product Kt^* retained, Equation (5.6) yields

$$\begin{aligned}
&\frac{dp}{dt} + \frac{2}{3} K\, t^*_{xy} = 0, \\
&\left(\frac{d}{dt} + \frac{1}{\tau}\right) t^*_{xy} + K(p + t^*_{yy}) = 0, \\
&\left(\frac{d}{dt} + \frac{1}{\tau}\right) t^*_{yy} + \frac{dp}{dt} = 0, \\
&\left(\frac{d}{dt} + \frac{1}{\tau}\right) t^*_{zz} + \frac{dp}{dt} = 0, \\
&\left(\frac{d}{dt} + \frac{1}{\tau}\right) t^*_{xz} = 0, \left(\frac{d}{dt} + \frac{1}{\tau}\right) t^*_{yz} = 0.
\end{aligned} \tag{5.8}$$

The first three equations can be solved for p, t^*_{xy} and t^*_{yy}, and the other components of $\mathbf{t}^*$ then determined from the remaining equations. Assuming a dependence on time as $\exp(\alpha t/\tau)$, we obtain a cubic equation for α:

$$\alpha(\alpha + 1)^2 = \frac{2}{3}(K\tau)^2.$$

There is one real root and a complex-conjugate pair which we denote by r and $x \pm iy$. The equation shows immediately that $r > 0$ and we also obtain

$$x = -1 - \frac{1}{2} r, \quad y = \sqrt{r\left(1 + \frac{3}{4} r\right)}$$

so the complex roots have a negative real part.

If the solution for the pressure is written as

$$p = A\, e^{rt/\tau} + B\, e^{(x+iy)t/\tau} + C\, e^{(x-iy)t/\tau} \tag{5.9}$$

then t^*_{xy} and t^*_{yy} are determined by the second and third of Equations (5.8) and the constants A, B and C can be evaluated in terms of the initial values of p, t^*_{xy} and t^*_{yy}. For large t the dominant parts are those which grow as $\exp(rt/\tau)$ and these are

$$\begin{aligned}
&p = A \exp(rt/\tau), \quad t^*_{xy} = -3rp/2K\tau, \\
&t^*_{yy} = t^*_{zz} = -\frac{1}{2} t^*_{xx} = -rp/(r + 1),
\end{aligned} \tag{5.10}$$

while t^*_{xz} and t^*_{yz} vanish for large t. The fluxes have the normal form; their values at time t are determined by the pressure and velocity gradient at the same time. The initial values of $\mathbf{t}^*$ can be chosen independently of the thermodynamic variables, but with any initial conditions the normal form is dominant for times which are large in comparison to the relaxation time τ.

If t is set equal to 0 in the asymptotic expressions, they do not take on the correct initial values, unless the fluxes already have the normal dependence on p and K at the initial time. The normal solution is a special solution and so cannot be forced to fit arbitrary initial data. If the normal solution is used for all times, then A must be adopted as the initial value for p rather than $p(0) = A + B + C$. With this adjustment of initial values to compensate for the ageing period, the normal solution agrees with the exact solution for large times.

For small K, $r \to \frac{2}{3}(K\tau)^2$ and t^*_{xy} satisfies the linear viscosity law while the remaining $\mathbf{t}^*$'s vanish. When K is large, $r \to [\frac{2}{3}(K\tau)^2]^{1/3}$; then $\mathbf{t}^*$ approaches diagonal form with $t^*_{yy} = t^*_{zz} = -\frac{1}{2}\, t^*_{xx} = -p$. The expansion of the constitutive equations in powers of K is the same as the Chapman-Enskog expansion; it can be shown from the solution to the cubic equation that the series converges for $K < \sqrt{2}/3\tau$ but for larger values a branch point destroys convergence. This simple analytic behavior is unlikely to be typical of more general situations but appears instead to be peculiar to the assumed conditions of uniform shear and the Maxwellian model.

The exponential growth exhibited by Equation (5.10) is not present in the solution to the linearized equations. Because of the analyticity of K (in a neighborhood of $K = 0$), it is legitimate to expand the solution in powers of K for any t, but the expansion does not converge uniformly in t. One may choose K small enough for the linear equations to hold initially, but as t increases the deviations from linearity eventually become dominant.

5.4. FOURIER TRANSFORM OF THE COLLISION OPERATOR

As shown by Bobylev, a Fourier transformation with respect to velocity puts the collision operator for Maxwellian molecules into a particularly convenient form. Let the Fourier transform of the distribution function be denoted by F:

$$F(\mathbf{x},\mathbf{k},t) = (2\pi)^{-3/2} \int d^3v \, \exp\,(i\mathbf{k}\cdot\mathbf{v})\, f(\mathbf{x},\mathbf{v},t).$$

We abbreviate the Fourier transformation by $\mathcal{F}$ so $F = \mathcal{F}f$. In the form (3.7) the collision operator is

$$J = \int d^3v_1 \int d^2e \, \Sigma(\mathbf{e}\cdot\hat{\mathbf{g}})\, [f(\mathbf{v}')\, f(\mathbf{v}_1') - f(\mathbf{v})\, f(\mathbf{v}_1)]$$

where the hat denotes a unit vector and Σ is given by

$$\Sigma = 4|\mathbf{e}\cdot\mathbf{g}|\,\sigma.$$

For Maxwellian molecules Σ is a function only of the angle between $\mathbf{e}$ and $\boldsymbol{g}$. Following the manipulations which led to Equation (3.18), we find for the transform of the collision operator.

$$\mathcal{F}J = (2\pi)^{-3/2}\int d^3v\, d^3v_1 \int d^2e\, \Sigma(\mathbf{e}\cdot\hat{\boldsymbol{g}}) f(\mathbf{v}) f(\mathbf{v}_1)\,[\exp(i\mathbf{k}\cdot\mathbf{v}') - \exp(i\mathbf{k}\cdot\mathbf{v})].$$

For $\mathbf{v}'$ we use Equation (3.11), $\mathbf{v}' = \mathbf{v} - \mathbf{e}\,\mathbf{e}\cdot\boldsymbol{g}$, which gives

$$\mathcal{F}J = (2\pi)^{-3/2}\int d^3v\, d^3v_1 \int d^2e\, \Sigma(\mathbf{e}\cdot\hat{\boldsymbol{g}})\, f(\mathbf{v})\, f(\mathbf{v}_1)$$
$$\exp(i\mathbf{k}\cdot\mathbf{v})[\exp(-i\mathbf{k}\cdot\mathbf{e}\,\mathbf{e}\cdot\boldsymbol{g}) - 1].$$

A unit vector $\mathbf{e}'$ such that $\mathbf{e}'\cdot\hat{\boldsymbol{g}} = \mathbf{e}\cdot\hat{\mathbf{k}}$, $\mathbf{e}'\cdot\hat{\mathbf{k}} = \mathbf{e}\cdot\hat{\boldsymbol{g}}$ can be obtained by rotating $\mathbf{e}$ about an axis perpendicular to $\boldsymbol{g}$ and $\mathbf{k}$. Changing variables from $\mathbf{e}$ to $\mathbf{e}'$, and then dropping the prime, we get

$$\mathcal{F}J = C[F]$$

where the operator C is given by

$$\begin{aligned} C[F] &= (2\pi)^{-3/2}\int d^2e\,\Sigma(\mathbf{e}\cdot\hat{\mathbf{k}})\int d^3v\, d^3v_1 f f_1 \\ &\quad\times \{\exp[i\mathbf{v}\cdot(\mathbf{k}-\mathbf{e}\mathbf{e}\cdot\mathbf{k})]\exp[i\mathbf{v}_1\cdot\mathbf{e}\mathbf{e}\cdot\mathbf{k}] - \exp[i\mathbf{k}\cdot\mathbf{v}]\} \\ &= (2\pi)^{3/2}\int d^2e\,\Sigma(\mathbf{e}\cdot\hat{\mathbf{k}})[F(\mathbf{k}-\mathbf{e}\mathbf{e}\cdot\mathbf{k})F(\mathbf{e}\mathbf{e}\cdot\mathbf{k}) - F(\mathbf{k})F(0)] \\ &= (2\pi)^{3/2}\int d^2e\,\Sigma(\mathbf{e}\cdot\hat{\mathbf{k}})[F(\mathbf{k}_\perp)F(\mathbf{k}_\parallel) - F(\mathbf{k})\,F(0)]. \end{aligned} \tag{5.11}$$

Here $\mathbf{k}_\parallel$ and $\mathbf{k}_\perp$ denote the components of $\mathbf{k}$ parallel and perpendicular to $\mathbf{e}$. The transformed Boltzmann equation is

$$\frac{\partial F}{\partial t} - i\nabla_{\mathbf{k}}\cdot\nabla F = C[F]. \tag{5.12}$$

The collision operator can be given an alternative form which is more transparent if the independent variable is written as $\boldsymbol{g}$ instead of $\mathbf{k}$:

$$C[F] = (2\pi)^{3/2}\int d\Omega\, \boldsymbol{g}\sigma\left[F\left(\frac{1}{2}(\boldsymbol{g}+\boldsymbol{g}')\right)F\left(\frac{1}{2}(\boldsymbol{g}-\boldsymbol{g}')\right) - F(\boldsymbol{g})F(0)\right].$$

Here the quantities $\boldsymbol{g}, \boldsymbol{g}'$ are related to one another, and to the "scattering angles," in the same way as if they were relative velocities before and after a collision rather than $\mathbf{k}$-vectors. There is considerable simplification since C is a two-dimensional integral over angles whereas J is a fivefold integral.

The same transformation can be carried through for the bilinear functional

$$J[f,h] = \int d^3v_1 \int d^2e\, \Sigma(\mathbf{e}\cdot\hat{\mathbf{g}})[f(\mathbf{v}')h(\mathbf{v}_1') - f(\mathbf{v})h(\mathbf{v}_1)].$$

We find for $C[F,H] = \mathcal{F}J[f,h]$

$$C[F,H] = (2\pi)^{3/2} \int d^2e\, \Sigma(\mathbf{e}\cdot\hat{\mathbf{k}})[F(\mathbf{k}_\parallel)H(\mathbf{k}_\perp) - F(\mathbf{k})H(0)]$$

where F,H are the Fourier transforms of f,h.

A Galilean transformation $\mathbf{v} \to \mathbf{v} - \mathbf{w}$ changes $F(\mathbf{k})$ according to $F(\mathbf{k}) \to \exp(-i\mathbf{w}\cdot\mathbf{k})F(\mathbf{k})$ and the collision operator is invariant,

$$C[e^{-i\mathbf{w}\cdot\mathbf{k}}F] = e^{-i\mathbf{w}\cdot\mathbf{k}}C[F].$$

Of course Galilean invariance holds for any force law, but for Maxwellian molecules the collision operator is invariant under a wider group of transformations which are called Bobylev symmetries. Since $k_\parallel^2 + k_\perp^2 = k^2$ we obtain

$$C[e^{-ak^2}F] = e^{-ak^2}C[F] \tag{5.13}$$

for any a. In the equilibrium case, this corresponds to a shift in the temperature. In addition, the collision operator C is invariant to a change in scale of $\mathbf{k}$; using

$$e^{\alpha\mathbf{k}\cdot\nabla_\mathbf{k}}F(\mathbf{k}) = F(e^\alpha\mathbf{k})$$

we can express the scale invariance by

$$C[e^{\alpha\mathbf{k}\cdot\nabla_\mathbf{k}}F] = e^{\alpha\mathbf{k}\cdot\nabla_\mathbf{k}}C[F]. \tag{5.14}$$

If the gas is spatially homogeneous, the symmetries (5.13), (5.14) extend to the Boltzmann equation, but otherwise they are violated by the convective term $-i\nabla_\mathbf{k}\cdot\nabla F$.

5.5. EIGENFUNCTIONS FOR THE LINEARIZED COLLISION OPERATOR

Let F_0 denote the Fourier transform of the Maxwellian f_0. To linearize we put $F = F_0(1 + \Psi)$ and expand, retaining only linear terms in Ψ. Then $CF = -F_0(\mathbf{k})F_0(0)L\Psi$ where L is given by

$$L\Psi = \int d^2e\, \Sigma(\mathbf{e}\cdot\hat{\mathbf{k}})[\Psi(0) + \Psi(\mathbf{k}) - \Psi(\mathbf{k}_\parallel) - \Psi(\mathbf{k}_\perp)]. \tag{5.15}$$

This operator is related to I by

$$F_0L\Psi = \mathcal{F}f_0I\Phi$$

where Ψ and Φ are related by

$$F_0\Psi = \mathcal{F} f_0 \Phi. \tag{5.16}$$

Note that L is the same no matter what Maxwellian is used in the linearization; this is a consequence of the invariance properties of C listed in the preceding section.

In the following for simplicity we use units with $m/2\kappa T = n = 1$. Then the Maxwellian is

$$f_0 = \pi^{-3/2} \exp(-v^2)$$

and its Fourier transform is

$$F_0 = (2\pi)^{-3/2} \exp(-k^2/4).$$

Consider the eigenvalue equation for L:

$$L\Psi = \lambda\Psi.$$

Rotational invariance of L implies that Ψ can be written in the form $h(k)Y_{lm}(\mathbf{k})$ where h depends on only the magnitude of $\mathbf{k}$ and $Y_{lm}(\mathbf{k})$ denotes the spherical harmonic of the angles θ, φ which determine the direction of $\mathbf{k}$. The eigenvalues are independent of m so an eigenvalue for a given l-value is (at least) $2l + 1$-fold degenerate. It follows from scale invariance that $h = Ak^r$ where A is a constant and the parameter r, which we assume to be real, must be non-negative in order for the first term in Equation (5.15) to exist. To determine the eigenvalues we take $m = 0$ and use $Y_{l0}(\theta,\varphi) = \sqrt{(2l + 1)/4\pi}\, P_l(\cos\theta)$ where P_l is the Legendre polynomial. In a coordinate system with z-axis along $\mathbf{k}$, $\Psi(\mathbf{k}) = A'k^r$ where $A' = A\sqrt{(2l + 1)/4\pi}$; in addition $\Psi(\mathbf{k}_\parallel) = A'(k\cos\chi)^r P_l(\cos\chi)$ where χ is the angle between $\mathbf{e}$ and $\mathbf{k}$, while $\Psi(\mathbf{k}_\perp) = A'(k\sin\chi)^r P_l(\sin\chi)$. The first term in L vanishes except for $r = l = 0$ when it is a constant, so we get

$$L\Psi = \Psi \int d^2e\, \Sigma(\mathbf{e}\cdot\hat{\mathbf{k}})[1 + \delta_{r0}\delta_{l0} - \cos^r\chi\, P_l(\cos\chi) - \sin^r\chi\, P_l(\sin\chi)].$$

This identifies the eigenvalues, denoted by $\lambda_l(r)$, as

$$\lambda_l(r) = \int d^2e\, \Sigma(\cos\chi)[1 + \delta_{r0}\delta_{l0} - \cos^r\chi\, P_l(\cos\chi) - \sin^r\chi\, P_l(\sin\chi)].$$

Alternative ways of writing this are

$$\begin{aligned}\lambda_l(r) &= 2\pi \int g\sigma \sin\theta\, d\theta \left[1 + \delta_{r0}\delta_{l0} - \cos^r\tfrac{1}{2}\theta\, P_l(\cos\tfrac{1}{2}\theta) - \sin^r\tfrac{1}{2}\theta\, P_l(\sin\tfrac{1}{2}\theta)\right] \\ &= \pi\alpha \int_0^\infty ds\, s^{-2} \left[1 + \delta_{r0}\delta_{l0} - \cos^r\tfrac{1}{2}\theta\, P_l(\cos\tfrac{1}{2}\theta) - \sin^r\tfrac{1}{2}\theta\, P_l(\sin\tfrac{1}{2}\theta)\right].\end{aligned}$$

The operator I has the same eigenvalues, while its eigenfunctions are obtained from Equation (5.16). To evaluate the Fourier transform, we use the expansion of a plane wave in spherical harmonics:

$$\exp(i\mathbf{k}\cdot\mathbf{v}) = 4\pi \sum_{l,m} i^l j_l(kv)\overline{Y}_{lm}(\mathbf{k})Y_{lm}(\mathbf{v}).$$

Here the bar denotes the complex conjugate and j_l is the spherical Bessel function:

$$j_l(\rho) = \sqrt{\pi/2\rho}\,J_{l+1/2}(\rho).$$

Denoting the eigenfunction of I by Φ, we get

$$f_0\Phi = U(v)Y_{lm}(\mathbf{v})$$

where

$$U(v) = (-i)^l A v^{-1/2}\int_0^\infty dk\,k^{r+3/2}J_{l+1/2}(kv)F_0(k).$$

The remaining integral can be evaluated by using

$$\int_0^\infty dt\,t^{\mu-1}J_\nu(at)e^{-t^2} =$$

$$\tfrac{1}{2}\,(a/2)^\nu\,\frac{\Gamma(\frac{1}{2}(\mu+\nu))}{\Gamma(\nu+1)}\,e^{-a^2/4}{}_1F_1(\tfrac{1}{2}(\nu-\mu)+1;\nu+1;\tfrac{1}{4}\,a^2)$$

where ${}_1F_1$ is the hypergeometric function. For large values of its argument the hypergeometric function has the asymptotic form

$${}_1F_1(a;c;z)\rightarrow\frac{\Gamma(c)}{\Gamma(a)}\,e^z z^{a-c}+\frac{\Gamma(c)}{\Gamma(c-a)}\,e^{-i\pi a}z^{-a}.$$

If it is required that Φ be square integrable with the Maxwellian as a weight, then the first term must drop out which requires $r = 2n + l$ where n is a non-negative integer. In this case ${}_1F_1$ is a Sonine, or Laguerre, polynomial:

$${}_1F_1(-n;l+\tfrac{3}{2};z) = n!\,\frac{\Gamma(l+\frac{3}{2})}{\Gamma(n+l+\frac{3}{2})}\,S^n_{l+1/2}(z).$$

To summarize, square integrability requires $r = 2n + l$ and the eigenfunctions of I are

$$\Phi_{nlm}(\mathbf{v}) = N_{nl}S^n_{l+1/2}(v^2)v^l Y_{lm}(\mathbf{v}) \tag{5.17}$$

where the constants N_{nl} may be chosen for convenience. The eigenfunctions Ψ_{nim} of L are

$$\Psi_{nlm} = A_{nl}k^{2n+l}Y_{lm}(\mathbf{k}),\quad A_{nl} = i^l\,\frac{1}{2^{2n+l}n!}\,N_{nl}. \tag{5.18}$$

Denoting the corresponding eigenvalues by $\lambda_{nl} = \lambda_l(2n + l)$, we get

$$\lambda_{nl} = \int d^2e \, \Sigma(\cos\chi)[1 + \delta_{n0}\delta_{l0} - \cos^{2n+l}\chi P_l(\cos\chi) - \sin^{2n+l}\chi P_l(\sin\chi)]. \tag{5.19}$$

In general there is no physical requirement that (Φ, Φ) be finite and so values of r other than $2n + l$ are admissible. Normalization of the distribution function requires that $f_0\Phi$ be integrable, and this is the case if $r \geq 0$. If the average energy is finite $v^2 f_0 \Phi$ must be integrable which requires either $r > 2$ or $r = 2n + l$; other restrictions may be imposed in particular circumstances. For many purposes the eigenvalues and eigenfunctions with $r = 2n + l$ are sufficient.

Equation (5.19) shows that $\lambda_{00} = \lambda_{01} = \lambda_{10} = 0$; the corresponding eigenfunctions are summational invariants. The cases $n = 1$, $l = 1$ and $n = 0$, $l = 2$ correspond to the functions S_i and T_{ij} which were previously shown to be eigenfunctions. The recursion relation $(2l + 1)\cos\chi P_l(\cos\chi) = (l + 1)P_{l+1}(\cos\chi) + lP_{l-1}(\cos\chi)$ for the Legendre polynomials implies a similar relation for the eigenvalues:

$$(2l + 1)\lambda_{nl} = (l + 1)\lambda_{n-1,l+1} + l\lambda_{n,l-1}.$$

With $n = 1, l = 1$ this confirms the relation (5.4) between λ_{11} and λ_{02}; $l = 0$ reveals the degeneracy $\lambda_{n0} = \lambda_{n-1,1}$.

A similar analysis can be applied to the collision operator for self-diffusion:

$$I_S\Phi = \int d^2v_1 g \, bdb \, d\varphi f_0(\mathbf{v}_1)[\Phi(\mathbf{v}) - \Phi(\mathbf{v}')].$$

The same functions Φ_{nlm} are also eigenfunctions of I_S but the eigenvalues are given by

$$\lambda_{nl}^S = \int d^2e \, \Sigma(\cos\chi)[1 - \cos^{2n+l}\chi P_l(\cos\chi)].$$

5.6. PROPERTIES OF THE EIGENFUNCTIONS

The functions Φ_{nlm} have a particular interest in kinetic theory as they occur in several contexts. This is not surprising since they are orthogonal polynomials in $\mathbf{v}$ with the Maxwellian as a weight. In this section some of their properties will be collected together.

The spherical harmonics are

$$Y_{lm}(\theta, \varphi) = \epsilon\left\{\frac{2l + 1}{4\pi}\frac{(l - |m|)!}{(l + |m|)!}\right\}^{1/2} P_l^m(\cos\theta)e^{im\varphi}$$

where $\epsilon = 1, m \leq 0, \epsilon = (-1)^m, m > 0$ and P_l^m is the associated spherical harmonic. For any vector $\mathbf{v}$ with polar angles θ, φ, we write $Y_{lm}(\mathbf{v}) = Y_{lm}(\theta, \varphi)$.

The spherical harmonics are the expression in polar coordinates of quantities which transform under rotations according to an irreducible representation of the rotation group. One may also use the irreducible tensors T^l which are equivalent to $v^l Y_{lm}$, but are expressed in Cartesian coordinates. The components of T^l, denoted by $T^l_{i_1 i_2 \ldots i_l}$, are homogeneous polynomials of degree l in the components of $\mathbf{v}$ with the properties that they are symmetric in the subscripts $i_1, \ldots, i_l$ and that no nonvanishing tensor of lower order can be formed by contraction. The first few are

$$T^0 = 1,$$

$$T^1_i = v_i,$$

$$T^2_{ij} = v_i v_j - \frac{1}{3}\delta_{ij} v^2,$$

$$T^3_{ijk} = v_i v_j v_k - \frac{1}{5} v^2(\delta_{ij} v_k + \delta_{ik} v_j + \delta_{jk} v_i).$$

The components of T^l can be expressed as linear combinations of $v^l Y_{lm}$ and vice versa. As one example, if v_m denotes the quantities

$$v_1 = -\frac{1}{\sqrt{2}}(v_x + i v_y), \quad v_0 = v_z, \quad v_{-1} = \frac{1}{\sqrt{2}}(v_x - i v_y)$$

then

$$v_m = \sqrt{(4\pi/3)}\; Y_{1m}(\mathbf{v}).$$

The eigenfunctions Φ_{nlm} could be written with $T^l_{ijk\ldots}$ in place of $v^l\, Y_{lm}$.

Sonine polynomials are defined by the expansion

$$(1 - t)^{-(p+1)} \exp\left(-\frac{xt}{1 - t}\right) = \sum_{n=0}^{\infty} S_p^n(x) t^n.$$

The terminology is quite variable; some authors refer to the S_p^n as Laguerre polynominals and the quantities $(-1)^n\, \Gamma(p + n + 1)\, S_p^n$ are called both Sonine and Laguerre polynomials. Our usage follows that of Chapman and Cowling.

The S_p^n are polynomials of degree n, and with any fixed p but $n = 0,1,2 \ldots$ provide a complete set for the expansion of functions which are square integrable on the interval $0, \infty$ with weight $e^{-x} x^p$. For $n = 0,1$ they are

$$S_p^0 = 1, \quad S_p^1 = p + 1 - x.$$

One can proceed to higher n-values by using the recursion formula

$$S_p^n = (2n + p - 1 - x)\, S_p^{n-1} - (n + p - 1)\, S_p^{n-2}.$$

Other useful relations are

$$x \frac{d}{dx} S_p^n + S_{p+1}^{n-1} = 0, \quad x \frac{d}{dx} S_p^n = nS_p^n - (n + p)\, S_p^{n-1},$$

$$S_p^n = S_{p+1}^n - S_{p+1}^{n-1}, \quad S_{p+1}^n = \sum_{r=0}^{n} S_p^{n-r}.$$

From these equations one can derive the differential equation

$$x \frac{d^2}{dx^2} S_p^n + (p + 1 - x) \frac{d}{dx} S_p^n + nS_p^n = 0.$$

In addition Sonine polynomials satisfy the orthogonality and normalization conditions

$$\int_0^\infty e^{-x}\, x^p\, S_p^n(x)\, S_p^q(x)\, dx = \delta_{nq} \frac{1}{n!} \Gamma(n + p + 1).$$

Two eigenfunctions of I with different values of l,m are orthogonal because of orthogonality of the spherical harmonics, and the preceding equation shows them to be orthogonal if the n-values are different. If the constants N_{nl} are given the values

$$N_{nl} = \left\{ 2\pi^{3/2} \frac{n!}{\Gamma(n + l + \frac{3}{2})} \right\}^{1/2}$$

then they are orthonormal,

$$(\Phi_{nlm}, \Phi_{pqr}) = \delta_{np}\, \delta_{lq} \delta_{mr}$$

where the scalar product is $(\Psi, \Phi) = \int d^3v\, f_0\, \overline{\Psi}\, \Phi$.

The Φ_{nlm} $(\mathbf{v})$ span the space of functions which are square integrable with weight $\exp(-v^2)$. They are polynomials of degree $2n + l$ in $\mathbf{v}$, that is, polynomials in each of the three components of $\mathbf{v}$ with the sum of the three degrees being $2n + l$. Any polynomial in $\mathbf{v}$ of degree r can be written as a linear combination of Φ_{nlm} $(\mathbf{v})$ with degrees $2n + l$ not exceeding r.

The functions Φ_{nlm} are closely related to the eigenfunctions of the Fokker-Planck operator obtained in Chapter 2. To see the relation, consider the eigenvalue problem for the energy of an oscillator in quantum mechanics (written in terms of $\mathbf{v}$):

$$\left(-\frac{1}{2} \nabla^2 + \frac{1}{2} v^2 \right) W = \lambda W. \tag{5.20}$$

The eigenfunctions and eigenvalues are

$$W_{\mathbf{n}} = \sqrt{f_0}\, B_{\mathbf{n}}\, H_{n_1}(v_1)\, H_{n2}(v_2)\, H_{n3}(v_3),$$

$$\lambda_{\mathbf{n}} = n_1 + n_2 + n_3 + \frac{3}{2}$$

where $\mathbf{n}$ denotes the three integers n_1, n_2, n_3 and H_n is the Hermite polynomial. If the constants $B_{\mathbf{n}}$ have the values

$$B_{\mathbf{n}} = (2^{n_1+n_2+n_3}\, n_1!\, n_2!\, n_3!)^{-1/2}$$

then the eigenfunctions are normalized according to

$$\int d^3v |W_{\mathbf{n}}|^2 = 1.$$

Equation (5.20) can also be solved in polar coordinates. If the eigenfunction is written in the form $W_{nlm} = R_{nl}(v)\, Y_{lm}(\mathbf{v})$ then R_{nl} satisfies

$$\frac{d^2}{dv^2} R_{nl} + \frac{2}{v}\frac{d}{dv} R_{nl} + \left[2\lambda - v^2 - \frac{l(l+1)}{v^2}\right] R_{nl} = 0.$$

This is solved by

$$R_{nl} = Cv^l \exp(-v^2/2)\, S^n_{l+1/2}(v^2)$$

and the eigenvalue is $\lambda = 2n + l + \frac{3}{2}$. Comparing with (5.17), we see that $W_{nlm} = \sqrt{f_0}\, \Phi_{nlm}$. For fixed λ the two sets of functions $W_{\mathbf{n}}$ and $\sqrt{f_0}\, \Phi_{nlm}$ must be the same. Hence we can expand $W_{\mathbf{n}}$ as

$$W_{\mathbf{n}} = \sqrt{f_0} \sum_{n,l,m} C^{nlm}_{\mathbf{n}}\, \Phi_{nlm}$$

where the coefficients $C^{nlm}_{\mathbf{n}}$ vanish except when $n_1 + n_2 + n_3 = 2n + l$. The product of Hermite polynomials transforms according to a reducible representation of the rotation group, and the preceding expansion corresponds to the reduction of the representation into its irreducible components.

To summarize the relation between the different sets of functions, the eigenfunctions F_{nlm} of the Fokker-Planck operator, W_{nlm} of the oscillator Hamiltonian, and Φ_{nlm} of the collision operator are related by

$$F_{nlm} = \sqrt{f_0}\, W_{nlm} = f_0\, \Phi_{nlm}.$$

The functions $F_{\mathbf{n}}$ and $W_{\mathbf{n}}$ are also eigenfunctions of Ω and H respectively; this is related to the degeneracy of the eigenvalues. The eigenvalues of I have a lower degree of degeneracy and the functions $\Phi_{\mathbf{n}} = F_{\mathbf{n}}/f_0$ are not eigenfunctions.

The oscillator eigenfunctions are eigenfunctions of the Fourier transform (since the oscillator Hamiltonian looks the same after the transformation):

$$\mathcal{F}\sqrt{f_0}\, \Phi_{nlm} = i^{2n+l} \sqrt{f_0}\, \Phi_{nlm}.$$

The expansion of $\mathcal{F}$ in its eigenfunctions is

$$\exp(i\mathbf{k}\cdot\mathbf{v}) = (2\pi)^{3/2}\sum_{nlm} i^{2n+l}\sqrt{f_0(k)}\,\overline{\Phi}_{nlm}(\mathbf{k})\sqrt{f_0(v)}\,\Phi_{nlm}(\mathbf{v}). \tag{5.21}$$

5.7. MOMENT EQUATIONS

Let f be expanded as

$$f = f_0 \sum_{nlm} M_{nlm}\,\Phi_{nlm}. \tag{5.22}$$

Then the coefficients are moments calculated with the Φ_{nlm}:

$$M_{nlm} = \int d^3v\,\Phi_{nlm}\,f.$$

Taking moments of the Boltzmann equation we obtain a set of equations for the moments:

$$\frac{\partial}{\partial t} M_{nlm} + (\mathbf{v}\cdot\nabla f)_{nlm} = J_{nlm}. \tag{5.23}$$

By using the recursion relations for the spherical harmonics and the Sonine polynomials, we can express the convective term $(\nabla\cdot\mathbf{v}f)_{nlm}$ in terms of moments of order $2n + l + 1$ and lower.

For the moments of the collision operator, we have

$$J_{nlm} = \int d^3v\,d^3v_1\,f(\mathbf{v})\,f(\mathbf{v}_1)\,Q_{nlm}(\mathbf{v},\mathbf{v}_1)$$

where

$$Q_{nlm}(\mathbf{v},\mathbf{v}_1) = \frac{1}{2}\int d^2e\,\Sigma[\Phi_{nlm}(\mathbf{v}') + \Phi_{nlm}(\mathbf{v}_1') - \Phi_{nlm}(\mathbf{v}) - \Phi_{nlm}(\mathbf{v}_1)].$$

As a function of $\mathbf{v}$ and $\mathbf{v}_1$ independently, Q_{nlm} can be expanded in the Φ_{nlm}:

$$Q_{nlm}(\mathbf{v},\mathbf{v}_1) = \sum C(nlm,pqr,stu)\,\Phi_{pqr}(\mathbf{v})\,\Phi_{stu}(\mathbf{v}_1). \tag{5.24}$$

Then J_{nlm} is given by an expansion in the moments

$$J_{nlm} = \sum C(nlm,pqr,stu)\,M_{pqr}M_{stu}. \tag{5.25}$$

The coefficients are symmetric, $C(nlm,pqr,stu) = C(nlm,stu,pqr)$, so each term in the sum occurs twice.

Equation (5.25) applies for any force law, but in the general case the expansion is an infinite series. For Maxwellian molecules the series truncates to a finite sum: Since Σ is independent of the relative velocity, Q_{nlm} is a polynomial of degree $2n + l$ in $\mathbf{v}, \mathbf{v}_1$ and consequently the expansion (5.24) contains only terms for which $2p + q + 2s + u = 2n + l$.

If $s = t = u = 0$ then using the orthogonality of the Φ_{nlm} and $\Phi_{000} = 1$ we obtain

$$C(nlm,pqr,000) = \int d^3v\, f_0(\mathbf{v})\, \overline{\Phi}_{pqr}(\mathbf{v}) \int d^3v_1\, f_0\,(\mathbf{v}_1)\, Q_{nlm}.$$

The $\mathbf{v}_1$-integral is just $-\frac{1}{2}I$ and since Φ_{nlm} is its eigenvector we get

$$C(nlm,pqr,000) = -\frac{1}{2}\lambda_{nl}\,\delta_{np}\,\delta_{lq}\,\delta_{mr}.$$

Equation (5.23) becomes

$$\left(\frac{\partial}{\partial t} + \lambda_{nl}\right) M_{nlm} + (\mathbf{v}\cdot\nabla f)_{nlm} = \sum C(nlm,pqr,stu)\, M_{pqr} M_{stu}$$

where now terms on the right with $pqr = nlm$ or $stu = nlm$ are omitted. In addition terms with $n = l = 0$, $n = 0$, $l = 1$, and $n = 1$, $l = 0$ drop out because the corresponding Φ_{nlm} are summational invariants. The right-hand side involves only moments of lower order than M_{nlm}, while moments of higher order occur only in the convective term. If the higher moments are assumed to be independent of position, this coupling drops out so the moments of low order satisfy a closed set of equations (as illustrated by the case of uniform shear); once the low-order moments are known, those of higher order can be worked out from the moment equations.

To determine the remaining coefficients we use the transformed Boltzmann equation (5.12). The Fourier transform of Equation (5.22) is

$$F(\mathbf{k}) = F_0(\mathbf{k}) \sum_{nlm} M_{nlm} \overline{\Psi}_{nlm}(\mathbf{k}) \tag{5.26}$$

and the corresponding expansion of the collision operator is

$$C = F_0(\mathbf{k}) \sum_{nlm} J_{nlm}\, \overline{\Psi}_{nlm}(\mathbf{k}) \tag{5.27}$$

where Ψ_{nlm} denotes the functions (5.18). If the expansion (5.26) is substituted into Equation (5.11) for C the result is

$$C = F_0(\mathbf{k})\, F_0(0) \sum M_{pqr} M_{stu} X_{pqr,stu} \tag{5.28}$$

where

$$X_{pqr,stu} = \frac{1}{2}\int d^2e\; \Sigma(\cos\chi)[\overline{\Psi}_{pqr}(\mathbf{k}_\perp)\,\overline{\Psi}_{stu}(\mathbf{k}_\parallel) + \overline{\Psi}_{stu}(\mathbf{k}_\perp)\,\overline{\Psi}_{pqr}(\mathbf{k}_\parallel)$$
$$- \overline{\Psi}_{pqr}(\mathbf{k})\,\overline{\Psi}_{stu}(0) - \overline{\Psi}_{stu}(\mathbf{k})\,\overline{\Psi}_{pqr}(0)].$$

The last terms vanish unless $pqr = 000$ or $stu = 000$ and since these cases have already been treated we retain only the first two terms. Using Equation (5.18) we get

$$X_{pqr,stu} = A_{pq} A_{st}\, k^{2(p+s)+q+t}\, Z_{pqr,stu}$$

where A_{pqr} denotes the normalization constant given in (5.18) and

$$Z_{pqr,stu} = \frac{1}{2}\int d^2e\, \Sigma(\cos\chi)\, [|\mathbf{k}_\perp|^{2p+q}|\mathbf{k}_\|| ^{2s+t}\,\overline{Y}_{qr}(\mathbf{k}_\perp)\,\overline{Y}_{tu}(\mathbf{e})$$

$$+ |\mathbf{k}_\perp|^{2s+t}|\mathbf{k}_\||^{2p+q}\,\overline{Y}_{tu}(\mathbf{k}_\perp)\,\overline{Y}_{qr}(\mathbf{e})].$$

To evaluate $Z_{pqr,stu}$ we transform the coordinate system by a rotation R such that the new z-axis lies along $\mathbf{k}$. The transformation of the spherical harmonics is

$$Y_{lm}(R\mathbf{e}) = \sum_{\mu=-l}^{l} D^l_{\mu m}(R)\, Y_{l\mu}(\mathbf{e})$$

where D^l denotes the l'th representation of the rotation group. If χ,φ are polar angles for $\mathbf{e}$ then the angles for $\mathbf{k}_\perp$ are $\frac{1}{2}\pi - \chi$, $\varphi + \pi$. We obtain

$$Z_{pqr,stu} = \frac{1}{2}(-1)^{r+u}\sum_{\mu\nu}\overline{D}^q_{\mu r}\,\overline{D}^t_{\nu u}\int d^2e\,\Sigma(\cos\chi)$$

$$\times\left[\sin^{2p+q}\theta\cos^{2s+t}\theta\, Y_{q\mu}(\tfrac{1}{2}\pi - \chi, \pi + \varphi)\, Y_{t\nu}(\chi,\varphi)\right.$$

$$\left. + \sin^{2s+t}\theta\cos^{2p+q}\theta\, Y_{t\nu}(\tfrac{1}{2}\pi - \chi, \pi + \varphi)\, Y_{q\mu}(\chi,\varphi)\right].$$

The φ-integration forces $\mu + \nu = 0$ and the product of representations is given by

$$\overline{D}^p_{\mu r}\,\overline{D}^t_{-\mu u} = \sum_l \left(\frac{4\pi}{2l+1}\right)^{1/2}(q,r,t,u|l,m)(q,\mu,t,-\mu|l,0)\,\overline{Y}_{lm}$$

where $(q,r,t,u|l,m)$ denotes the Clebsch-Gordon coefficients for the rotation group. The labels are restricted by the rules of addition of angular momenta: q,t and l must form the sides of a triangle and $m = r + u$. This yields

$$Z_{pqr,stu} = \sum_l \frac{1}{\sqrt{2l+1}}\,\overline{Y}_{lm}(\mathbf{k})(q,r,t,u|l,m)\, I(l,p,q,s,t)$$

where

$$I(l,p,q,s,t) = 2\pi^{3/2}\sum_\mu (q,\mu,t,-\mu|l,0)\int_0^\pi d\chi\sin\chi\,\Sigma(\cos\chi)$$

$$\times\left[\sin^{2p+q}\chi\cos^{2s+t}\chi\, Y_{q\mu}(\tfrac{1}{2}\pi - \chi, 0)\, Y_{t,\mu}(\chi,0)\right.$$

$$\left. + \sin^{2s+t}\chi\cos^{2p+q}\chi\, Y^t_\mu(\tfrac{1}{2}\pi - \chi, 0)\, Y_{q,\mu}(\chi, 0)\right].$$

By substituting the preceding results into Equation (5.28) and comparing with the expansion (5.27), the coefficients $C(nlm,pqr,stu)$ can be identified as

$$C(nlm,pqr,stu) = (2\pi)^{-3/2}\frac{1}{\sqrt{2l+1}}(A_{pq}A_{st}/A_{nl})(q,r,t,u|l,m)\ I(l,p,q,s,t).$$

To summarize the restrictions on the labels, it is necessary that $2p + q + 2s + t = 2n + l$, $r + u = m$, and l,q,t form a triangle; in other cases the coefficients vanish.

The moment equations simplify considerably if the gas is isotropic. Then M_{nlm} vanishes except for $l = m = 0$. Denoting the moments for this case by M_n, we obtain

$$\left(\frac{\partial}{\partial t} + \lambda_{n0}\right) M_n = \sum_{p=2}^{n-2} C_{np} M_p M_{n-p}.$$

The coefficients are found to be

$$C_{np} = (1/4\sqrt{\pi})\left[\frac{(2n+1)!}{(2p+1)!\,(2n-2p+1)!}\right]^{1/2}$$
$$\times \int_0^{\pi} d\chi\ \sin\chi\ [\sin^{2p}\chi\ \cos^{2(n-p)}\chi + \cos^{2p}\chi\ \sin^{2(n-p)}\chi].$$

It should be recalled that the density n has been dropped from the Maxwellian; when reintroduced it appears as a factor in the C-coefficients and in the eigenvalues λ_{nl}.

5.8. BKW SOLUTION

Let F denote the particular function

$$F = (2\pi)^{-3/2}[1 + (\alpha - 1)k^2]\exp(-\alpha k^2).$$

A straightforward calculation gives

$$C[F] = (2\pi)^{-3/2}\lambda(\alpha - 1)^2 k^4 \exp(-\alpha k^2)$$

where

$$\lambda = \int d^2e\ \Sigma(\cos\chi)\sin^2\chi\cos^2\chi = \frac{1}{3}\lambda_{02}.$$

If α depends on time then

$$\frac{\partial F}{\partial t} = (2\pi)^{-3/2}(1 - \alpha)\frac{d\alpha}{dt}k^4\exp(-\alpha k^2).$$

Consequently, if α satisfies

$$\frac{d\alpha}{dt} = \lambda(1 - \alpha)$$

then F is a solution to the Boltzmann equation (without spatial variations). If α is taken to be 1 when $t = 0$ then

$$\alpha(t) = 1 - \exp(-\lambda t). \tag{5.29}$$

In terms of the velocity distribution function, the solution is

$$f(\mathbf{v}, t) = (2\pi\alpha)^{-3/2} \frac{1}{2\alpha} \left[5\alpha - 3 + \frac{1 - \alpha}{2\alpha} v^2 \right] \exp(-v^2/4\alpha).$$

Since negative values of f are physically unacceptable, it is necessary that $5\alpha \geq 3$ which means that the time in (5.29) must satisfy $\lambda t \geq \ln \frac{5}{2}$.

This solution was found independently by Bobylev, and Krook and Wu. It has quite a special form but is the only known exact solution, with closed form, to the Boltzmann equation for a gas not in equilibrium.

Notes on Chapter 5

5.3 The solution to the moment equations for shear flow is given in Ikenberry and Truesdell [1956], Truesdell [1956], Galkin [1958, 1964, 1966], and Truesdell and Muncaster [1980]. The solution for heat flow under a uniform temperature gradient has been obtained by Asmolov, Makashev, and Nosik [1980]. In the latter case there is no time dependence and the functional relation between heat flow and temperature gradient is linear no matter how large the gradient.

5.4 See Bobylev [1975a].

5.5 The eigenvalue problem for the linearized collision operator for Maxwellian molecules was solved by Wang Chang and Uhlenbeck [1970], and Mott-Smith [1954]. See also Waldmann [1958]. The eigenvalue problem for a mixture is treated in Sirovich [1966]. The argument here is due to Bobylev [1975a]. Extensive numerical results are given in Alterman, Frankowski, and Pekeris [1962], for both Maxwellian molecules and hard spheres.

We mention here a model of the Boltzmann equation due to Bhatnagar, Gross, and Krook [1954], Welander [1954], and Gross and Jackson [1959], commonly called the BGK model. In the model the collision operator has only a single (or, in more elaborate versions, only a few) eigenvalues in addition to the vanishing eigenvalue for the summational invariants. The resulting simplification is such that it is possible to obtain many results in analytical form. For applications of the BGK model, see Cercignani [1969].

5.6 The Maxwellian eigenfunctions have been applied extensively to other force laws, such as hard spheres. Chapman and Cowling [1970] work out the collision operator for the Sonine-polynomial generating function and thereby evaluate matrix elements of the collision operator. See also Mott-Smith [1954], Pekeris, Alterman, Finkelstein, and Frankowski [1962], Ford [1968], and Foch and Ford [1970].

5.7 The coefficients C_{nlm} are worked out by Ikenberry and Truesdell [1956] for orders $n + l \leq 5$. The coefficients in arbitrary order were evaluated by Ernst [1979] for the isotropic case and by Hendriks and Nieuwenhuizen [1982] in general. The moment equations for both Maxwellian and non-Maxwellian molecules are treated in Grad [1949], Mott-Smith [1954], and Vestner and Waldmann [1977].

5.8 The BKW solution is obtained in Bobylev [1975b, 1976, 1979, 1980], and Krook and Wu [1976, 1977]. We note here an argument of Nikol'skii [1964a, b] which applies to hard spheres rather than Maxwellian molecules: The collision operator for hard spheres has the scaling property $J[f(\lambda \mathbf{v})] = \lambda^{-4} J[f(\mathbf{v})]$. It follows that if $f_0(\mathbf{v}, t)$ is a solution to the spatially homogeneous Boltzmann equation, then a solution with spatial variations is given by $f(\mathbf{x}, \mathbf{v}, t) = f_0((\mathbf{v}t - \mathbf{x})/\tau, \chi(t))$, where τ is a constant and $d\chi/dt = (\tau/t)^4$. Unfortunately, for hard spheres a solution f_0 is known only at equilibrium and the Nikol'skii argument then yields just a special case of the solution discussed in Section 3.9.

For reviews of the literature related to the BKW solution, see Ernst [1981, 1983], and Hoare [1984].

6

THE LINEARIZED BOLTZMANN EQUATION

6.1. INTRODUCTION

The main obstacle to a complete analysis of the Boltzmann equation is its nonlinearity. However, there is an interesting class of processes with small amplitude for which the Boltzmann equation can be reduced to a linear equation. The resulting linearized Boltzmann equation will be discussed in this chapter.

The linearized Boltzmann equation has a normal solution, which could be obtained by linearizing the full normal solution. However, if a linear process such as sound propagation is the matter of interest, it is more efficient to linearize the equation at the beginning. In addition known methods of linear analysis provide a basis for dealing with questions such as the existence and uniqueness of solutions and the convergence of the Chapman-Enskog expansion. One particular set of questions to be dealt with concerns the linearized collision operator I and the integral equations whose solutions determine the transport coefficients.

Of course one should be cautious in generalizing lessons learned from the linearized equation to the full nonlinear Boltzmann equation. An obvious defect of linearization is the absence of the heating which occurs when there are dissipative processes; the case of uniform shear discussed in the preceding chapter demonstrates the qualitative difference which can occur between solutions to the linear and nonlinear equations. Solutions to the linearized equation can, however, be useful in analyzing nonlinear effects and in addition the "normal modes" of the linearized Boltzmann equation are needed in the analysis of the ring operator, which will be discussed in Chapter 12.

At the Navier-Stokes approximation there is a linear dependence on the gradients of the thermodynamic variables. This local linearity is to be contrasted with the global linearity at issue here. To obtain the linearized Boltzmann equation, we will expand about the Maxwellian for a strict equilibrium state, rather than about the local Maxwellian.

6.2. LINEARIZED NAVIER-STOKES EQUATIONS

Before treating the linearized Boltzmann equation, we first discuss the linearized hydrodynamic equations. Consider a process with space- and time-dependent thermodynamic variables $T(\mathbf{x},t)$, $\rho(\mathbf{x},t)$ and $\mathbf{u}(\mathbf{x},t)$. If the deviation from equilibrium is small, these differ by small amounts from constant values which characterize an equilibrium reference state. We may take the latter to be at rest since this can always be brought about by a Galilean transformation. Hence, we write

$$T(\mathbf{x},t) = T + \delta T(\mathbf{x},t), \quad \rho(\mathbf{x},t) = \rho + \delta\rho(\mathbf{x},t), \quad \mathbf{u}(\mathbf{x},t) = \delta\mathbf{u}(\mathbf{x},t).$$

Through terms which are linear in the small variations the equation of continuity becomes

$$\frac{\partial \delta\rho}{\partial t} + \rho \nabla \cdot \delta\mathbf{u} = 0.$$

Similarly, the equation of motion (1.25) yields

$$\rho \frac{\partial \delta\mathbf{u}}{\partial t} + \nabla \delta p + \nabla \cdot \delta\mathbf{t}^* = 0$$

and the temperature equation (1.29) reduces to

$$\frac{\partial \delta T}{\partial t} + T\left(\frac{\partial p}{\partial \epsilon}\right)_\rho \nabla \cdot \delta\mathbf{u} + (\rho C_v)^{-1} \nabla \cdot \delta\mathbf{s}^* = 0.$$

All quantities except the small variations are to be evaluated at the equilibrium reference state.

For $\mathbf{s}^*$ and $\mathbf{t}^*$ we will use the constitutive equations (1.35). This results in the linearized Navier-Stokes equations:

$$\frac{\partial \delta\rho}{\partial t} + \rho \nabla \cdot \delta\mathbf{u} = 0,$$

$$\frac{\partial \delta T}{\partial t} + T\left(\frac{\partial p}{\partial \epsilon}\right)_\rho \nabla \cdot \delta\mathbf{u} - (\lambda/\rho C_v)\, \nabla^2 \delta T = 0,$$

$$\rho \frac{\partial \delta\mathbf{u}}{\partial t} + \nabla \delta p - \eta \nabla^2 \delta\mathbf{u} - \left(\frac{1}{3}\eta + \eta'\right) \nabla\nabla \cdot \delta\mathbf{u} = 0.$$

We only consider the initial value problem in infinite space, for which a convenient method of solution is by the Fourier-Laplace transform. Thus, we suppose $\delta\rho$, δT and $\delta\mathbf{u}$ to depend on position and time according to $\exp(-\zeta t - i\mathbf{k}\cdot\mathbf{x})$ which yields the set of algebraic equations

$$\zeta\delta\rho + \rho i\mathbf{k}\cdot\delta\mathbf{u} = 0,$$

$$\zeta\delta T + T\left(\frac{\partial p}{\partial\epsilon}\right)_{\rho} i\mathbf{k}\cdot\delta\mathbf{u} + (\lambda/\rho C_v)\,k^2\,\delta T = 0,$$

$$(\zeta\rho - \eta k^2)\,\delta\mathbf{u} + i\mathbf{k}\,\delta p - \left(\frac{1}{3}\eta + \eta'\right)\mathbf{k}\,\mathbf{k}\cdot\delta\mathbf{u} = 0.$$

In terms of the components of $\delta\mathbf{u}$ parallel and perpendicular to $\mathbf{k}$, denoted by $\delta u_{\parallel}$ and $\delta\mathbf{u}_{\perp}$ respectively, the equations become

$$\zeta\delta\rho + \rho ik\,\delta u_{\parallel} = 0,$$

$$\zeta\delta T + T\left(\frac{\partial p}{\partial\epsilon}\right)_{\rho} ik\,\delta u_{\parallel} + (\lambda/\rho C_v)k^2\,\delta T = 0,$$

$$(\zeta - \bar{\nu}k^2)\,\delta u_{\parallel} + ik\,\delta p = 0,$$

$$(\zeta - \nu k^2)\,\delta\mathbf{u}_{\perp} = 0$$

where $\nu = \eta/\rho$ is the kinematic viscosity and $\bar{\nu} = (\frac{4}{3}\eta + \eta')/\rho$. Using $\delta\rho$ and δT as independent variables, we make the replacement $\delta p = (\partial p/\partial T)_{\rho}\,\delta T + (\partial p/\partial\rho)_T\,\delta\rho$. The resulting equations can be written in matrix form as

$$A\,\xi = \zeta\xi \tag{6.1}$$

where ξ has five components which depend linearly on $\delta\rho$, δT, $\delta u_{\parallel}$ and the two components of $\delta\mathbf{u}_{\perp}$. We will use the "normal coordinates" defined in Equation (1.17):

$$\xi_{\alpha} = \left\{\left(1/\sqrt{TK_T}\right)\frac{\delta\rho}{\rho},\ \sqrt{\rho C_v}\,\frac{\delta T}{T},\ \sqrt{\rho/T}\,\delta u_{\parallel},\ \sqrt{\rho/T}\,\delta u_{\perp}^1,\ \sqrt{\rho/T}\,\delta u_{\perp}^2\right\}$$

where the superscript denotes the two components of $\delta\mathbf{u}_{\perp}$. In this representation the matrix A takes a form which we write as

$$A = -(ikc/\sqrt{\gamma})\,A_1 + k^2 A_2$$

where the matrices A_1, A_2 are given by

$$A_1 = \begin{bmatrix} 0 & 0 & 1 & 0 & 0 \\ 0 & 0 & \sqrt{\gamma-1} & 0 & 0 \\ 1 & \sqrt{\gamma-1} & 0 & 0 & 0 \\ 0 & 0 & 0 & 0 & 0 \\ 0 & 0 & 0 & 0 & 0 \end{bmatrix}, \qquad A_2 = \begin{bmatrix} 0 & 0 & 0 & 0 & 0 \\ 0 & \gamma D_T & 0 & 0 & 0 \\ 0 & 0 & \bar{\nu} & 0 & 0 \\ 0 & 0 & 0 & \nu & 0 \\ 0 & 0 & 0 & 0 & \nu \end{bmatrix}$$

Here $\gamma = C_p/C_v$ while c and D_T are the sound speed and thermal diffusivity,

$$c = \sqrt{\gamma/\rho K_T}, \; D_T = \gamma/\rho C_p.$$

In the derivation use has been made of Equation (1.13) for the difference in specific heats.

When the linearization is done on higher-order hydrodynamic equations such as the Burnett equations, A contains third and higher powers of k.

Equation (6.1) shows that ζ is an eigenvalue of A. Since A is a 5×5 matrix, it has five eigenvalues and eigenvectors. It can be seen by inspection that νk^2 is a degenerate eigenvalue with the eigenvectors being the transverse modes ξ_4, ξ_5. The remaining three are solutions to the secular equation which works out to

$$\zeta^3 - \zeta^2 k^2(\bar{\nu} + \gamma D_T) + \zeta k^2(c^2 + \gamma \bar{\nu} D_T k^2) - D_T c^2 k^4 = 0.$$

We solve this equation by an expansion in powers of k. The five eigenvalues including the two already noted are, through terms of second order in k,

$$\zeta_1 = D_T k^2, \; \zeta_{2,3} = \pm ikc + \Gamma k^2, \; \zeta_4 = \zeta_5 = \nu k^2. \tag{6.2}$$

Here Γ is the sound damping constant, given by

$$\Gamma = \frac{1}{2}[\bar{\nu} + (\gamma - 1) D_T].$$

The five eigenvectors of A can be displayed as the columns of a 5×5 matrix. Denoting this matrix by R, we write it as an expansion in k,

$$R = R_0 + (ikD_T/c) R_1 + \dots \tag{6.3}$$

Then R_0 is found to be

$$R_0 = (1/\sqrt{2\gamma}) \begin{bmatrix} \sqrt{2(\gamma-1)} & 1 & 1 & 0 & 0 \\ -\sqrt{2} & \sqrt{\gamma-1} & \sqrt{\gamma-1} & 0 & 0 \\ 0 & -\sqrt{\gamma} & \sqrt{\gamma} & 0 & 0 \\ 0 & 0 & 0 & \sqrt{2\gamma} & 0 \\ 0 & 0 & 0 & 0 & \sqrt{2\gamma} \end{bmatrix}$$

and R_1 is given by

$$R_1 = \sqrt{(\gamma - 1)/2}\, R_0 \times \begin{bmatrix} 0 & 1 & -1 & 0 & 0 \\ -1 & 0 & \alpha & 0 & 0 \\ 1 & -\alpha & 0 & 0 & 0 \\ 0 & 0 & 0 & 0 & 0 \\ 0 & 0 & 0 & 0 & 0 \end{bmatrix}$$

Here α is the abbreviation

$$\alpha = [1/\sqrt{8(\gamma - 1)}]\,[\gamma - 1 - (\overline{\nu}/D_T)].$$

The columns of R, reading from left to right, are the eigenvectors of A corresponding to the eigenvalues $\zeta_1, \ldots, \zeta_5$ respectively. It can be verified that $R^T = R^{-1}$ where the superscript T denotes the transpose, so R is an orthogonal matrix. The transformation $R^T A R$ converts A into diagonal form.

With the aid of the eigenvectors of A, linear combinations of $\delta\rho$, δT and $\delta\mathbf{u}$ can be constructed which have a simple exponential dependence on the time. These hydrodynamic normal modes are the elements of the column matrix

$$\phi = R^{-1}\,\xi.$$

In the lowest approximation R can be replaced by R_0 and the elements of ϕ are

$$\phi_1 = \sqrt{\frac{\gamma - 1}{\gamma T K_T}}\,\frac{\delta\rho}{\rho} - \sqrt{\frac{\rho C_v}{\gamma}}\,\frac{\delta T}{T},$$

$$\phi_2 = \frac{1}{\sqrt{2\gamma T K_T}}\,\frac{\delta\rho}{\rho} + \sqrt{\frac{\gamma - 1}{2\gamma}\,\rho C_v}\,\frac{\delta T}{T} - \sqrt{\frac{\rho}{2T}}\,\delta u_\parallel,$$

$$\phi_3 = \frac{1}{\sqrt{2\gamma T K_T}}\,\frac{\delta\rho}{\rho} + \sqrt{\frac{\gamma - 1}{2\gamma}\,\rho C_v}\,\frac{\delta T}{T} + \sqrt{\frac{\rho}{2T}}\,\delta u_\parallel,$$

$$\phi_4 = \sqrt{\rho/T}\,\delta u_\perp^1,$$

$$\phi_5 = \sqrt{\rho/T}\,\delta u_\perp^2.$$

To see what processes the normal modes correspond to, consider first the case when only ϕ_1 is present. The vanishing of the other ϕ's gives

$$\delta\mathbf{u} = 0, \quad \frac{\delta\rho}{\rho} = -\sqrt{(\gamma - 1) T K_T\,\rho C_v}\,\frac{\delta T}{T}.$$

The last equation can be reduced to $\delta\rho = (\partial\rho/\partial T)_p\,\delta T$ which implies $\delta p = 0$. Thus the normal mode ϕ_1 corresponds to the transfer of energy at constant pressure and vanishing velocity, that is, to heat conduction.

The modes ϕ_2 and ϕ_3 correspond to sound propagation. The frequency $\omega = i\zeta$ is related to k by

$$\omega = \pm ck + i\Gamma k^2.$$

The first term describes wave propagation with speed c while the second describes damping due to thermal conduction and viscosity. The sound propagation is longitudinal since $\mathbf{u}$ is parallel to $\mathbf{k}$, and the condition $\phi_1 = 0$ gives

$$\frac{\delta\rho}{\rho} = \sqrt{T K_T\,\rho C_v/(\gamma - 1)}\,\frac{\delta T}{T}.$$

This reduces to $\delta\rho = (\partial p/\partial T)_S \delta T$ which shows that sound propagation is adiabatic, to the approximation considered.

The modes ϕ_4 and ϕ_5 are viscous shear modes, with the velocity being transverse to $\mathbf{k}$ and the density and temperature being constant. Since the eigenvalue is real, the shear modes are purely diffusive; in this respect they are similar to the heat mode but are to be contrasted with the propagating sound modes.

6.3. LINEARIZATION OF THE BOLTZMANN EQUATION

To linearize the Boltzmann equation we write the distribution function as

$$f = f_0(1 + \Phi) \tag{6.4}$$

where f_0 is a strict Maxwellian. Assuming the deviation Φ from equilibrium to be small, we retain only linear terms and the result is the linearized Boltzmann equation:

$$\frac{\partial \Phi}{\partial t} + \mathbf{v}\cdot\nabla\Phi + nI\Phi = 0. \tag{6.5}$$

Here I is the linearized collision operator obtained in Chapter 4. We recall the basic properties of I that it is symmetric and non-negative:

$$(h, Ik) = (Ih, k), \; (h, Ih) \geq 0 \tag{6.6}$$

where the scalar product is as defined in Equation (4.19).

One can also linearize about a nonequilibrium distribution. Suppose the conditions at the boundary of the gas are such that it is maintained in a nonequilibrium steady state. The distribution function f_S satisfies

$$\mathbf{v}\cdot\nabla f_S + J[f_S] = 0$$

together with the appropriate boundary conditions. For small deviations from the steady state we write

$$f = f_S(1 + \Phi)$$

and neglect quadratic terms in Φ. The resulting equation is

$$\frac{\partial \Phi}{\partial t} + \mathbf{v}\cdot\nabla\,\Phi + I_S\,\Phi + \Phi\mathbf{v}\cdot\nabla \ln f_S = 0$$

where the operator I_S is defined by

$$f_S\, I_S \Phi = \int d^3v_1\, g\, b db\, d\varphi\, [f_S f_{S1}(\Phi + \Phi_1) - f_S' f_{S1}'(\Phi' + \Phi_1')].$$

The behavior of small perturbations about a steady state is a matter of considerable interest but here only Equation (6.5) will be treated.

6.4. EIGENVALUE PROBLEM

Suppose Φ to have an exponential dependence on position and time, $\Phi(\mathbf{x},\mathbf{v},t) = g(\mathbf{v})\exp(-\zeta t - i\mathbf{k}\cdot\mathbf{x})$. Then we obtain

$$B\,g = \zeta\,g \tag{6.7}$$

where B denotes the operator

$$B = nI - i\mathbf{k}\cdot\mathbf{v} \tag{6.8}$$

The solutions will be required to satisfy $(g,g) < \infty$ and this determines an eigenvalue problem. However, this condition is more stringent than is dictated by physical conditions and so may exclude some solutions of physical interest.

For $\mathbf{k} = 0$ the summational invariants are eigenfunctions with eigenvalue 0. Since there are five independent summational invariants, this eigenvalue is fivefold degenerate. For nonvanishing $\mathbf{k}$ we expect five independent solutions which reduce to summational invariants when $\mathbf{k}$ approaches 0. Our main interest is in these solutions, which will be labeled by a subscript α:

$$B\,g_\alpha = \zeta_\alpha g_\alpha,\ \alpha = 1,\ldots,5. \tag{6.9}$$

Before constructing the solution to the preceding equation, we first consider some of its general properties. The operator B is not self-adjoint, and its Hermitian conjugate (for real $\mathbf{k}$) is

$$B^\dagger = nI + i\mathbf{k}\cdot\mathbf{v}. \tag{6.10}$$

Let J denote the operation of complex conjugation, so we have $B^\dagger = JBJ$. In addition let τ denote the operator which reverses the direction of the velocity,

$$\tau\,g(\mathbf{v}) = g(-\mathbf{v}),\ \tau^2 = 1. \tag{6.11}$$

Then B satisfies

$$\tau B\tau = B,\ \tau J\,B\,\tau J = B. \tag{6.12}$$

Equation (6.9) implies that

$$B^\dagger\,\bar{g}_\alpha = \bar{\zeta}_\alpha\,\bar{g}_\alpha.$$

Hence the function $h_\alpha = \bar{g}_\alpha$ is an eigenfunction of $B^\dagger$ with $\bar{\zeta}_\alpha$ as its eigenvalue. The h_α also reduce to summational invariants when $\mathbf{k}$ vanishes. It follows from $(h_\alpha,\ Bg_\beta) = (B^\dagger h_\alpha, g_\beta)$ that $(\zeta_\alpha - \zeta_\beta)(h_\alpha, g_\beta) = 0$. Thus h_α and g_β are orthogonal if $\zeta_\alpha \neq \zeta_\beta$. If there is degeneracy so $\zeta_\alpha = \zeta_\beta$ for some α, β then if necessary linear combinations can be formed such that h_α and g_β are orthogonal. In any case we may assume

$$(h_\alpha, g_\beta) = 0,\ \alpha \neq \beta$$

and may choose a normalization so that

$$(h_\alpha, g_\beta) = \delta_{\alpha\beta}. \tag{6.13}$$

It need not be the case that $(g_\alpha, g_\beta) = \delta_{\alpha\beta}$ because of the absence of Hermitian symmetry for B.

It follows from Equation (6.12) that

$$B(\tau \bar{g}_\alpha) = \bar{\zeta}_\alpha (\tau \bar{g}_\alpha)$$

so $\bar{\zeta}_\alpha$ is an eigenvalue of B as well as of $B^\dagger$. Since there is an odd number (five), at least one of the eigenvalues must be real with the remainder forming complex-conjugate pairs. As it turns out, the members of one pair are equal and therefore real; this degeneracy is due to invariance under rotations and reflections of $\mathbf{v}$ in the plane perpendicular to $\mathbf{k}$. In summary, there are three real eigenvalues, two of which are equal, and one complex-conjugate pair.

The property (6.6) of I shows that

$$\mathrm{Re}(g_\alpha, B g_\alpha) = n \,\mathrm{Re}\,(g_\alpha, I g_\alpha) \geq 0$$

for real $\mathbf{k}$. Since $(g_\alpha, B g_\alpha) = \zeta_\alpha (g_\alpha, g_\alpha)$ and (g_α, g_α) is real, we see that any eigenvalue has a non-negative real part:

$$\mathrm{Re}\,\zeta_\alpha \geq 0.$$

The inequality is strict except in the special case $\mathbf{k} = 0$, $\zeta_\alpha = 0$, with g_α being a summational invariant.

6.5. PERTURBATION SOLUTION

The solutions to Equation (6.7) will be constructed by perturbation theory with the wavenumber being the expansion parameter; this is the same as the Chapman-Enskog expansion for states near equilibrium. To avoid complications due to more than one perturbation parameter, we restrict $\mathbf{k}$ to a single direction, say the x-axis. Then the equation to be solved is

$$(nI - ikv_x)\, g_\alpha = \zeta_\alpha g_\alpha.$$

We expand in powers of k:

$$\zeta_\alpha = \zeta_\alpha^{(0)} - ik\,\zeta_\alpha^{(1)} + \ldots, \; g_\alpha = g_\alpha^{(0)} - ik\, g_\alpha^{(1)} + \ldots$$

Since we are concerned with the perturbation of a vanishing eigenvalue, $\zeta_\alpha^{(0)} = 0$ so $g_\alpha^{(0)}$ is a summational invariant. After substituting the expansion into the equation and equating the coefficients of each power of k, we find the sequence of equations

$$nIg_\alpha^{(1)} + v_x g_\alpha^{(0)} = \zeta_\alpha^{(1)} g_\alpha^{(0)},$$

$$nIg_\alpha^{(2)} + v_x g_\alpha^{(1)} = \zeta_\alpha^{(2)} g_\alpha^{(0)} + \zeta_\alpha^{(1)} g_\alpha^{(1)}, \quad \text{etc.}$$

These equations can be solved successively for the coefficients in the expansion. The detailed calculations are straightforward but here only the results will be given. Through first-order terms in k the eigenfunctions are

$$
\begin{aligned}
g_1 &= \sqrt{\frac{2}{5}}\left[\frac{\epsilon}{\kappa T} - \frac{5}{2} + \frac{ik}{p}\left(\mathcal{S}_x - \theta\eta\, v_x\right)\right],\\
g_{2,3} &= \sqrt{\frac{2}{15}}\left\{\frac{\epsilon}{\kappa T} + \frac{ik}{p}\left[\mathcal{S}_x + \frac{1}{2}\eta\left(1 - \frac{1}{2}\theta\right) v_x\right]\right\}.\\
&\mp \sqrt{\frac{m}{2\kappa T}}\left\{v_x + \frac{ik}{p}\left[\mathcal{T}_{xx} - \frac{2}{15}\eta\left(1 - \frac{7}{2}\theta\right)\frac{\epsilon}{\kappa T} - \theta\eta\right]\right\}, \qquad (6.14)\\
g_4 &= \sqrt{m/\kappa T}\left(v_y + \frac{ik}{p}\mathcal{T}_{xy}\right),\\
g_5 &= \sqrt{m/\kappa T}\left(v_z + \frac{ik}{p}\mathcal{T}_{xz}\right).
\end{aligned}
$$

Here $\epsilon = \frac{1}{2}mv^2$ and we have used the dimensionless parameter $\theta = 2\lambda m/5\kappa\eta$ while $\mathcal{S}_i$, $\mathcal{T}_{ij}$ are the quantities defined by Equations (4.16) and (4.17). The normalization of the g_α has been adjusted to agree with Equation (6.13). The eigenvalues to second order in k are the same as (6.2) with the parameters having values for a monatomic gas. In particular the thermal diffusivity is $D_T = 2\lambda/5n\kappa$, $c = \sqrt{5\kappa T/3m}$ is the sound speed, and the damping constant Γ is given by

$$\Gamma = (2\eta/3\rho)\left[1 + \frac{1}{2}\theta\right].$$

To compare the eigenfunctions with those obtained from the Navier-Stokes equations, it is necessary to calculate the thermodynamic variables from the functions (6.14). With f having the form (6.4), the deviations of the thermodynamic variables from their equilibrium values are

$$\delta n = \int d^3v f_0 \Phi = n\,(1,\Phi), \quad \delta T = \frac{2}{3\kappa}(\epsilon, \Phi) - T(1, \Phi), \quad \delta \mathbf{u} = (\mathbf{v}, \Phi).$$

Suppose for example that only the first normal mode is present so $\Phi \propto g_1$. Then except for a multiplicative constant

$$\delta n = -\sqrt{\frac{2}{5}}, \quad \delta T = \sqrt{\frac{2}{5}}(T/n), \quad \delta \mathbf{u} = 0.$$

The corresponding values of the ξ's are

$$\xi_1 = -\sqrt{2\kappa/5n}, \quad \xi_2 = \sqrt{3\kappa/5n}, \; \xi_3 = \xi_4 = \xi_5 = 0.$$

Except for normalization, this is the same as the corresponding eigenvector displayed in the matrix R_0, that is, the first column of R_0. The remaining columns of R_0 are obtained in a similar way from $g_2, \ldots, g_5$.

We list here the contributions to the eigenfunctions and eigenvalues obtained in the next order, corresponding to the linearized Burnett equations. The results for the eigenfunctions, normalized to satisfy Equation (6.13), are

$$g_1^{(2)} = -\sqrt{\frac{2}{5}\frac{1}{np}}\left\{\frac{2\lambda}{5\kappa}\mathcal{T}_{xx} - A_{xx} - (8\theta - 9\theta^2 - 3\theta_2 - 4\omega_3)\frac{\eta^2}{15m\kappa T}\epsilon - (\theta^2 + \theta_2)\frac{\eta^2}{2m}\right\},$$

$$g_{2,3}^{(2)} = -\sqrt{\frac{2}{15}\frac{1}{np}}\left\{\frac{5\kappa T}{2m}\mathcal{T}'_{xx} - \left(1 - \frac{1}{2}\theta\right)\frac{\eta}{2m}\mathcal{T}_{xx} - A_{xx}\right.$$

$$\left. + \frac{\eta^2}{30m}\left(1 + 17\theta - \frac{59}{4}\theta^2 + 16\theta_2 - 12\omega_3\right)\frac{\epsilon}{\kappa T} - (6\theta - 5\theta^2 - 4\omega_3 + 4\theta_2)\frac{\eta^2}{4m}\right\}$$

$$\pm \sqrt{\frac{m}{2\kappa T}\frac{1}{n^2 m}}\left\{-B_{xxx} + \frac{2}{3}\mathcal{S}'_x + \left(1 - \frac{7}{2}\theta\right)\frac{2\eta}{15\kappa T}\mathcal{S}_x\right.$$

$$\left. + \left(1 - \theta + \frac{25}{4}\theta^2 + 10\omega_2\right)\frac{\eta^2}{30\kappa T}v_x\right\},$$

$$g_4^{(2)} = \sqrt{\frac{m}{\kappa T}\frac{1}{n^2 m}}\left(B_{xyx} - \frac{\eta^2\omega_2}{4\kappa T}v_y\right),$$

$$g_5^{(2)} = \sqrt{\frac{m}{\kappa T}\frac{1}{n^2 m}}\left(B_{xzx} - \frac{\eta^2\omega_2}{4\kappa T}v_z\right).$$

The parameters θ_2, ω_2 and ω_3 are the same as defined in Chapter 4. The remaining quantities are determined by the integral equations

$$IA_{ij} = v_i\,\mathcal{S}_j - (2\lambda/3\kappa)\delta_{ij}\left(\epsilon - \frac{3}{2}\kappa T\right),$$

$$IB_{ijk} = \mathcal{T}_{ij}v_k - \eta\left(\delta_{ik}\,v_j + \delta_{jk}\,v_i - \frac{2}{3}\delta_{ij}\,v_k\right),$$

$$I\mathcal{S}'_i = \mathcal{S}_i, \quad I\mathcal{T}'_{ij} = \mathcal{T}_{ij}$$

plus the conditions that they be orthogonal to the summational invariants.

The eigenvalues ζ_1, ζ_4 and ζ_5 have no terms which are third order in k while

$$\zeta_2^{(3)} = -\zeta_3^{(3)} = (c\eta^2/30n^2 m\kappa T)(-8\,\omega_3 + 4\,\theta_2 + 4\,\theta - 7\,\theta^2 - 4 + 10\omega_2).$$

The perturbation expansion for sound propagation has been worked out by Foch to the fifth order in k. He compared the theoretical results with experimental data for absorption and dispersion and found good agreement over a wide range of conditions.

6.6. POLYNOMIAL SOLUTIONS

The linearized Boltzmann equation has some special solutions which are polynomials in $\mathbf{x}$ and t. These can be obtained from the perturbation expansion as follows. The function

$$g_\alpha(\mathbf{k},\mathbf{v}) \exp\,[-i\mathbf{k}\cdot\mathbf{x} - \zeta_\alpha t] \tag{6.15}$$

is a solution to Equation (6.5) for any values of $\mathbf{k}$ within the radius of convergence of the series. Its derivatives with respect to $\mathbf{k}$ are also solutions, as are the derivatives at $\mathbf{k} = 0$:

$$\Phi(\mathbf{x},t) = \frac{\partial}{\partial k_i}\frac{\partial}{\partial k_j} \cdots g_\alpha(\mathbf{k},\mathbf{v}) \exp\,[-i\mathbf{k}\cdot\mathbf{x} - \zeta_\alpha t]\,|_{k=0}. \tag{6.16}$$

It is necessary to determine ζ_α and g_α to higher order in $\mathbf{k}$ the more derivatives are taken, but the result in any order is a polynomial in $\mathbf{x}$ and t.

Examples are readily worked out from the expressions already obtained for g_α and ζ_α. In particular if only one derivative is taken in Equation (6.16) one finds for $\alpha = 1$

$$\Phi_1 = x\left(\epsilon - \frac{5}{2}\kappa T\right) - n^{-1}(\mathcal{S}_x - \theta\eta v_x). \tag{6.17}$$

This describes heat flow with a constant temperature gradient. Similarly for $\alpha = 2$ the result is

$$\Phi_2 = -(x + ct)\left(\frac{2}{5}c\epsilon - v_x\kappa T\right)$$
$$+\frac{2}{5}(c/n)\left[\mathcal{S}_x + \frac{1}{2}\left(1 - \frac{1}{2}\theta\right)\eta v_x\right] - (1/nm)\left[\kappa T\mathcal{T}_{xx} - \frac{2}{15}\left(1 - \frac{7}{2}\theta\right)\eta\epsilon - \theta\eta\right],$$

which corresponds to uniform gradients of the temperature and of the longitudinal component of velocity. For $\alpha = 4$ we get

$$\Phi_4 = xv_y - (1/nm)\mathcal{T}_{xy}$$

which describes uniform shear.

6.7. PROPERTIES OF THE LINEARIZED COLLISION OPERATOR

In this section we consider in detail certain mathematical properties of the operator I, including the nature of its spectrum and the related question of the existence of solutions to the integral equation

$$IX = \chi. \tag{6.18}$$

For brevity in this and the following section we use units with $n = m/2\kappa T = 1$; then the Maxwellian is $f_0 = \pi^{-3/2}\exp(-v^2)$.

It will be assumed that the cross section is bounded, $\sigma(g,\theta) \leq \sigma_0$. In classical mechanics this requires that the potential vanish for distances greater than a finite range. In quantum mechanics it is only necessary that the potential decrease for large r faster than r^{-2} and be suitably well behaved for finite r.

If the cross section is bounded, or even integrable so that the total cross section σ_T exists, then I can be split into two parts

$$I = \nu - K \tag{6.19}$$

where ν amounts to multiplication by the collision frequency $\nu(v)$:

$$\nu(v) = \int d^3v_1\, g\, \sigma_T(g)\, f_0(\mathbf{v}_1).$$

The remaining operator K is given by

$$K\Phi = \int d^3v_1\, d\Omega\, g\, \sigma\, f_0(\mathbf{v}_1)[\Phi(\mathbf{v}') + \Phi(\mathbf{v}_1') - \Phi(\mathbf{v}_1)].$$

The collision frequency is given in Equation (3.29) for the case that the total cross section is constant. It has a minimum ν_0 at $v = 0$ and increases monotonically, eventually becoming linear in v. In quantum theory a similar behavior is found for hard spheres; although σ_T is not constant, it approaches a nonvanishing constant for large g and this results in a linear dependence of ν on v for large v. Such a behavior of the collision frequency occurs more generally if the particles repel one another strongly at short distances. If they are only weakly repulsive, or attractive, at short distances, the cross section vanishes as g goes to infinity. Then ν may increase without bound, approach a constant, or approach zero as v increases, depending on which of the factors g or σ_T outweighs the other. Potentials are referred to as hard or soft depending on whether ν has values between a minimum and infinity, or between 0 and a maximum; a hard-core potential is one for which $\nu(v)$ increases linearly in v for large v.

Now consider the operator K, whose last term will be denoted by K_1:

$$K_1\Phi = \int d^3v_1 g\, \sigma_T(g)\, f_0\; \Phi(\mathbf{v}_1).$$

We write this as

$$K_1\Phi = \int d^3v_1\, f_0(\mathbf{v}_1)\, k_1(\mathbf{v}, \mathbf{v}_1)\, \Phi(\mathbf{v}_1)$$

where

$$k_1(\mathbf{v}, \mathbf{v}_1) = |\mathbf{v} - \mathbf{v}_1|\; \sigma_T(|\mathbf{v} - \mathbf{v}_1|). \tag{6.20}$$

Let the remainder of K be denoted by K_2, which we write in the form

$$\begin{aligned} K_2\Phi &= \int d^3v_1\, d\Omega\, g\, \sigma\, f_0(\mathbf{v}_1)[\Phi(\mathbf{v}') + \Phi(\mathbf{v}_1')] \\ &= \int d^3v_1\, d^3v_0\, d\Omega\, g\, \sigma\, f_0(\mathbf{v}_1)\Phi(\mathbf{v}_0)[\delta(\mathbf{v}_0 - \mathbf{v}') + \delta(\mathbf{v}_0 - \mathbf{v}_1')] \\ &= \int d^3v_0\, f_0(\mathbf{v}_0)\, k_2(\mathbf{v}, \mathbf{v}_0)\Phi(\mathbf{v}_0) \end{aligned}$$

where the function k_2 is given by

$$f_0(\mathbf{v}_0)k_2(\mathbf{v}, \mathbf{v}_0) = \int d^3v_1\, d\Omega\, g\, \sigma\, f_0(\mathbf{v}_1)[\delta(\mathbf{v}_0 - \mathbf{v}') + \delta(\mathbf{v}_0 - \mathbf{v}_1')]. \tag{6.21}$$

To reduce the preceding expression we use the form Equation (3.14) for the collision operator. In this form the first term of (6.21) is

$$\frac{1}{2}m\int d^3v_1\, d^3v'\, d^3v_1'\, \sigma\, f_0(\mathbf{v}_1)\, \delta(\mathbf{v}_0 - \mathbf{v}')\, \delta(\boldsymbol{G} - \boldsymbol{G}')\, \delta(E - E'). \tag{6.22}$$

After setting $\mathbf{v}' = \mathbf{v}_0$ we use the second δ-function to evaluate $\mathbf{v}_1$ at $\mathbf{v}_1 = \mathbf{y} + \mathbf{v}_1'$, where $\mathbf{y} = \mathbf{v}_0 - \mathbf{v}$. The expression (6.22) becomes

$$4m\int d^3v_1'\, \sigma\, f_0(\mathbf{y} + \mathbf{v}_1')\, \delta(E - E'). \tag{6.23}$$

The energy difference is

$$E - E' = \frac{1}{2}m(v^2 + v_1^2 - v'^2 - v_1'^2) = m\mathbf{y}\cdot(\mathbf{v}_1' - \mathbf{v}). \tag{6.24}$$

The component of $\mathbf{v}_1'$ parallel to $\mathbf{y}$ can be integrated out with the help of the δ-function and we get for (6.23)

$$(4/y)\int d^2z\, \sigma(g, \theta)\, f_0(\mathbf{y} + \mathbf{v}_\parallel + \mathbf{z}). \tag{6.25}$$

Here $\mathbf{v}_\parallel$ is the component of $\mathbf{v}$ parallel to $\mathbf{y}$ and $\mathbf{z}$ denotes the component of $\mathbf{v}_1'$ normal to $\mathbf{y}$. The arguments g, θ can be expressed in terms of $\mathbf{v}$, $\mathbf{v}_0$, and $\mathbf{z}$ as follows. For g we have

$$g = \mathbf{v} - \mathbf{v}_1 = \mathbf{v} - \mathbf{y} - \mathbf{v}_1' = \mathbf{v}_\perp - \mathbf{y} - \mathbf{z}$$

where $\mathbf{v}_\perp$ is the component of $\mathbf{v}$ normal to $\mathbf{y}$. The last step follows because, in view of Equation (6.24), conservation of energy forces the components of $\mathbf{v}$ and $\mathbf{v}_1'$ parallel to $\mathbf{y}$ to be the same. Since both $\mathbf{z}$ and $\mathbf{v}_\perp$ are perpendicular to $\mathbf{y}$, we get

$$g = \sqrt{y^2 + (\mathbf{z} - \mathbf{v}_\perp)^2}.$$

For $\mathscr{g}'$ we have

$$\mathscr{g}' = \mathbf{v}' - \mathbf{v}_1' = \mathbf{v}_0 - \mathbf{v}_1' = \mathbf{y} + \mathbf{v}_\perp - \mathbf{z}.$$

Consequently the scattering angle is determined by

$$\cos\theta = \frac{\mathscr{g}\cdot\mathscr{g}'}{\mathscr{g}^2} = \frac{(\mathbf{z} - \mathbf{v}_\perp)^2 - y^2}{(\mathbf{z} - \mathbf{v}_\perp)^2 + y^2}.$$

These results simplify if the origin of the $\mathbf{z}$-plane is shifted by $\mathbf{v}_\perp$. The expression (6.25) becomes

$$(4/y)\int d^2z\,\sigma(\mathscr{g},\theta)\,f_0(\mathbf{v}_0 + \mathbf{z})$$

where $\mathscr{g}$, θ are now given by

$$\mathscr{g} = \sqrt{y^2 + z^2}, \quad \cos\theta = \frac{z^2 - y^2}{z^2 + y^2}. \tag{6.26}$$

The remaining contribution to k_2 can be reduced in a similar way and the result for both contributions is

$$f_0(\mathbf{v}_0)\,k_2(\mathbf{v},\mathbf{v}_0) = (4/y)\int d^2z\,f_0(\mathbf{v}_0 + \mathbf{z})[\sigma(\mathscr{g},\theta) + \sigma(\mathscr{g},\pi - \theta)]$$

where $\mathscr{g}$ and θ are determined by Equation (6.26). From this result it is seen that $k_2(\mathbf{v}, \mathbf{v}_0)$ is non-negative.

The preceding integral can easily be worked out if σ is constant, as is the case for classical hard spheres. Choosing the z-axis of the coordinate system to be along $\mathbf{y}$, we get

$$f_0(\mathbf{v}_0)\,k_2(\mathbf{v},\mathbf{v}_0) =$$

$$(8\sigma/y)\pi^{-3/2}\int_{-\infty}^{\infty}dz_x\int_{-\infty}^{\infty}dz_y\,\exp(-[(v_{0x} + z_x)^2 + (v_{0y} + z_y)^2 + v_{0z}{}^2])$$

$$= (8\sigma/y\sqrt{\pi})\,\exp(-v_{0z}^2).$$

For an arbitrary orientation of the coordinate system, this becomes

$$f_0(\mathbf{v}_0)\,k_2(\mathbf{v},\mathbf{v}_0) = (8\sigma/y\sqrt{\pi})\,\exp[-(\mathbf{v}_0\cdot\mathbf{y})^2/y^2].$$

For later use we list an equivalent form

$$\sqrt{f_0(\mathbf{v}_0)f_0(\mathbf{v})}\,k_2(\mathbf{v},\mathbf{v}_0) = (8\sigma/y\sqrt{\pi})\,\exp\left(-\frac{1}{4}[y^2 + (\hat{\mathbf{y}}\cdot\mathbf{V})^2]\right) \tag{6.27}$$

where $\mathbf{V} = \mathbf{v}_0 + \mathbf{v}$ and $\hat{\mathbf{y}}$ is a unit vector in the $\mathbf{y}$ direction. From this form it is seen that K_2 is symmetric, $k_2(\mathbf{v},\mathbf{v}_0) = k_2(\mathbf{v}_0,\mathbf{v})$, which also follows more generally from the fact that I, ν and K_1 are symmetric.

The function $\nu(v)$ is real and continuous and it follows that the operator ν is self-adjoint. Its spectrum consists of the values of the function $\nu(v)$, which is a segment of the positive real axis.

Regarding K, we prove a theorem due to Carleman that it is a compact operator. For K_1 we use the criterion that an integral operator is compact if its kernel is square integrable (with, in the present case, the weight function f_0). Replacing the cross section by its bound σ_0 we find from (6.20)

$$\int d^3v\, d^3v_1\, f_0(\mathbf{v})\, f_0(\mathbf{v}_1)|k_1(\mathbf{v},\mathbf{v}_1)|^2 \leq (4\pi\sigma_0)^2 \int d^3v\, d^3v_1\, f_0(\mathbf{v})\, f_0(\mathbf{v}_1)|\mathbf{v} - \mathbf{v}_1|^2.$$

It is straightforward to evaluate the integral and the result is

$$\int d^3v\, d^3v_1\, f_0(\mathbf{v})\, f_0(\mathbf{v}_1)|k_1(\mathbf{v},\mathbf{v}_1)|^2 \leq 48\ \pi^2\sigma_0^2.$$

Thus K_1 is compact. We have also obtained an upper bound for its norm:

$$\|K_1\| \leq 4\sqrt{3}\,\pi\sigma_0.$$

For hard spheres this can be expressed in terms of the collision frequency as

$$\|K_1\| \leq \frac{1}{2}\sqrt{3\pi}\,\nu_0. \tag{6.28}$$

The proof for K_2 is somewhat longer and we first show that it is bounded. For this purpose we use the lemma that if an integral operator K has a kernel $k(\mathbf{v},\mathbf{v}_0)$ satisfying

$$\int d^3v\, \sqrt{f_0(\mathbf{v})f_0(\mathbf{v}_0)}|k(\mathbf{v},\mathbf{v}_0)| \leq C, \qquad \int d^3v_0\sqrt{f_0(\mathbf{v})f_0(\mathbf{v}_0)}|k(\mathbf{v},\mathbf{v}_0)| \leq C,$$

then it has the bound $\|K\| \leq C$. From Equation (6.27) we get

$$\int d^3v_0\, \sqrt{f_0(\mathbf{v})f_0(\mathbf{v}_0)}\, k_2(\mathbf{v},\mathbf{v}_0) \leq (8\sigma_0/\sqrt{\pi})\int d^3y\, y^{-1}\exp(-y^2/4) \leq 64\sqrt{\pi}\,\sigma_0.$$

The symmetry of K_2 then yields the bound

$$\|K_2\| \leq 64\sqrt{\pi}\,\sigma_0.$$

For classical hard spheres the preceding bound is $\|K_2\| \leq 8\nu_0$ which is to be compared with (6.28). The sum of the two bounds is a bound for K:

$$\|K\| \leq \left(8 + \frac{1}{2}\sqrt{3\pi}\right)\nu_0. \tag{6.29}$$

A slightly better estimate is needed for the proof of compactness. In Equation (6.27) substitute $\mathbf{V} = \mathbf{y} + 2\mathbf{v}$ and take polar coordinates for $\mathbf{y}$ with

z-axis along $\mathbf{v}$. Setting $\mathbf{v}\cdot\mathbf{y} = vy\cos\theta = vyw$ we get

$$\int d^3v_0\sqrt{f_0(\mathbf{v})\,f(\mathbf{v}_0)}\;v\,k_2(\mathbf{v},\mathbf{v}_0)$$

$$\leq 16\sqrt{\pi}\,\sigma_0\;v\int_0^{\infty} dy\;y\int_{-\infty}^{\infty} dw\;\exp\left\{-\frac{1}{4}[y^2 + (y + 2v\,w)^2]\right\} \leq 32\,\pi\,\sigma_0.$$

We next use $v_0 \leq v + y$, the previous estimate, and Equation (6.27) to obtain

$$\int d^3v_0\sqrt{f_0(\mathbf{v})\,f_0(\mathbf{v}_0)}\;v_0\,k_2(\mathbf{v},\mathbf{v}_0)$$

$$\leq v\int d^3v_0\sqrt{f_0(\mathbf{v})\,f_0(\mathbf{v}_0)}\;k_2(\mathbf{v},\mathbf{v}_0) + (8\sigma_0/\sqrt{\pi})\int d^3y\;\exp\,(-y^2/4) \leq 96\,\pi\,\sigma_0.$$

The lemma then shows that the operator with kernel $v\,k_2(\mathbf{v},\mathbf{v}_0)$ is bounded by $96\,\pi\,\sigma_0$. Now we introduce an operator K_2^R with kernel

$$\begin{aligned} k_2^R(\mathbf{v},\mathbf{v}_0) &= k_2(\mathbf{v},\mathbf{v}_0),\; v \leq R,\\ &= 0,\quad v \geq R. \end{aligned}$$

Then K_2^R is compact since its kernel, as is readily verified, is square integrable. The difference of the two kernels satisfies

$$\begin{aligned} k_2(\mathbf{v},\mathbf{v}_0) - k_2^R(\mathbf{v},\mathbf{v}_0) &= 0,\quad v \leq R,\\ &= k_2(\mathbf{v},\mathbf{v}_0),\; v \geq R \end{aligned}$$

which implies

$$|k_2(\mathbf{v},\mathbf{v}_0) - k_2^R(\mathbf{v},\mathbf{v}_0)| \leq (v/R)\,k_2(\mathbf{v},\mathbf{v}_0)$$

and consequently

$$\|K_2 - K_2^R\| \leq R^{-1}\,\|vK_2\| \leq 96\,\pi\,\sigma_0/R.$$

Taking the limit $R \to \infty$ we obtain K_2 as the uniform limit of compact operators from which it follows that K_2 is compact.

The preceding proof is essentially the one given by Carleman; although his argument was provided for classical hard spheres it requires only a change in wording to apply to any bounded cross section. It had previously been shown by Hecke for hard spheres that $\nu^{-1}K$ is compact and this relative compactness is sufficient to prove the following main results regarding the spectrum of I. Grad has obtained the strongest result, that $(K_2)^3$ has a square-integrable kernel (which implies that K_2 is compact) if the cross section satisfies

$$\sigma(g,\theta) < C(1 + g^{\epsilon-2}) \tag{6.30}$$

where C and ϵ are positive constants with $\epsilon < 1$.

It follows from the compactness of K that I is self-adjoint, since it is the sum of the self-adjoint operator ν and the bounded self-adjoint operator $-K$.

Since I has been shown to be non-negative, its spectrum must lie along the positive real axis (including 0); we know already that 0 is an eigenvalue. Additional information about the spectrum is provided by Weyl's theorem that the essential spectrum (consisting of the continuous spectrum, limit points of the point spectrum, and eigenvalues with infinite multiplicity) of a closed operator is unchanged by the addition of a compact operator. It follows that I has an essential spectrum which coincides with the (continuous) spectrum of ν. For hard potentials this ranges from a minimum ν_0 to ∞; in this case there are no limit points of the spectrum lying below ν_0 and in particular the eigenvalue 0 is separated by a finite distance from any other point in the spectrum. For soft potentials, with $\nu(v)$ ranging from a maximum down to 0, the essential spectrum covers the same region so the eigenvalue 0 is not isolated.

For hard potentials, the nature of the spectrum assures that the Fredholm alternative applies to the integral equation (6.18): For an inhomogeneous term χ in the Hilbert space, the integral equation has a solution in the Hilbert space if and only if χ is orthogonal to the solutions to the homogeneous equation (that is, to the summational invariants). In the cases of interest the inhomogeneous terms were shown to be orthogonal to the summational invariants and so we are assured of the existence of solutions (for hard potentials); furthermore, the transport coefficients such as thermal conductivity and viscosity are finite since they are given by scalar products of functions with finite norm.

The situation is different when the potential is soft. Since the essential spectrum extends down to the origin the solution to the integral equation may or may not have finite norm, depending on the nature of the singularity. As a consequence, the transport coefficients may not be finite.

There can be additional eigenvalues of I other than 0. For example, for hard potentials there may be eigenvalues lying between 0 and ν_0. That such eigenvalues exist for classical hard spheres can be shown with the aid of the variational principle of Chapter 4. Using calculations given there we find that

$$(S_i, I\,S_i)[(S_j, S_j)]^{-1} = \frac{8}{15}\sqrt{2}\,\nu_0 = 0.754\,\nu_0.$$

Since S_i is orthogonal to the summational invariants, the eigenvalue 0 does not enter into the left-hand side. Consequently, there must be one or more nonzero eigenvalues less than ν_0 in order to pull the average of I below the start of the continuous spectrum at ν_0. It has been shown that in fact there are infinitely many such eigenvalues with a limit point at ν_0.

The preceding description of the spectrum of I applies to potentials which satisfy Grad's condition (6.30) so that the separation (6.19) with compact K can be made. In classical mechanics if the force does not strictly vanish beyond a finite range, the total cross section is infinite; incident particles undergo some scattering no matter how far they pass from the target. Of course with an infinite total cross section, the separation (6.19) is not possible. Maxwell's force law falls into this category but the eigenvalues of I for this case were obtained in the preceding chapter. There it was seen that I has a pure point spectrum with the only limit point lying at infinity; such an operator is said to have compact resolvent. It has been shown by Pao that this is fairly typical; for a large class of forces with infinite range, I has compact resolvent.

6.8. PROPERTIES OF B

Consider next the operator B defined in Equation (6.8). Again we assume Grad's condition (6.30) to be satisfied so that I can be written in the form (6.19) with compact K. To avoid the complications of dealing with more than one perturbation parameter, we again take $\mathbf{k}$ to be in a fixed direction, say along the x-axis. The function $-ikv_x + \nu$ is continuous and multiplication by it defines a closed operator with dense domain consisting of those functions Φ such that $v_x\Phi$ and $\nu\Phi$ have finite norms; the adjoint operator consists of multiplication by $+ikv_x + \nu$. Since B is obtained by adding a bounded self-adjoint operator, it follows that B is closed and has dense domain, and its formal conjugate $B^\dagger = ikv_x + I$ is its adjoint.

Considering I as an unperturbed operator and $-ikv_x$ as a perturbation, we now show that for hard-core potentials the perturbation is analytic at $k = 0$. According to a theorem of Rellich, this is the case if the perturbation is relatively bounded, meaning that there exist constants a, b such that

$$\|v_x\Phi\| \le a\|\Phi\| + b\|I\Phi\| \tag{6.31}$$

for all Φ in the domain of I. For nonvanishing cross section, say $\sigma(g, \theta) \ge \epsilon > 0$, and for hard-core potentials with $\nu(v)$ increasing linearly in v for large v, we can bound v by $\nu(v)$, say $v \le b\,\nu(v)$. Since $|v_x| \le v$, we have

$$\|v_x\Phi\| \le b\|\nu\Phi\|.$$

By the triangle inequality,

$$\|\nu\Phi\| = \|(I + K)\Phi\| \le \|I\Phi\| + \|K\Phi\| \le \|I\Phi\| + c\|\Phi\| \tag{6.32}$$

where c is the norm of the bounded operator K. Consequently,

$$\|v_x\Phi\| \le b[c\|\Phi\| + \|I\Phi\|]$$

and taking $a = bc$ we obtain the desired inequality (6.31).

For classical hard spheres, (6.31) can be formulated in specific terms as follows. The collision frequency is given by Equation (3.29):

$$\nu(v) = \frac{1}{2}\nu_0 \left\{ e^{-v^2} + (2v + v^{-1}) \int_0^v e^{-t^2}\, dt \right\}.$$

The function

$$F(v) = e^{-v^2} + (2v + v^{-1}) \int_0^v e^{-t^2}\, dt - \sqrt{\pi}\, v$$

has a derivative

$$F'(v) = v^{-1} e^{-v^2} + (2 - v^{-2}) \int_0^v e^{-t^2}\, dt - \sqrt{\pi}$$

whose vanishing implies that

$$F(v) = 2v^{-1} \int_0^v e^{-t^2}\, dt,$$

that is, the derivative can vanish only where the function is positive. Since $F(0) = 2$, $F(\infty) = 0$, the function is non-negative and we get the inequality

$$v \leq 2\nu(v)/(\sqrt{\pi}\, \nu_0).$$

Combining this with (6.32) and using the bound (6.29) for K, we obtain (6.31) with

$$a = \sqrt{3} + 16\pi^{-1/2}, \quad b = 2/(\sqrt{\pi}\, \nu_0).$$

Analyticity of the perturbation means that for each eigenvalue and eigenfunction of I there exist, for sufficiently small k, an eigenvalue and eigenfunction of B which are analytic in k in a region including $k = 0$. Thus, for hard-core interactions the linearized Chapman-Enskog expansion is convergent for sufficiently small k. It is also a consequence of analyticity that points in the spectrum of B do not move, for small k, far from their positions at $k = 0$. Consequently, the eigenvalues $\zeta_\alpha(k)$ for the normal solution, which for $k = 0$ are separated by a finite distance from the rest of the spectrum, remain separated and lie further to the left (that is, with smaller real parts) than other points in the spectrum. The normal solution, therefore, decays with time more slowly than other solutions, which is the reason for ageing to hydrodynamics.

The continuous spectrum is also perturbed analytically. However, it is possible to locate the essential spectrum without reference to a perturbation expansion. Since K is compact, the essential spectrum of B coincides with the set of values taken on by the function $-ikv_x + \nu$. For hard atoms this is a region, roughly hyperbolic in shape, which extends from ν_0 (for $v = 0$) to ∞ (for $v \to \infty$); for soft atoms it reaches from ν_0 down to the imaginary axis. In the latter case the normal solution, if it exists at all, is embedded in the essential spectrum so there is no ageing to hydrodynamics in the same sense as for hard atoms. For hard, but nonhard-core, atoms the proof of analyticity of the perturbation fails, but a normal solution may nevertheless exist and be separated from the rest of the spectrum.

The condition (6.31) means that collisions, described by I, dominate over the free-streaming described by $\mathbf{v}\cdot\nabla$; this is referred to as collision-dominated flow. A different approach must be used for rapidly varying processes. It is unlikely that the perturbation expansion converges if k is large; the radius of convergence should have a value roughly equal to $1/\ell$ where ℓ is the mean free path. Even if the normal solution could be obtained by analytic continuation from small k-values, it need not be the dominant solution for long times. As a general rule one expects the perturbation expansion to fail at a "level crossing" (which may occur at a complex value of k) where different points in the spectrum come together; the normal solution then becomes mixed with other solutions. (Such a singularity need not occur if the operator is normal at a level crossing, so the rule does not contradict the analyticity of B at $k = 0$.)

Analyticity of the perturbation is evidently limited to $\mathbf{k}$-values within a bounded region; the distribution function must have compact support in $\mathbf{k}$-space. The nonlinear collision operator contains a product of two distribution functions evaluated at the same position $\mathbf{x}$; in $\mathbf{k}$-space the product is nonvanishing in a region obtained by combining the two individual $\mathbf{k}$-vectors. Thus, in general, solutions to the nonlinear Boltzmann equation do not have compact support in $\mathbf{k}$-space, so their Chapman-Enskog expansion is not expected to be convergent. The case of uniform shear discussed in the previous chapter is unusual because the polynomial dependence on position corresponds in $\mathbf{k}$-space to derivatives of finite order evaluated at $\mathbf{k} = 0$, as in Equation (6.16). In other cases it appears that stronger nonlinearities may require slower spatial variations in order for the Chapman-Enskog expansion to provide a good approximation. It is noteworthy that the higher approximations have been used successfully in the analysis of data on sound propagation (which is linear), whereas the Burnett equations have not demonstrated a superiority over the Navier-Stokes equations in treating strongly nonlinear processes such as shocks.

The preceding analysis sheds no light on the nature of B in those cases for which the total cross section is infinite. For Maxwellian molecules a straightforward computation using the known eigenfunctions of I shows that

$$\|v_z\Phi_{nlm}\|^2 = \frac{3n + 2l + 3}{(2l + 1)(2l + 3)}(l + m + 1)(l - m + 1) + \frac{n + 1}{(2l + 1)(2l - 1)}(l + m)(l - m).$$

For $m = 0$ we get

$$\|v_z \Phi_{nl0}\|^2 = \frac{3n + 2l + 3}{(2l + 1)(2l + 3)} (l + 1)^2 + \frac{n + 1}{(2l + 1)(2l - 1)} l^2$$
$$\geq n + 1 + \tfrac{1}{2} l$$

However, it has been shown by Mott-Smith that for large n, l the eigenvalues of I increase only as $(n + l)^{1/4}$ so the Rellich criterion is not satisfied. This indicates some nonanalyticity at $k = 0$, but its detailed nature is not known.

6.9. INITIAL AND BOUNDARY VALUE PROBLEM

In this section we consider the initial and boundary value problem for the linearized Boltzmann equation. The discussion follows the analysis of Scharf, based on the theory of semi-groups.

Let T^t, $t \geq 0$ denote a one-parameter family of linear operators on a Hilbert space. It is called a semi-group if it satisfies the group combination law $T^t T^s = T^{t+s}$ and contains the identity, $T^0 = 1$. The group combination law implies that T^t is continuous in t if it is continuous at $t = 0$, or $\lim_{t\to 0} T^t = 1$ where the limit is meant in the strong sense. It is called a contraction semi-group if $\|T^t\| \leq 1$. The generator A is defined by

$$A\Phi = \lim_{t \to 0} \frac{1}{t} (T^t - 1)\Phi$$

where the limit is strong; the domain of A consists of those Φ for which the limit exists.

An operator A is said to be dissipative if its domain $\mathfrak{D}(A)$ is dense and

$$\text{Re}(\Phi, A\Phi) \leq 0$$

for all Φ in $\mathfrak{D}(A)$. It can be shown that the generator of a continuous contraction semi-group is dissipative. This and the converse are contained in

Theorem 1: An operator A is the generator of a continuous contraction semi-group if and only if it is closed and dissipative, and the adjoint $A^\dagger$ is dissipative.

The next theorem provides for the existence of a solution to the initial value problem.

Theorem 2: Let A satisfy the conditions of Theorem 1, and let $h(t)$, as an element of the Hilbert space, be continuously differentiable for $t \geq 0$. Then the equation

$$\frac{d\Phi(t)}{dt} = A\Phi(t) + h(t)$$

with the initial condition $\lim_{t\to 0} \Phi(t) = \Phi_0$ has a unique solution. The solution can be written as

$$\Phi(t) = T^t \Phi_0 + \int_0^t T^{t-s} h(s)\, ds$$

where T^t is the semi-group generated by A and $\Phi(t)$, $t > 0$, is in $\mathfrak{D}(A)$.

In applications to the linearized Boltzmann equation, we set

$$\frac{d\Phi}{dt} = A\Phi$$

where

$$A\Phi = -\mathbf{v}\cdot\nabla\Phi - I\Phi.$$

Consider first a spatially unbounded gas. The appropriate Hilbert space is $\mathcal{H} = \mathcal{L}^2(\mathbf{x}) \otimes \mathcal{L}^2_{f_0}(\mathbf{v})$ with scalar product

$$(\Phi, \Psi) = \int d^3v\, d^3x\, f_0(\mathbf{v})\overline{\Phi}\Psi$$

where for brevity the density in f_0 has been set equal to 1. We assume the total cross section to be finite so I can be separated into the two parts ν and $-K$. The part K (which is really $1 \otimes K$) is not compact in the product space but is still bounded. As the domain for A we take functions Φ which are absolutely continuous in $\mathbf{x}$ and for which $(\mathbf{v}\cdot\nabla + \nu)\Phi$ lies in $\mathcal{H}$; the last condition is equivalent to $\mathbf{v}\cdot\nabla\Phi$ and $\nu\Phi$ each belonging to $\mathcal{H}$. The operator A is then densely defined, closed, and has adjoint

$$A^\dagger\Phi = \mathbf{v}\cdot\nabla\Phi - I\Phi. \tag{6.33}$$

It only remains to show that A and $A^\dagger$ are dissipative and this is immediate:

$$\text{Re}(\Phi, A\Phi) = \text{Re}(\Phi, A^\dagger\Phi) = -\text{Re}(\Phi, I\Phi) \leq 0.$$

Thus, A is the generator of a continuous contraction semi-group. This guarantees the existence of a unique solution to the initial value problem, given by Theorem 2 with $h = 0$:

$$\Phi(t) = T^t\Phi_0.$$

Since $A^\dagger$ is dissipative and has dissipative adjoint, it also generates a semi-group which is the adjoint $(T^t)^\dagger$.

Before discussing the existence theory for a gas with boundaries, we first consider the boundary conditions one would intuitively expect to impose. For atoms incident on a boundary, the velocity satisfies $\mathbf{v}\cdot\mathbf{n} \geq 0$ where $\mathbf{n}$ is the outward-drawn normal to the boundary. The boundary affects the distribution function f for emitted atoms, that is, those with $\mathbf{v}\cdot\mathbf{n} \leq 0$. For example, with specular reflection from a fixed boundary, the normal component of $\mathbf{v}$ is reversed and the boundary condition is

$$f(\mathbf{v}) = f(\mathbf{v} - 2\mathbf{n}\,\mathbf{n}\cdot\mathbf{v}). \tag{6.34}$$

(The dependence on position has been suppressed, but of course in statements of boundary conditions $\mathbf{x}$ must lie on the boundary.)

At the other extreme, the boundary may act as a thermal reservoir, with definite values of the thermodynamic variables. Let f_B denote the Maxwellian

$$f_B(\mathbf{v}) = n_B(m/2\pi\kappa T_B)^{3/2}\exp[-m(\mathbf{v} - \mathbf{u}_B)^2/2\kappa T_B] \tag{6.35}$$

where n_B, T_B and $\mathbf{u}_B$ are thermodynamic variables characterizing the boundary. The boundary condition is $f(\mathbf{v}) = f_B(\mathbf{v})$ for $\mathbf{v}\cdot\mathbf{n} \leq 0$, which is referred to as complete accommodation. Actually specification of the preceding Maxwellian corresponds to a boundary which is a particle reservoir as well as a thermal reservoir; it emits or absorbs particles in such a way as to maintain the density at the assigned value. If however the boundary is impenetrable, the density is determined by the incident distribution, as follows. The condition that there be no net flux of atoms in or out of the boundary is

$$\int d^3v\, f(\mathbf{v})\,\mathbf{v}\cdot\mathbf{n} = 0$$

which can be written

$$\int_+ d^3v\, f(\mathbf{v})\mathbf{v}\cdot\mathbf{n} = -\int_- d^3v\, f(\mathbf{v})\,\mathbf{v}\cdot\mathbf{n} = -\int_- d^3v\, f_B(\mathbf{v})\,\mathbf{v}\cdot\mathbf{n}$$

where the subscripts $+,-$ denote the ranges of velocity with $\mathbf{v}\cdot\mathbf{n} \geq 0$ and $\mathbf{v}\cdot\mathbf{n} \leq 0$, respectively. For simplicity, we suppose the wall to have no motion normal to itself so $\mathbf{n}\cdot\mathbf{u}_B = 0$. The integral over the Maxwellian can then be carried out to give

$$n_B = \sqrt{2\pi m/\kappa T_B}\int_+ d^3v\, f(\mathbf{v})\,\mathbf{v}\cdot\mathbf{n}$$

and the boundary condition becomes

$$f(\mathbf{v}) = n_B^{-1} f_B(\mathbf{v})\int_+ d^3v'\, f(\mathbf{v}')\,\mathbf{v}'\cdot\mathbf{n}, \qquad \mathbf{v}\cdot\mathbf{n} \leq 0.$$

In practice the extremes of specular reflection and complete accommodation may not give an adequate description and sometimes a linear combination of the two is used. More generally one may assume a linear relation between the emitted and incident distributions with the form

$$f(\mathbf{v}) = \int_{+} d^3v'\, S(\mathbf{v},\mathbf{v}')\, f(\mathbf{v}'), \qquad \mathbf{v}\cdot\mathbf{n} \le 0 \tag{6.36}$$

where $S(\mathbf{v},\mathbf{v}')$ describes the scattering of atoms by the boundary.

The boundary conditions previously stated apply to the full distribution function, but in application to the linearized Boltzmann equation must be translated into conditions on Φ. This is immediate in the case of specular reflection when Equation (6.34) yields

$$\Phi(\mathbf{v}) = \Phi(\mathbf{v} - 2\mathbf{n}\,\mathbf{n}\cdot\mathbf{v}).$$

In case of complete accommodation it is necessary to linearize the emitted Maxwellian (6.35) about the Maxwellian f_0 used to obtain the linearized Boltzmann equation. This yield

$$\Phi(\mathbf{v}) = \frac{\delta n}{n} + \frac{m}{\kappa T}\mathbf{v}\cdot\delta\mathbf{u} + \left(\frac{1}{2}mv^2 - \frac{3}{2}\kappa T\right)\frac{\delta T}{\kappa T^2}, \qquad \mathbf{v}\cdot\mathbf{n} \le 0. \tag{6.37}$$

Here n, T and $\mathbf{u}$ denote the parameters in f_0 and δn, δT, $\delta\mathbf{u}$ are the deviations from them of the parameters in f_B. For an impenetrable wall, δn is again fixed by the condition that the normal component of the flux vanish, or

$$\int d^3v\, f_0(\mathbf{v})\, \Phi(\mathbf{v})\mathbf{v}\cdot\mathbf{n} = 0.$$

Dropping the condition $\mathbf{n}\cdot\mathbf{u}_B = 0$ we obtain in place of (6.37)

$$\begin{aligned}\Phi(\mathbf{v}) &= \left(\frac{1}{2}mv^2 - 2\kappa T\right)\frac{\delta T}{\kappa T^2} + \frac{1}{\kappa T}(m\mathbf{v} - \sqrt{2\pi m\kappa T}\mathbf{n})\cdot\delta\mathbf{u} \\ &\quad + \frac{1}{n}\sqrt{2\pi m\kappa T}\int_{+} d^3v'\, f_0(\mathbf{v}')\, \Phi(\mathbf{v}')\, \mathbf{v}'\cdot\mathbf{n}, \qquad \mathbf{v}\cdot\mathbf{n} \le 0.\end{aligned} \tag{6.38}$$

If the general stochastic condition (6.36) is linearized, the result is

$$\begin{aligned}\Phi(\mathbf{v}) &= f_0^{-1}(\mathbf{v})\int_{+} d^3v'\, S(\mathbf{v},\mathbf{v}')\, f_0(\mathbf{v}') - 1 \\ &\quad + f_0^{-1}(\mathbf{v})\int_{+} d^3v'\, S(\mathbf{v},\mathbf{v}')\, f_0(\mathbf{v}')\, \Phi(\mathbf{v}'), \qquad \mathbf{v}\cdot\mathbf{n} \le 0.\end{aligned}$$

In all cases, the boundary condition for Φ has the form

$$\Phi(\mathbf{v}) = \mathcal{I}(\mathbf{v}) + \int_{+} d^3v'\, R(\mathbf{v},\mathbf{v}')\, f_0(\mathbf{v}')\, \Phi(\mathbf{v}'), \qquad \mathbf{v}\cdot\mathbf{n} \le 0 \tag{6.39}$$

where $\mathcal{I}$ is a prescribed inhomogeneous term.

For a gas with boundaries, we use the Hilbert space $\mathcal{H}_V$ with scalar product

$$(\Phi, \Psi)_V = \int d^3v\, f_0(\mathbf{v}) \int_V d^3x\, \overline{\Phi}\Psi$$

where the spatial integration extends over the volume V occupied by the gas. This volume need not be finite; we could consider, for example, a half-space bounded by an infinite plane, or the region between two parallel planes, and so forth.

Consider the operator A acting on functions Φ which are absolutely continuous in $\mathbf{x}$ and are such that $\mathbf{v}\cdot\nabla\Phi$ and $\nu\Phi$ lie in $\mathcal{H}_V$. For such Φ we have

$$\text{Re}(\Phi,\Phi)_V = -\frac{1}{2}\int d^3v f_0(\mathbf{v}) \int |\Phi|^2 \mathbf{v}\cdot d\mathbf{A} - (\Phi, I\Phi)_V \le -\frac{1}{2}\int d^3v f_0(\mathbf{v}) \int |\Phi|^2 \mathbf{v}\cdot d\mathbf{A}$$

where $d\mathbf{A}$ is an element of area on the boundary. Clearly A will be dissipative if

$$\int d^3v f_0(\mathbf{v}) \int |\Phi|^2 \mathbf{v}\cdot d\mathbf{A} \ge 0, \quad \Phi \text{ in } \mathcal{D}(A)$$

and we wish to determine the conditions which will insure that this is the case. The physically interesting boundary conditions, as previously discussed, do not involve correlations between different points of the boundary and so it is natural to require

$$\int d^3v f_0(\mathbf{v}) |\Phi|^2 \mathbf{v}\cdot\mathbf{n} \ge 0, \quad \Phi \text{ in } \mathcal{D}(A)$$

for all points $\mathbf{x}$ on the boundary. We express this in a different form as follows: Let $\mathbf{v}_+$ and $\mathbf{v}_-$ denote velocities in the outward and inward directions, satisfying $\mathbf{v}_+\cdot\mathbf{n} \ge 0$, $\mathbf{v}_- = \mathbf{v}_+ - 2\,\mathbf{n}\,\mathbf{n}\cdot\mathbf{v}_+$. Then the inequality is equivalent to

$$\int_+ d^3v f_0(\mathbf{v})\mathbf{v}\cdot\mathbf{n}\,(|\Phi_+|^2 - |\Phi_-|^2) \ge 0 \tag{6.40}$$

where $\Phi_\pm = \Phi(\mathbf{v}_\pm)$. Let the emitted distribution be obtained from the incident distribution according to

$$\Phi_- = R\Phi_+. \tag{6.41}$$

The operator R characterizes the boundary and corresponds to the second term in (6.39). To express the inequality (6.40) in a convenient form, we introduce a Hilbert space $\mathcal{H}'$ associated with each point $\mathbf{x}$ on the boundary, with the scalar product

$$(\Phi, \Psi)' = \int_+ d^3v f_0(\mathbf{v})\mathbf{v}\cdot\mathbf{n}\,\overline{\Phi}\,\Psi.$$

Then (6.40) can be written

$$(\Phi,\Phi)' - (R\Phi,R\Phi)' \geq 0$$

which is to say $\|R\| \leq 1$, or R is a contraction in the space $\mathcal{H}'$.

For the domain of A we take functions Φ which are absolutely continuous in $\mathbf{x}$, have $\mathbf{v}\cdot\nabla\Phi$ and $\nu\Phi$ in $\mathcal{H}_V$, and satisfy the boundary condition (6.41) with R being a contraction. Then A is dissipative. We must also determine $A^\dagger$ and show that it is dissipative. In order for $(\Psi, A\Phi)$ to be a continuous linear functional of Φ in $\mathcal{D}(A)$ it is first of all necessary that Ψ be absolutely continuous in $\mathbf{x}$ and have $\mathbf{v}\cdot\nabla\Psi$ and $\nu\Psi$ in $\mathcal{H}_V$. For such Ψ we can integrate by parts to obtain

$$\begin{aligned}(\Psi,A\Phi) &= (A\dagger\Psi,\Phi) - \int d^3v f_0(\mathbf{v})\int \overline{\Psi}\,\Phi\,\mathbf{v}\cdot d\mathbf{A}\\ &= (A^\dagger\Psi,\Phi) - (\Psi_+,\Phi_+)' + (\Psi_-,\Phi_-)'.\end{aligned}$$

For Ψ in $\mathcal{D}(A^\dagger)$ it is necessary that the last two terms cancel,

$$(\Psi_+,\Phi_+)' = (\Psi_-,\Phi_-)' = (\Psi_-,R\Phi_+)' = (R^\dagger\Psi_-,\Phi_+)'.$$

Consequently, Ψ must satisfy the adjoint boundary condition

$$\Psi_+ = R^\dagger\Psi_- \tag{6.42}$$

where $R^\dagger$ is the adjoint of R in the space $\mathcal{H}'$.

Thus $A^\dagger$ is the operator (6.33) on a domain of functions which are absolutely continuous in $\mathbf{x}$, have $\mathbf{v}\cdot\nabla\Phi$ and $\nu\Phi$ in $\mathcal{H}_V$, and satisfy the boundary condition (6.42). If R is a contraction, then so is its adjoint and an argument like the preceding one shows that $A^\dagger$ is dissipative. Hence we have proved the existence theorem: The linearized Boltzmann equation for a gas with boundaries, with boundary conditions and domain $\mathcal{D}(A)$ as previously described, has a unique solution.

The boundary condition (6.41) is necessarily linear since the domain $\mathcal{D}(A)$ must be a linear space. In order to include inhomogeneous boundary conditions, it is necessary to transfer the inhomogeneity from the boundary conditions to the Boltzmann equation. To do this, let χ be chosen to satisfy the inhomogeneous part of the boundary condition, that is

$$\chi(\mathbf{v}) = \mathcal{I}(\mathbf{v}), \quad \mathbf{v}\cdot\mathbf{n}\leq 0.$$

This limits χ only on the boundary. In the interior we choose it so that

$$h(t) = -\frac{\partial\chi}{\partial t} - \mathbf{v}\cdot\nabla\chi - I\chi$$

is continuously differentiable in t. Then $\Psi = \Phi - \chi$ satisfies

$$\frac{\partial\Psi}{\partial t} = A\Psi + h,$$

with homogeneous boundary conditions. The proof of the existence theorem goes through as before, and the solution can be expressed in the form given in Theorem 2. This is particularly simple if the boundary conditions are time independent, as we may then choose χ to be the initial value Φ_0 and take $\Psi_0 = 0$. Theorem 2 yields

$$\Phi(t) = \Phi_0 + \int_0^t ds\, T^{t-s} h = \Phi_0 - \int_0^t ds\, T^{t-s}[\mathbf{v} \cdot \nabla \Phi_0 + I\Phi_0].$$

The quantity in brackets is not necessarily $A\Phi_0$ since Φ_0 need not be in $\mathcal{D}(A)$.

It remains to show that R is a contraction for the particular boundary conditions previously discussed. This is immediate for specular reflection since then $R = 1$. The boundary condition (6.38) can be put in the form

$$R\Phi_+ = \int_+ d^3v' f_0(\mathbf{v}')\Phi(\mathbf{v}')\mathbf{v}' \cdot \mathbf{n} \left[\int_+ d^3v' f_0(\mathbf{v}')\mathbf{v}' \cdot \mathbf{n} \right]^{-1} = (1, \Phi)'/(1,1)'$$

so R is a contraction:

$$(R\Phi_+, R\Phi_+)' = |(1,\Phi)'|^2/(1,1)' \leq (\Phi, \Phi)'.$$

Thus, we have obtained the results, due to Scharf, that the boundary conditions considered, namely specular reflection, complete accommodation, and the stochastic boundary condition if the corresponding R is a contraction, each leads to a well-posed problem with a unique solution.

6.10. GREEN'S FUNCTIONS

With the aid of Green's functions, the solution to the linearized Boltzmann equation can be expressed in a form which makes the boundary values explicit. Let $G(\mathbf{x}, \mathbf{v}, \mathbf{x}_0, \mathbf{v}_0, t)$ satisfy the linearized Boltzmann equation

$$\left(\frac{\partial}{\partial t} + \mathbf{v} \cdot \nabla + I \right) G(\mathbf{x}, \mathbf{v}, \mathbf{x}_0, \mathbf{v}_0, t) = 0,$$

the initial condition

$$f_0(\mathbf{v}_0)G(\mathbf{x}, \mathbf{v}, \mathbf{x}_0, \mathbf{v}_0, 0) = \delta(\mathbf{x} - \mathbf{x}_0)\delta(\mathbf{v} - \mathbf{v}_0),$$

and the homogeneous boundary condition (6.41)

$$G_- = RG_+, \quad \mathbf{x} \text{ on boundary}$$

where

$$G_\pm(\mathbf{x}, \mathbf{v}, \mathbf{x}_0, \mathbf{v}_0, t) = G(\mathbf{x}, \mathbf{v}_\pm, \mathbf{x}_0, \mathbf{v}_0, t), \ \mathbf{x} \text{ on boundary}.$$

The solution to the preceding equations can be written

$$f_0(\mathbf{v}_0)G(\mathbf{x}, \mathbf{v}, \mathbf{x}_0, \mathbf{v}_0, t) = T^t\, \delta(\mathbf{x} - \mathbf{x}_0)\delta(\mathbf{v} - \mathbf{v}_0)$$

where T^t acts on the delta functions as functions of $\mathbf{x}, \mathbf{v}$ with $\mathbf{x}_0, \mathbf{v}_0$ held fixed. The solution Φ to the linearized Boltzmann equation can be expressed in terms of G by

$$\begin{aligned}\Phi(\mathbf{x}, \mathbf{v}, t) &= (T^t\Phi)(\mathbf{x}, \mathbf{v}) \\ &= \int d^3v_0 d^3x_0 f_0(\mathbf{v}_0) G(\mathbf{x}, \mathbf{v}, \mathbf{x}_0, \mathbf{v}_0, t)\Phi(\mathbf{x}_0, \mathbf{v}_0).\end{aligned}$$

The Green's function satisfies the principle of reciprocity, which can be shown as follows. Define the adjoint function $G^\dagger$ by

$$\left(\frac{\partial}{\partial t} - \mathbf{v} \cdot \nabla + I\right) G^\dagger(\mathbf{x}, \mathbf{v}, \mathbf{x}_0, \mathbf{v}_0, t) = 0$$

$$f_0(\mathbf{v}_0) G^\dagger(\mathbf{x}, \mathbf{v}, \mathbf{x}_0, \mathbf{v}_0, 0) = \delta(\mathbf{x} - \mathbf{x}_0)\delta(\mathbf{v} - \mathbf{v}_0),$$

$$G^\dagger_+ = R^\dagger G^\dagger_-, \quad \mathbf{x} \text{ on boundary.}$$

Then $G^\dagger$ can be represented by

$$f_0(\mathbf{v}_0) G^\dagger(\mathbf{x}, \mathbf{v}, \mathbf{x}_0, \mathbf{v}_0, t) = (T^t)^\dagger \delta(\mathbf{x} - \mathbf{x}_0)\delta(\mathbf{v} - \mathbf{v}_0)$$

and we have

$$((T^t)^\dagger\Phi)(\mathbf{x}, \mathbf{v}) = \int d^3x_0\, d^3v_0 f_0(\mathbf{v}_0) G^\dagger(\mathbf{x}, \mathbf{v}, \mathbf{x}_0, \mathbf{v}_0, t)\Phi(\mathbf{x}_0, \mathbf{v}_0).$$

It then follows from the definition of the adjoint plus the fact that G is real that

$$G^\dagger(\mathbf{x}, \mathbf{v}, \mathbf{x}_0, \mathbf{v}_0, t) = G(\mathbf{x}_0, \mathbf{v}_0, \mathbf{x}, \mathbf{v}, t). \tag{6.43}$$

On the other hand we have

$$\tau(\mathbf{v} \cdot \nabla + I)\tau = -\mathbf{v} \cdot \nabla + I \tag{6.44}$$

where τ is the operator (6.11) which reverses the direction of the velocity. Insofar as the boundary conditions are concerned, if $\Phi_- = R\Phi_+$ then $\tau\Phi$ satisfies $(\tau\Phi)_+ = (\tau R \tau)(\tau\Phi)_-$. If we assume

$$\tau R \tau = R^\dagger \tag{6.45}$$

then τ changes the boundary conditions to their adjoint so $\tau A \tau = A^\dagger$, implying that

$$\tau T^t \tau = (T^t)^\dagger.$$

This in turn shows that

$$G^\dagger(\mathbf{x}, \mathbf{v}, \mathbf{x}_0, \mathbf{v}_0, t) = G(\mathbf{x}, -\mathbf{v}, \mathbf{x}_0, -\mathbf{v}_0, t)$$

and using (6.43) we obtain the desired statement of reciprocity:

$$G(\mathbf{x}, \mathbf{v}, \mathbf{x}_0, \mathbf{v}_0, t) = G(\mathbf{x}_0, -\mathbf{v}_0, \mathbf{x}, -\mathbf{v}, t) \tag{6.46}$$

Basically reciprocity is due to invariance of the atomic equations of motion to time reversal, which at the level of the Boltzmann equation manifests itself in Equations (6.44) and (6.45).

To obtain the solution to a problem with inhomogeneous boundary conditions, we use Equation (6.46) to write the equation for G as

$$\left(-\frac{\partial}{\partial t_0} - \mathbf{v}_0\cdot\nabla_0 + I_0\right) G(\mathbf{x},\mathbf{v},\mathbf{x}_0,\mathbf{v}_0,t - t_0) = 0, \tag{6.47}$$

where the subscript on ∇_0, I_0 indicates that they operate on G as a function of $\mathbf{x}_0, \mathbf{v}_0$ with $\mathbf{x}, \mathbf{v}$ held fixed. Let Φ be a solution to

$$\left(\frac{\partial}{\partial t_0} + \mathbf{v}_0\cdot\nabla_0 + I_0\right) \Phi_0(\mathbf{x}_0,\mathbf{v}_0,t_0) = 0 \tag{6.48}$$

with inhomogeneous boundary conditions. Multiply (6.47) by $f_0\Phi$, (6.48) by f_0G, subtract and integrate to obtain

$$\int_0^t dt_0 \int d^3x_0\, d^3v_0 f_0(\mathbf{v}_0) \left[\left(\frac{\partial}{\partial t} + \mathbf{v}_0\cdot\nabla_0\right) G(\mathbf{x},\mathbf{v},\mathbf{x}_0,\mathbf{v}_0,t - t_0)\Phi(\mathbf{x}_0,\mathbf{v}_0,t_0)\right.$$

$$\left. + G(\mathbf{x},\mathbf{v},\mathbf{x}_0,\mathbf{v}_0,t - t_0)I_0\Phi(\mathbf{x}_0,\mathbf{v}_0,t_0) - \Phi(\mathbf{x}_0,\mathbf{v}_0,t_0)I_0G(\mathbf{x},\mathbf{v},\mathbf{x}_0,\mathbf{v}_0,t - t_0)\right] = 0.$$

The last two terms cancel because of the symmetry of I. On carrying out the time integration in the first term and using the initial condition for G, we get

$$\begin{aligned}\Phi(\mathbf{x},\mathbf{v},t) &= \int d^3x_0\, d^3v_0 f_0(\mathbf{v}_0) G(\mathbf{x},\mathbf{v},\mathbf{x}_0,\mathbf{v}_0,t)\Phi(\mathbf{x}_0,\mathbf{v}_0,0)\\ &\quad - \int_0^t dt_0 \int d^3v_0 \int d\mathbf{A}_0\cdot\mathbf{v}_0 G(\mathbf{x},\mathbf{v},\mathbf{x}_0,\mathbf{v}_0,t - t_0)\Phi(\mathbf{x}_0,\mathbf{v}_0,t_0)\end{aligned} \tag{6.49}$$

where $d\mathbf{A}_0$ is an element of area on the boundary at $\mathbf{x}_0$. In the last term we use the adjoint boundary condition for G to obtain

$$\begin{aligned}\int d\mathbf{A}_0\cdot\int d^3v_0 f_0(\mathbf{v}_0)\mathbf{v}_0 G\Phi &= \int d\mathbf{A}_0\cdot\int_+ d^3v_0 f_0(\mathbf{v}_0)\mathbf{v}_0(G_+\Phi_+ - G_-\Phi_-)\\ &= \int d\mathbf{A}_0\cdot\int_+ d^3v_0 f_0(\mathbf{v}_0)\mathbf{v}_0[(R^\dagger G)_-\Phi_+ - G_-\Phi_-]\\ &= \int d\mathbf{A}_0\cdot\int_+ d^3v_0 f_0(\mathbf{v}_0)\mathbf{v}_0 G_-(R\Phi_+ - \Phi_-).\end{aligned}$$

Thus (6.49) asks for $\Phi_- - R\Phi_+$ which is just the prescribed inhomogeneous term $\mathcal{J}$, so

$$\Phi(\mathbf{x},\mathbf{v},t) = \int d^3x_0 d^3v_0 f_0(\mathbf{v}_0) G(\mathbf{x},\mathbf{v},\mathbf{x}_0,\mathbf{v}_0,t)\Phi(\mathbf{x}_0,\mathbf{v}_0,0)$$
$$- \int_0^t \int d\mathbf{A}_0 \cdot \int_+ d^3v_0 \mathbf{v}_0 G(\mathbf{x},\mathbf{v},\mathbf{x}_0,\mathbf{v}_0,t - t_0)\mathcal{J}(\mathbf{x}_0,\mathbf{v}_0,t_0).$$

This expresses the solution Φ in terms of its initial values and the prescribed boundary values.

The Green's function can be obtained by variational differentiation of Φ with respect to its initial value:

$$f_0(\mathbf{v}_0)G(\mathbf{x},\mathbf{v},\mathbf{x}_0,\mathbf{v}_0,t) = \delta\Phi(\mathbf{x},\mathbf{v},t)/\delta\Phi(\mathbf{x}_0,\mathbf{v}_0,0)$$

which can also be written as

$$f_0(\mathbf{v})G(\mathbf{x},\mathbf{v},\mathbf{x}_0,\mathbf{v}_0,t) = \delta f(\mathbf{x},\mathbf{v},t)/\delta f(\mathbf{x}_0,\mathbf{v}_0,0))|_{f=f_0}.$$

Higher derivatives, such as

$$\delta^2 f(\mathbf{x},\mathbf{v},t)/\delta f(\mathbf{x}_0,\mathbf{v}_0,0)\delta f(\mathbf{x}_1,\mathbf{v}_1,0)|_{f=f_0}$$

can be calculated from the full (i.e., nonlinearized) Boltzmann equation, and such Green's functions would include nonlinear effects.

Notes on Chapter 6

Discussions of the linearized Boltzmann equation are given in Wang Chang and Uhlenbeck [1970], and Cercignani [1969]. For the linearized Enskog equation, see Gross and Wisnivesky [1968], and Kamgar-Parsi, Cohen, and de Schepper [1987].

6.1 The derivation of the linearized hydrodynamic equations from the linearized Boltzmann equation, including the effects of the ageing period, is discussed in Sirovich [1963a, b, c], Ernst [1970], Bixon, Dorfman, and Mo [1971], Ellis and Pinsky [1975], and Pinsky [1976]. Solutions to the Boltzmann equation for the Lorentz gas (which is linear) are discussed in Hauge [1970].

6.2 See Titulaer and Felderhof [1980] and Felderhof and Titulaer [1980]. The solution to the linearized Navier-Stokes equations has been discussed by de Schepper, van Rijs, and Cohen [1985] over the full range of values for **k**.

6.5 For the description of sound on the basis of the linearized Boltzmann equation, see Wang Chang and Uhlenbeck [1970], Pekeris, Alterman, Finkelstein, and Frankowski [1962], Sirovich and Thurber [1963], Buckner and Ferziger [1966b], Grad [1966], Foch [1967a], Foch and Uhlenbeck [1967b], Foch and Ford [1970], and Ford [1973]. The perturbation solution (6.14) is worked out in McLennan [1966b].

6.6 The polynomial solutions were obtained for Maxwellian molecules by Grad [1963a] and for other force laws by McLennan [1965].

6.7 See Hecke [1922]. Proofs that K is compact are given in Carleman [1957], Grad [1963b], and Dorfman [1963]. Extensive numerical results on the eigenvalues and eigenfunctions of I for hard spheres are given in Alterman, Frankowski, and Pekeris [1962]. Kuscer and Williams [1967] show that the eigenvalues have a limit point at ν_0. Forces with infinite range are discussed in Pao [1974a, b].

6.8 Analyticity of the perturbation is shown in McLennan [1965]. For the Rellich theorem, see Kato [1980]. The spectrum of B is discussed in McLennan [1966b], Skvortsov [1966, 1967], Arsen'ev [1968], Sirovich and Thurber [1969], and Illner and Kuscer [1979]; for similar results in the neutron diffusion problem, see Corngold [1964, 1968]. The growth of the eigenvalues of I for large n, l is discussed in Mott-Smith [1954].

6.9 See Scharf [1967, 1969]. The theorems on semi-groups are discussed in, for example, Yoshida [1980] and Pazy [1983]. Existence and uniqueness for stationary flow are treated in Guiraud [1968, 1972].

6.10 See Case [1969].

7

THE GENERALIZED BOLTZMANN EQUATION

7.1. INTRODUCTION

In this chapter we consider the generalization of the Boltzmann equation to include effects which become important when the density is increased. The Boltzmann equation is limited to low density primarily because it incorporates only binary collisions. Hence, the main task in proceeding to higher density is to include collisions in clusters of three particles, four particles, and so forth.

The derivation of the Boltzmann equation given in Chapter 3 serves the purpose for which it was intended but gives little indication of the direction in which to proceed when it is desired to go beyond the low-density approximation. As described in the following, a systematic approach can be developed on the basis of a sequence of equations known as the BBGKY hierarchy; this sequence itself can be derived from the Liouville equation. Eventually a generalized Boltzmann equation is obtained in which corrections to the collision operator are expressed as a power series in the density. The term in this series following that corresponding to Boltzmann's equation involves the dynamics of a cluster of three particles and is called the triple collision operator; the next term depends on the dynamics of the four-body problem, and so on.

Clearly the dependence of the collision operator on density leads to a density dependence of the transport coefficients. There are also density-dependent contributions to the thermal conductivity and viscosity, referred to as collisional transfer, which are due to the spatial extent of the force between particles; one particle can hand energy and momentum over to another across a distance of the order of the force range. Collisional transfer shows up in the form of the collision operator, which even for binary collisions differs from that of Boltzmann in that the two distribution functions are not evaluated at the same position. In addition the fluxes of energy and momentum contain terms depending on the potential energy which lead to collisional transfer.

Actually it turns out that the expansion of the collision operator in powers of the density is not convergent. The spatial extent of a binary collision is limited by the force range, but the sequence of events which occurs in a higher-order collision operator can, depending on the initial conditions, stretch out over long distances. Fortunately in the three-body problem, the region of phase space leading to such extended collisions shrinks to small volumes and the triple collision operator is well defined. However, four-body (and more) collisions have a long-distance behavior such that the corresponding collision operator does not exist, at least on a domain which is large enough to include functions of physical interest. As a result the transport coefficients are not analytic functions of the density. In this chapter the main interest will be in the binary and triple collision operators, and we will proceed formally with the density expansion as if it were well defined. (In two dimensions the divergence occurs one order sooner, so the triple collision operator is not defined.)

7.2. DISTRIBUTION FUNCTIONS AND THE LIOUVILLE EQUATION

Consider a system of N identical particles, enclosed in a volume V. The system will be described by classical mechanics, with the particles interacting in pairs by a central conservative force. To exclude bound states, only repulsive forces will be considered. In addition the discussion is limited to short-ranged forces; the Coulomb interaction between charged particles requires a separate treatment.

Let $\mathbf{q}_\alpha, \mathbf{p}_\alpha$ denote the position and momentum of the α'th particle, referred to rectangular Cartesian coordinates. The six components of $\mathbf{q}_\alpha, \mathbf{p}_\alpha$ will be denoted by X_α. Let $U_{\alpha\beta} = U(\mathbf{q}_\alpha - \mathbf{q}_\beta)$ be the potential energy of interaction between particles α and β. For central forces, U is a function only of the distance $r_{\alpha\beta} = |\mathbf{q}_\alpha - \mathbf{q}_\beta|$.

The Hamiltonian for the system is

$$H_N = K_N + U_N \tag{7.1}$$

where K_N and U_N are the total kinetic and potential energies,

$$K_N = \sum_\alpha (p_\alpha^2/2m), \quad U_N = \frac{1}{2}\sum_{\alpha\beta} U_{\alpha\beta}. \tag{7.2}$$

Here m is the mass of a particle and the sums extend from 1 to N. Terms with $\alpha = \beta$ are not included as they would correspond to an interaction of a particle with itself; rather than indicating this explicitly, we adopt the convention that $U_{\alpha\alpha} = 0$. The factor $\frac{1}{2}$ is needed so that the interaction between a given pair of particles is counted only once.

Statistical mechanics is introduced through a Gibbs ensemble, with an associated probability density on the phase space of the N-particle system. Let F_N be the distribution function, defined so that $F_N\, dX_1..dX_N$ is the probability that the system has its coordinates and momenta in the indicated ranges. F_N is a function of all X_α and of the time t. Its definition requires the normalization

$$\int dX_1...dX_N\, F_N = 1 \tag{7.3}$$

where dX denotes $d^3q\, d^3p$. The components of momentum range from $-\infty$ to ∞ while the coordinates are restricted to the volume V.

Any distribution function must, from its definition, be non-negative. In addition it follows from the identity of particles that F_N is a symmetric function of the X_α.

If A is a function of the X_α its average, denoted by $\langle A \rangle$, is given by

$$\langle A \rangle = \int dX_1...dX_N\, F_N\, A.$$

The macroscopic variables are interpreted as averages, and they are then functions (or functionals) of any parameters such as temperature, chemical potential, and so on, which occur in F_N.

The time dependence of F_N is determined by the Liouville equation

$$\partial F_N/\partial t + \{F_N, H_N\} = 0$$

where $\{,\}$ denotes the Poisson bracket

$$\{A, B\} = \sum_\alpha \left(\frac{\partial A}{\partial \mathbf{q}_\alpha} \cdot \frac{\partial B}{\partial \mathbf{p}_\alpha} - \frac{\partial B}{\partial \mathbf{q}_\alpha} \cdot \frac{\partial A}{\partial \mathbf{p}_\alpha} \right).$$

We assume the potential energy to be sufficiently regular that the Liouville equation can be solved for $F_N(t)$ for all t in terms of its initial value.

The Liouville equation holds in the interior of the volume V to which the gas is confined. The walls of the container will be represented by the boundary condition that if $\mathbf{q}_\alpha$ lies on a wall, then F_N is an even function of the normal component of $\mathbf{p}_\alpha$. An alternative procedure is to include a term in the Liouville equation to represent collisions with the walls.

The full distribution function F_N contains more information than will normally be given, or of interest. Reduced distribution functions F_s, $s =$

$1,\ldots,N-1$, are defined by

$$F_s(1,\ldots,s) = V^s \int dX_{s+1}\ldots dX_N\, F_N(1,\ldots,N). \tag{7.4}$$

For brevity the particle labels $1,\ldots,N$ are used to denote the phases $X_1,\ldots,X_N$; when the time is listed as an argument it will be separated by a semicolon as in $F_s(1,\ldots,s;t)$. The selected set of particles in F_s has been taken to be those numbered $1,\ldots,s$ but of course the definition applies for any set of s particles. The F_s, $s < N$ are not normalized in the manner of Equation (7.3) for F_N, but satisfy instead

$$\int dX_1\ldots dX_s\, F_s(1,\ldots,s) = V^s, \quad s = 1,\ldots,N-1. \tag{7.5}$$

In the discussion of the Boltzmann equation, velocity was used as the independent variable rather than momentum, so $f\,d^3v$ corresponds to $F_1\,d^3p$. The two distribution functions also differ in their normalization, as $\int d^3v\,d^3q\,f = N$ while $\int dX\,F_1 = V$. Here the F_s are normalized as shown because the density then occurs in a convenient way as an expansion parameter.

The reduced distribution functions of low order are sufficient to determine the averages of most quantities of physical interest, which are usually sums of functions of variables referring to one or a pair of particles. Suppose A is a sum of one-particle functions:

$$A = \sum_\alpha a_\alpha.$$

An example is the total momentum, for which $a_\alpha = \mathbf{p}_\alpha$. Because of the identity of particles, each term in the sum has the same average so $\langle A\rangle$ can be obtained from F_1:

$$\langle A\rangle = \int dX_1\ldots dX_N\, Na_1\, F_N = (N/V)\int dX_1\, a_1\, F_1(1).$$

Similarly, the average of a sum over pairs can be calculated from F_2. Let B be such a sum,

$$B = \frac{1}{2}\sum_{\alpha\neq\beta} b_{\alpha\beta}.$$

We assume each term to be symmetric in the two particles, $b_{\alpha\beta} = b_{\beta\alpha}$ and the factor $\frac{1}{2}$ has been included so any term is counted only once. The average of B is

$$\langle B\rangle = \frac{1}{2}\int dX_1\ldots dX_N\, N(N-1)b_{1,2}\, F_N$$

$$= [N(N-1)/2V^2]\int dX_1\, dX_2\, b_{1,2}\, F_2\,(1,2).$$

7.3. THE BBGKY HIERARCHY

The F_s satisfy a coupled set of equations which can be derived from the Liouville equation. Integrating over $dX_{s+1}...dX_N$ and multiplying by V^s, we obtain a result which can be written as

$$\partial F_s/\partial t + \{F_s, H_s\} = V^s \int dX_{s+1}...dX_N\{H_N - H_s, F_N\}. \qquad (7.6)$$

Here H_s is the Hamiltonian for the subset of s particles, and is given by equations like (7.1) and (7.2) but with the sums running only from 1 to s. In the difference $H_N - H_s$ the kinetic energies for particles $1,...,s$ cancel. The kinetic energy of one of the remaining particles, say the $(s + 1)$'st, gives a contribution

$$V^s \int dX_{s+1}...dX_N(1/m)\mathbf{p}_{s+1} \cdot \partial F_N/\partial \mathbf{q}_{s+1}.$$

The integration over $\mathbf{q}_{s+1}$ can be converted to a surface integral at the walls and the integral over $\mathbf{p}_{s+1}$ then vanishes because of the boundary condition. The potential energy of a pair in the set $s + 1,...,N$, say $s + 1, s + 2$, yields

$$-V^s \int dX_{s+1}...dX_N \left(\frac{\partial F_N}{\partial \mathbf{p}_{s+1}} \cdot \frac{\partial U_{s+1,s+2}}{\partial \mathbf{q}_{s+1}} + \frac{\partial F_N}{\partial \mathbf{p}_{s+2}} \cdot \frac{\partial U_{s+1,s+2}}{\partial \mathbf{q}_{s+2}}\right)$$

$$= - V^s \int dX_{s+1}...dX_N \left(\frac{\partial}{\partial \mathbf{p}_{s+1}} - \frac{\partial}{\partial \mathbf{p}_{s+2}}\right) \cdot F_N \frac{\partial U_{s+1,s+2}}{\partial \mathbf{q}_{s+1}}.$$

We assume F_N to vanish for infinite values of momentum, so this expression vanishes after the integration over $\mathbf{p}_{s+1}$ and $\mathbf{p}_{s+2}$ is carried out. In (7.6) we are left with potentials $U_{\alpha\beta}$ where α is in the set $1,...,s$ and β in $s + 1,...,N$. The $N - s$ choices for β give the same contribution and we take $\beta = s + 1$ as a typical value. The remaining integral over $dX_{s+2}...dX_N$ can be written in terms of F_{s+1} and Equation (7.6) reduces to

$$\partial F_s/\partial t + \{F_s, H_s\} = [(N - s)/V] \int dX_{s+1} \{U^{(s+1)}, F_{s+1}\} \qquad (7.7)$$

Here $U^{(s+1)}$ denotes the sum

$$U^{(s+1)}(\mathbf{q}_1, ..., \mathbf{q}_{s+1}) = \sum_{\alpha=1}^{s} U(\mathbf{q}_\alpha - \mathbf{q}_{s+1}). \qquad (7.8)$$

Note that $U^{(s+1)}$, $s > 1$, is symmetric under an interchange of particles in the set $1,...,s$ but not under an interchange involving particle number $s + 1$; $U^{(2)}$ is the same as the pair potential U.

The sequence of equations (7.7), with s running from 1 to N, is referred to as the BBGKY hierarchy, the letters denoting the authors Bogoliubov, Born and Green, Kirkwood, and Yvon. The hierarchy is equivalent to the

Liouville equation; indeed the equation with $s = N$ is exactly the Liouville equation.

We now take the thermodynamic limit. This is a limit in which the volume V is increased without bound while the density $n = N/V$ is held constant. The hierarchy becomes

$$\partial F_s/\partial t + \{F_s, H_s\} = n\int dX_{s+1}\{U^{(s+1)},\ F_{s+1}\},\ s = 1,2,\ldots \tag{7.9}$$

The constant n is to be distinguished from the local density, which may vary from point to point. Henceforth we will use the hierarchy in this limiting form.

For finite volume Equation (7.4) implies

$$\int dX_s\, F_s(1,\ldots,s) = V\, F_{s-1}\,(1,\ldots,s-1).$$

Hence a given distribution function determines all others of lower order. In the thermodynamic limit neither this relation nor the normalization condition (7.5) can be maintained, but a rough equivalent is the absence of statistical correlations among particles which are far apart in space. If for example the s'th particle is moved far from the others we expect $F_s(1,..,s) \to F_{s-1}(1,.., s-1)\, F_1(s)$, in asymptotic agreement with the preceding relation.

7.4. RELATION OF THE HIERARCHY TO THE KINETIC EQUATION

The Poisson bracket with H_1 is $\{F_1, H_1\} = \mathbf{v}\cdot\nabla F_1$ where $\mathbf{v} = \mathbf{p}/m$ is the velocity. Hence, the first equation of the hierarchy can be written

$$\left(\frac{\partial}{\partial t} + \mathbf{v}\cdot\nabla\right) F_1 = nJ \tag{7.10}$$

where

$$J(1) = \int dX_2\{U(1,2),\ F_2(1,2)\}. \tag{7.11}$$

On the left-hand side Equation (7.10) is the same as the Boltzmann equation, but it is not a closed equation for F_1 since F_2 occurs in J.

The simplest closure assumption is $F_2 = 0$ which gives the Knudsen equation, but this is tenable only when the density is so low that interactions can be neglected altogether. Another possibility is to neglect statistical correlations between the particles,

$$F_2(1,2) = F_1(1)\, F_1(2), \tag{7.12}$$

which is called the assumption of molecular chaos. (Actually "molecular chaos" has the more general connotation that each $F_s, s > 1$, is a product of

F_1's.) When the Poisson bracket in J is worked out, this yields

$$\partial F_1/\partial t + \mathbf{v}\cdot\nabla F_1 + \nabla_{\mathbf{p}}\cdot\mathcal{F} F_1 = 0 \tag{7.13}$$

where $\nabla_{\mathbf{p}}$ denotes the divergence in momentum space and

$$\mathcal{F}(1) = -n\frac{\partial}{\partial\mathbf{q}_1}\int dX_2\, F_1(2)\, U(1,2).$$

Thus $\mathcal{F}(1)$ is the total force exerted on particle 1 by the other particles and appears in the equation in the same way as would an external force. It is determined in a self-consistent manner by integrating over F_1, which in turn is determined from Equation (7.13). Equation (7.13) is called the Vlasov equation and is used for an elementary description of plasmas, since the self-consistent field gives a good account of the long-range Coulomb force. It does not provide an adequate description of short range forces and will not be considered further here.

If (7.12) is satisfied at an initial time, then after a period of interaction each X_1, X_2 depends on both of their initial values so F_2 does not factorize into a product of F_1's; molecular chaos cannot be maintained over a period of time because interactions develop correlations. However, it was shown by Bogoliubov that molecular chaos can be used in a limiting sense to construct a time-independent functional, whereby F_2 at a given time is determined by F_1 at the same time. The construction of the time-independent functional will be carried out in detail (to a certain order of approximation), but the method can be briefly summarized as follows. Let $\mathbf{F} = \{F_1, F_2, \ldots\}$ denote the set of reduced distribution functions. If they are specified at an initial time t_0 then their values at time t are determined by the hierarchy. We express this formally in terms of the solution operator $\mathcal{S}(t)$ to the hierarchy by

$$\mathbf{F}(t) = \mathcal{S}(t_0 - t)\mathbf{F}(t_0).$$

Let the assumption of molecular chaos be imposed at the initial time,

$$F_s(1,\ldots,s,t_0) = F_1(1,t_0)\cdots F_1(s,t_0). \tag{7.14}$$

This means that $\mathbf{F}(t_0)$ is a functional of $F_1(t_0), \mathbf{F}(t_0) = \mathbf{F}[F_1(t_0)]$. By solving the initial value problem backwards, we may express $F_1(t_0)$ in terms of the F_s's at time $t, F_1(t_0) = (\mathcal{S}(t - t_0)\mathbf{F}(t))_1$. This yields

$$\mathbf{F}(t) = \mathcal{S}(t_0 - t)\mathbf{F}[(\mathcal{S}(t - t_0)\mathbf{F}(t))_1].$$

We now have F_s at time t depending on F_1 at the same time but the functional relation explicitly involves the initial time t_0. The next step is to take the limit $t_0 \to -\infty$. Provided the limit exists, it yields the time-independent functional

$$\mathbf{F} = \lim_{\tau\to-\infty} \mathcal{S}(\tau)\mathbf{F}[(\mathcal{S}(-\tau)\mathbf{F})_1]. \tag{7.15}$$

The functional dependence expressed by the preceding equation is consistent (at least formally) with the hierarchy; that is, it is maintained with time. Correlations are present [since (7.15) does not have the form (7.14)] but they are allowed to develop in a way which is consistent with the dynamics. Once the time-independent functional for F_2 is known, it can be substituted into Equation (7.11) to make J into a collision operator acting on F_1. The first hierarchy equation becomes a closed equation for F_1, which is called the kinetic equation or the generalized Boltzmann equation.

In practice the time-independent functional can be constructed only by means of an expansion in powers of the density. The collision operator J is then also expressed as a density expansion,

$$J = J_2 + nJ_3 + \dots \tag{7.16}$$

As indicated by the notation, J_2 is the binary collision operator, J_3 the triple collision operator, and so forth. Except in the spatially homogeneous case, J_2 is not the same as Boltzmann's collision operator, and to distinguish them the latter will be denoted by J_B. Of course this expansion does not make the density dependence completely explicit, since F_1 itself is in general a function of the density.

7.5. STREAMING OPERATORS

The expansion (7.16) will be developed in the following, but first we introduce a set of operators which are used repeatedly. Let $S_s(t)$ be the operator on functions $A(X_1, \dots, X_s)$ which replaces $X_1, \dots, X_s$ by the values $X_1(t), \dots, X_s(t)$ to which they evolve in t seconds under the action of H_s:

$$S_s(t)\, A(X_1, \dots, X_s) = A(X_1(t), \dots, X_s(t)).$$

The $S_s(t)$ are called streaming operators and satisfy

$$\frac{\partial}{\partial t} S_s(t)A = \{S_s(t)A, H_s\}. \tag{7.17}$$

They also have the group properties

$$S_s(0) = 1, \quad S_s(t)S_s(u) = S_s(t+u), \quad S_s^{-1}(t) = S_s(-t).$$

Since the Hamiltonian is a constant of the motion, it satisfies

$$S_s(t)\{A, H_s\} = \{S_s(t)\, A, H_s\}.$$

The streaming operator for the noninteracting motion will be denoted by $S_0(t)$; the number of particles will not be indicated since it can usually be determined from context. S_0 is a product of $S_1(t)$'s for the particles involved

$$S_0(1, \dots, s; t) = S_1(1;t) \cdots S_1(s;t)$$

and has the equation of motion

$$\frac{\partial}{\partial t} S_0(t)\, A \;=\; \{S_0(t)\, A, K\}$$

where K is the kinetic energy.

It is sometimes useful to write the streaming operators in an expontial form,

$$S_s(t) \;=\; \exp\,(t\mathscr{L}_s). \tag{7.18}$$

$\mathscr{L}_s$ is called the Liouville operator and satisfies

$$\mathscr{L}_s A \;=\; \{A, H_s\}.$$

The corresponding form for S_0 is

$$S_0(t) \;=\; \exp(t\mathscr{L}_0) \tag{7.19}$$

where $\mathscr{L}_0$ satisfies

$$\mathscr{L}_0 A \;=\; \{A, K\}.$$

According to Liouville's theorem the change of variables $X_1, \ldots, X_s \to X_1(t), \ldots, X_s(t)$ where the time dependence is governed by H_s, has unit Jacobian. Hence, we have

$$\int dX_1 \ldots dX_s\, S_s(t)\, A \;=\; \int dX_1 \ldots dX_s\, A.$$

This equation also holds if S_s is replaced by S_0.

Let $S_s^X(t)$ denote the mapping of a point $X = \{X_1, \ldots, X_s\}$ in the s-particle phase space to $X(t) = \{X_1(t), \ldots, X_s(t)\}$, so $X(t) = S_s^X(t)X$. The streaming operators can be represented by $S_s(t)A(X) = A(S_s^X(t)X)$. When two streaming operators are applied in succession, an inversion of the order of the mappings is to be noted, for example,

$$S_s(t)S_0(t')A(X) \;=\; S_s(t)A(S_0^X(t')X) \;=\; A(S_0^X(t')S_s^X(t)X).$$

7.6. BINARY COLLISION OPERATOR

To calculate the time-independent functional in detail, it is necessary to solve the initial value problem for the hierarchy. This is done by a perturbation procedure in which the right-hand side of the hierarchy is treated as a small quantity. For the purpose of deriving the binary collision operator, it is sufficient to obtain the time-independent functional to lowest order. We then neglect the right-hand side altogether, so F_1 and F_2 satisfy

$$\frac{\partial}{\partial t}F_1 + \{F_1, H_1\} = 0, \quad \frac{\partial}{\partial t}F_2 + \{F_2, H_2\} = 0.$$

The solution for F_2, with molecular chaos imposed at time t_0, is

$$F_2(1,2;t) = S_2(t_0 - t)\, F_2(1,2;t_0) = S_2(t_0 - t)\, F_1(1;t_0)\, F_1(2;t_0).$$

In addition F_1 satisfies

$$F_1(t_0) = S_1(t - t_0)F_1(t)$$

so

$$F_2(1, 2; t) = S_2(t_0 - t)S_0(t - t_0)F_1(1; t)F_1(2; t)$$

The time-independent functional, to lowest order in the density, is obtained by taking the limit $t_0 \rightarrow -\infty$. Let an operator Ω be defined by

$$\Omega = \lim_{\tau \rightarrow \infty} S_2(-\tau)S_0(\tau).$$

Then

$$F_2(1, 2) = \Omega\, F_1(1)F_1(2)$$

where the argument t has been dropped since all F's are evaluated at the same time. Bogoliubov's binary collision operator is now obtained by using (7.11):

$$J_2 = \int dX_2\{U(1, 2), \Omega\, F_1(1)F_2(2)\}.$$

Frequently the Poisson bracket with the potential energy is abbreviated as θ,

$$\theta A(1,2) = \{U(1,2), A(1,2)\} = \frac{\partial U(1,2)}{\partial \mathbf{q}_1} \cdot \left(\frac{\partial}{\partial \mathbf{p}_1} - \frac{\partial}{\partial \mathbf{p}_2}\right) A(1,2). \tag{7.20}$$

Then J_2 takes the form

$$J_2 = \int dX_2 \theta\, \Omega\, F_1(1)F_1(2). \tag{7.21}$$

The operator Ω is called the wave operator. Actually there are two wave operators, defined by

$$\Omega_{\pm} = \lim_{\tau \rightarrow \mp\infty} S_2(\tau)S_0(-\tau) \tag{7.22}$$

and related by time reversal. The one of interest at the moment is Ω_+, which we continue to denote by Ω.

The group property of the streaming operators implies

$$S_2(t)\Omega = \lim_{\tau \rightarrow \infty} S_2(t - \tau)S_0(\tau) = \lim_{\tau \rightarrow \infty} S_2(-\tau)S_0(t + \tau) = \Omega S_0(t)$$

which is called the intertwining property of the wave operator. Using the exponential forms (7.18) and (7.19), we get an equivalent expression of intertwining

$$\mathcal{L}_2\Omega = \Omega\mathcal{L}_0$$

which yields

$$\theta\Omega = (\mathcal{L}_0 - \mathcal{L}_2)\Omega = \mathcal{L}_0\Omega - \Omega\mathcal{L}_0.$$

Changing variables to relative and mass-center coordinates, we get

$$\mathcal{L}_0 = m^{-1}(\mathbf{p}_1 \cdot \nabla_1 + \mathbf{p}_2 \cdot \nabla_2) = \boldsymbol{g} \cdot \nabla_r + \boldsymbol{G} \cdot \nabla_c$$

where $\boldsymbol{g} = \mathbf{v}_1 - \mathbf{v}_2$, $\boldsymbol{G} = \frac{1}{2}(\mathbf{v}_1 + \mathbf{v}_2)$ are the relative and center-of-mass velocities and ∇_r, ∇_c denote derivatives with respect to the corresponding coordinates. Since the total momentum is conserved by both the free and interacting motions, Ω has no effect on the mass-center position or velocity so

$$\theta\Omega = \boldsymbol{g} \cdot \nabla_r \Omega - \Omega\, \boldsymbol{g} \cdot \nabla_r.$$

The resulting form for J_2 is

$$J_2 = \int dX_2[\boldsymbol{g} \cdot \nabla_r \Omega - \Omega\, \boldsymbol{g} \cdot \nabla_r] F_1(1) F_1(2). \tag{7.23}$$

In the spatially homogeneous case this reduces to

$$J_2 = \int dX_2\, \boldsymbol{g} \cdot \nabla_r \Omega\, F_1(\mathbf{p}_1) F_1(\mathbf{p}_2).$$

To evaluate the integral we use cylindrical polar coordinates with z-axis parallel to $\boldsymbol{g}$. The radial variable is the same as the impact parameter b so

$$J_2 = \int d^3p_2 \int b\,db\,d\varphi \int_{-\infty}^{\infty} dz \boldsymbol{g} \frac{d}{dz} \Omega\, F_1(\mathbf{p}_1) F_1(\mathbf{p}_2)$$

$$= \int d^3p_2 \int \boldsymbol{g}\; b\,db\,d\varphi [\Omega F_1(\mathbf{p}_1) F_1(\mathbf{p}_2)|_{z=\infty} - \Omega F_1(\mathbf{p}_1) F_1(\mathbf{p}_2)|_{z=-\infty}].$$

In this case S_0 has no effect so Ω is the same as the long-time limit of $S_2(-\tau)$. If z is negative there is no interaction during the receding motion so the last term in brackets in the preceding equation is the same as $F_1(\mathbf{p}_1)F_1(\mathbf{p}_2)$. In the first term $S_2(-\tau)$ takes the particles back through a collision and converts $\mathbf{p}_1, \mathbf{p}_2$ to their values $\mathbf{p}_1', \mathbf{p}_2'$ before the collision. The resulting form for J_2 is just the Boltzmann collision operator.

When F_1 depends on position, the binary collision operator differs from the Boltzmann collision operator because of the difference in position of the two colliding particles. To analyze this effect, we represent Ω in terms of the corresponding operator on phase space. Let X denote a point in the two-particle phase space, and define the wave operator Ω^X by

$$\Omega^X = \lim_{\tau \to \infty} S_2^X(\tau) S_0^X(-\tau).$$

(When the two wave operators are distinguished as in (7.22), the one above is Ω_+^X.) If the potential energy does not support bound states, it can be shown that

$$(\Omega^X)^{-1} = \lim_{\tau\to\infty} S_0^X(\tau)S_2^X(-\tau).$$

Then we have

$$\Omega\ A(X) = \lim_{\tau\to\infty} A(S_0^X(\tau)S_2^X(-\tau)X) = A((\Omega^X)^{-1}X).$$

Let $(\mathbf{r}^+, \boldsymbol{g}^+)$ denote the relative position and velocity obtained from the action of $(\Omega^X)^{-1}$:

$$(\mathbf{r}^+, \boldsymbol{g}^+) = (\Omega^X)^{-1}(\mathbf{r}, \boldsymbol{g}).$$

If A is a function of $(\mathbf{r}, \boldsymbol{g})$ then

$$\theta\Omega\ A(\mathbf{r}, \boldsymbol{g}) = \boldsymbol{g}\cdot\nabla_r\ A(\mathbf{r}^+, \boldsymbol{g}^+) - \boldsymbol{g}^+\cdot\nabla_{r^+}\ A(\mathbf{r}^+, \boldsymbol{g}^+). \tag{7.24}$$

The intertwining property for Ω^X implies

$$S_0^X(t)(\Omega^X)^{-1} = (\Omega^X)^{-1}\ S_2^X(t)$$

from which we get

$$S_0^X(t)(\mathbf{r}^+, \boldsymbol{g}^+) = (\Omega^X)^{-1}\ S_2^X(t)(\mathbf{r}, \boldsymbol{g}).$$

Differentiation with respect to t yields

$$\boldsymbol{g}\cdot\nabla_r\mathbf{r}^+ = \boldsymbol{g}^+.$$

If this is used in (7.24) it is seen that the dependence of A on $\mathbf{r}^+$ is unaffected by the differentiation so

$$\theta\Omega A = \boldsymbol{g}\cdot\nabla_r\ A(\mathbf{R}, \boldsymbol{g}^+)|_{\mathbf{R}=\mathbf{r}^+}.$$

Here $\mathbf{R}$ is a parameter which is held fixed in the differentiation then set equal to $\mathbf{r}^+$. To proceed further it is necessary to solve the equations of motion to obtain $\mathbf{r}^+, \boldsymbol{g}^+$ as functions of $\mathbf{r}, \boldsymbol{g}$. The collision operator depends on the trajectory of the particles throughout the course of the collision and not just on limiting initial and final values. A simple explicit form can be obtained only for hard spheres, which will be considered separately.

The scale of variation governing transport is the mean free path, which at low density is large in comparison to the force range. Thus, the spatial extent of a collision has no effect on the low-density limit of the transport coefficients but must be taken into account when density-dependent corrections to the transport coefficients are calculated.

7.7. TRIPLE COLLISION OPERATOR

To evaluate the time-independent functional to higher order in the density, it is necessary to obtain the solution to the hierarchy to higher order. To put the hierarchy into a suitable form we use $S_s(t)$ as an "integrating factor" to obtain

$$\begin{aligned}\frac{\partial}{\partial t} S_s(t)F_s(t) &= S_s(t)\left[\frac{\partial F_s(t)}{\partial t} + \{F_s(t), H_s\}\right] \\ &= S_s(t)n\int dX_{s+1}\{U^{(s+1)}, F_{s+1}(t)\}.\end{aligned}$$

Integrating we get

$$S_s(t)F_s(t) = F_s(0) + n\int_0^t dt_1 S_s(t_1)\int dX_{s+1}\{U^{(s+1)}, F_{s+1}(t_1)\} \tag{7.25}$$

or

$$F_s(t) = S_s(-t)F_s(0) + n\int_0^t dt_1\int dX_{s+1}S_s(t_1 - t)\{U^{(s+1)}, F_{s+1}(t_1)\}.$$

The leading term is the first approximation which was used in the preceding section. We iterate to generate the next approximation

$$\begin{aligned}F_s(t) &= S_s(-t)F_s(0) \\ &\quad + n\int_0^t dt_1\int dX_{s+1}\, S_s(t_1 - t)\{U^{(s+1)}, S_{s+1}(-t_1)F_{s+1}(0)\} + \dots\end{aligned}$$

The integral over time can be carried out with the aid of the identity

$$\begin{aligned}&\int_0^t dt_1\int dX_{s+1}\, S_s(t_1 - t)\,\{U^{(s+1)}, S_{s+1}(-t_1)F_{s+1}(0)\} \\ &= \int_0^t dt_1\int dX_{s+1}\, S_s(t_1 - t)\{H_{s+1} - H_s, S_{s+1}(-t_1)F_{s+1}(0)\} \\ &= \int_0^t dt_1 \frac{\partial}{\partial t_1}\int dX_{s+1}S_s(t_1 - t)S_{s+1}(-t_1)F_{s+1}(0) \\ &= \int dX_{s+1}[S_{s+1}(-t) - S_s(-t)]\, F_{s+1}(0)\end{aligned}$$

which yields

$$F_s(t) = S_s(-t)F_s(0) + n\int dX_{s+1}[S_{s+1}(-t) - S_s(-t)]F_{s+1}(0) + \dots \tag{7.26}$$

This can be written in a slightly different form by the change of variables

$X_{s+1} \to X_{s+1}(-t)$. Writing out the particle labels for clarity, we get

$$
\begin{aligned}
F_s(1, \ldots, s; t) &= S_s(1, \ldots, s; -t)F_s(1, \ldots, s; 0) \\
&\quad + n\int dX_{s+1}[S_{s+1}(1, \ldots, s+1; -t) - S_s(1, \ldots, s; -t)S_1(s+1; -t)] \\
&\qquad \times F_{s+1}(1, \ldots, s+1; 0) + \ldots
\end{aligned}
$$

If the F_s are given at an initial time t_0 rather than 0, then the argument $-t$ of the streaming operators is replaced by $t_0 - t$. Written out in detail, the results for F_1 and F_2 through terms linear in the density are

$$
\begin{aligned}
F_1(1;t) &= S_1(1;t_0 - t)F_1(1;t_0) + n\int dX_2[S_2(1,2;t_0 - t) \\
&\qquad - S_0(1,2;t_0 - t)]F_2(1,2;t_0), \\
F_2(1,2;t) &= S_2(1,2;t_0 - t)F_2(1,2;t_0) + n\int dX_3[S_3(1,2,3;t_0 - t) \\
&\qquad - S_2(1,2;t_0 - t)S_1(3;t_0 - t)]F_3(1,2,3;t_0).
\end{aligned}
\tag{7.27}
$$

Imposing the condition of molecular chaos at time t_0 we get

$$
\begin{aligned}
F_1(1; t) &= S_1(1; t_0 - t)F_1(1; t_0) \\
&\quad + n\int dX_2[S_2(1, 2; t_0 - t) - S_0(1, 2; t_0 - t)]F_1(1; t_0)F_1(2; t_0), \\
F_2(1, 2; t) &= S_2(1, 2; t_0 - t)F_1(1; t_0)F_1(2; t_0) \\
&\quad + n\int dX_3[S_3(1, 2, 3; t_0 - t) \\
&\quad - S_2(1, 2; t_0 - t)S_1(3; t_0 - t)]F_1(1; t_0)F_1(2; t_0)F_1(3; t_0).
\end{aligned}
$$

It is necessary to solve the first of these for $F_1(t_0)$ in terms of $F_1(t)$. Through terms of first order in density, the result is

$$
\begin{aligned}
F_1(1; t_0) = S_1(1; t - t_0)\Big\{ F_1(1; t) - n\int dX_2[S_2(1, 2; t_0 - t) \\
- S_0(1, 2; t_0 - t)]S_0(1, 2; t - t_0)F_1(1; t)F_1(2; t)\Big\}.
\end{aligned}
$$

This is to be substituted into the equation for F_2, and a straightforward calculation gives

$$
\begin{aligned}
F_2(1, 2; t) &= S_2(1, 2; t_0 - t)S_0(1, 2; t - t_0)F_1(1; t)F_1(2; t) \\
&\quad + n\int dX_3\{S_3(1,2,3;t_0 - t)S_0(1,2,3;t - t_0) - S_2(1,2;t_0 - t)S_0(1,2;t - t_0) \\
&\qquad \times [S_2(1,3;t_0 - t)S_0(1,3;t - t_0) \\
&\qquad + S_2(2,3;t_0 - t)S_0(2,3;t - t_0) - 1]\}F_1(1;t)F_1(2;t)F_1(3;t).
\end{aligned}
$$

We next take the limit $t_0 \to \infty$ and obtain for the time-independent functional

$$F_2(1, 2) = \Omega(1, 2)F_1(1)F_1(2) + n\int dX_3\{\Omega^{(3)}(1, 2, 3) - \Omega(1, 2)[\Omega(1, 3) + \Omega(2, 3) - 1]\}F_1(1)F_1(2)F_1(3). \tag{7.28}$$

Here $\Omega^{(3)}$ denotes the three-body wave operator, defined by

$$\Omega^{(3)} = \lim_{\tau \to \infty} S_3(-\tau)S_0(\tau).$$

The triple collision operator is obtained by substituting (7.28) into (7.11) with the result

$$J_3 = \int dX_2 dX_3\{U^{(3)}(1, 2, 3), \Omega^{(3)}(1, 2, 3) - \Omega(1, 2)[\Omega(1, 3) + \Omega(2, 3) - 1]\}F_1(1)F_1(2)F_1(3).$$

The triple collision operator is much more complex than the binary collision operator, since it involves the dynamics of the three-body problem. Detailed calculations of the contributions of three-body collisions to the transport coefficients have been carried out only for hard spheres.

7.8. THE ENSKOG EQUATION

For a moderately dense gas of hard spheres, Enskog obtained a simple generalization of the Boltzmann equation which takes into account the difference in positions of two colliding particles, and also includes an approximate description of three-body effects. Enskog's equation, particularly with certain phenomenological modifications, gives a good account of transport under some conditions which even approach liquid densities.

The hierarchy cannot be applied without modification to hard spheres, which do not have a potential energy. The binary collision operator in the form (7.23) can be worked out for hard spheres and if F_1 depends on position, the result includes Enskog's correction for the separation of the two particles. However, we will start with the hierarchy for hard spheres, which is interesting in itself.

Hard spheres move freely except on contact, where they make perfectly elastic collisions. (There is some ambiguity if three or more hard spheres come into contact simultaneously, but we will neglect this possibility.) There is no penetration into the overlapping region $r = |\mathbf{q}_\alpha - \mathbf{q}_\beta| < D$ where D is the hard-sphere diameter and $\mathbf{q}_\alpha, \mathbf{q}_\beta$ are the coordinates of a pair. Although the potential energy is not defined, one can think of it as the limit of "soft spheres" whose potential $U(r)$ vanishes for $r > D$ and for $r < D$ has a large and constant value. The distribution function must be such as to exclude the overlapping region. The equilibrium distribution function, for example, contains a factor $\exp(-U_N/\kappa T)$ (where U_N is the total potential energy) and this factor is seen, by the interpretation as the limit of soft spheres, to vanish in the overlapping region and elsewhere to be equal to 1.

Consider a system of two hard spheres, and let X denote the coordinates and momenta of both particles. If X is a nonoverlapping configuration, then the equations of motion determine the configuration $X(t) = S^X(t)X$ to which it evolves in time t. A collision will occur if the relative position $\mathbf{r} = \mathbf{q}_1 - \mathbf{q}_2$ and relative velocity $\boldsymbol{g} = \mathbf{v}_1 - \mathbf{v}_2$ are such that the line $\mathbf{r} + \boldsymbol{g}t$ intersects the sphere $r = D$ for some t. This is the case if the impact parameter

$$b = \sqrt{r^2 - (\mathbf{r}\cdot\hat{\boldsymbol{g}})^2},$$

where $\hat{\boldsymbol{g}}$ is the unit vector along $\boldsymbol{g}$, satisfies $b \leq D$. In general there are two points of intersection located at

$$D\mathbf{e}_{\pm} = \mathbf{r} - \hat{\boldsymbol{g}}[\mathbf{r}\cdot\hat{\boldsymbol{g}} \pm \sqrt{D^2 - b^2}]$$

where $\mathbf{e}_{\pm}$ are unit vectors. The collision will occur for positive t if the particles are approaching one another, or $\mathbf{r}\cdot\boldsymbol{g} < 0$, and the point of contact is then at $D\mathbf{e}_+$; similarly, a collision occurs for negative t at $D\mathbf{e}_-$ if $\mathbf{r}\cdot\boldsymbol{g} > 0$. After the collision, the relative velocity is changed to

$$\boldsymbol{g}' = \boldsymbol{g} - 2\mathbf{e}\,\mathbf{e}\cdot\boldsymbol{g}$$

where $\mathbf{e}$ is $\mathbf{e}_+$ or $\mathbf{e}_-$ depending on whether the collision occurs at a positive or negative time. The discontinuity on the sphere introduces an ambiguity which will be resolved by defining $S^X(t), t \geq 0$ as the limit of $S^X(t - s)$ when s approaches zero from above. This means that if the particles are just brought into contact at time t, then $S^X(t)$ does not change their velocities. If X is such that the particles are in contact and approaching, then $\lim_{t\to 0} S^X(t)X \neq X$ since the velocities are changed into those after the collision; however, $S^X(0) = 1$ so $S^X(t)$ is not continuous in t. For negative times we define $S^X(t)$ so that $S^X(-t) = \lim_{s\to 0} S^X(-t + s)$; then $S^X(-t) = \tau S^X(t)\tau$ where τ is the time-reversal operator: $\tau f(\mathbf{q},\mathbf{p}) = f(\mathbf{q}, -\mathbf{p})$. Note however that $S^X(-t)$ is not the inverse of $S^X(t)$.

To define the operator $S(t)$ on functions, let the function f vanish in the overlapping region. If X is nonoverlapping so is $S^X(t)X$ and so the definition $(S(t)f)(X) = f(S^X(t)X)$ makes sense; furthermore, the function $S(t)f$ vanishes in the overlapping region. Let W be the characteristic function for the nonoverlapping region ($W(X) = 1$ if X is nonoverlapping and 0 otherwise). Then $S(t)Wf$ is defined even if f does not vanish for overlapping points, and satisfies $S(t)Wf = WS(t)f$.

Now consider the derivation of the first equation of the hierarchy. Since the potential energy is not defined, it is necessary to proceed somewhat differently. Except when two or more particles come in contact, the motion of the system is the same as for free particles and so F_N satisfies

$$\partial F_N/\partial t + \sum_\alpha \mathbf{v}_\alpha \cdot \nabla_\alpha F_N = 0.$$

Using the definition of F_1 we obtain

$$\partial F_1/\partial t = -\mathbf{v}_1 \cdot \nabla_1 F_1 - n\int dX_2 \mathbf{v}_2 \cdot \nabla_2 F_2. \tag{7.29}$$

Here the integral extends over the nonoverlapping region, and with the divergence theorem can be converted to an integral over the surface where the two particles are in contact. This surface moves with particle 1; to deal with a fixed surface we first work in the frame of reference in which particle 1 is at rest, then transform back to the lab frame. With $\mathbf{v}_1 = 0$, Equation (7.29) becomes

$$\partial F_1/\partial t = \int d^3p_2 \int d\mathbf{A} \cdot \mathbf{v}_2 F_2$$

where $d\mathbf{A}$ is an element of area on the surface $r = D$, taken with its normal in the direction from $\mathbf{q}_1$ to $\mathbf{q}_2$. Transforming back to the original frame of reference, we get

$$\partial F_1/\partial t + \mathbf{v}_1 \cdot \nabla_1 F_1 = \int d^3p_2 \int d\mathbf{A} \cdot \boldsymbol{g} F_2.$$

The element of area on the sphere can be written

$$d\mathbf{A} = D^2 d^2e\, \mathbf{e}$$

where $\mathbf{e}$ is the unit vector such that $\mathbf{q}_2 = \mathbf{q}_1 + D\mathbf{e}$. This yields

$$\frac{\partial F_1(\mathbf{q}_1, \mathbf{p}_1)}{\partial t} + \mathbf{v}_1 \cdot \nabla_1 F_1(\mathbf{q}_1, \mathbf{p}_1) = D^2 \int d^3p_2 \int d^2e\, \mathbf{e} \cdot \boldsymbol{g}\, F_2(\mathbf{q}_1, \mathbf{p}_1, \mathbf{q}_1 + D\mathbf{e}, \mathbf{p}_2)$$

which is the first equation of the hierarchy for hard spheres.

To express F_2 as a functional of F_1, we impose the molecular-chaos condition a time s before the collision. The hard-sphere interaction is instantaneous and so we may take s to be small; the long-time limiting procedure used earlier is not needed. To obtain F_2 at the time of collision, we must shift forward by s seconds which gives

$$F_2(1, 2) = \lim_{s \to 0} S(-s) F_1(1) F_1(2).$$

If the two particles are approaching, or $\mathbf{e} \cdot \boldsymbol{g} \leq 0$, then $S(-s)$ has no effect in the limit. If they are receding, or $\mathbf{e} \cdot \boldsymbol{g} \geq 0$, it converts the momenta $\mathbf{p}_1, \mathbf{p}_2$ into those before the collision, $\mathbf{p}_1', \mathbf{p}_2'$. We divide the region of integration over d^2e into the two parts with $\mathbf{e} \cdot \boldsymbol{g}$ positive and negative and make the change of variables $\mathbf{e} \to -\mathbf{e}$ in the latter. This yields

$$\partial F_1/\partial t + \mathbf{v}_1 \cdot \nabla_1 F_1 = nD^2 \int d^3p_2 \int_+ d^2e \, \mathbf{e} \cdot \boldsymbol{g}$$
$$\times \{F_1(\mathbf{q}_1, \mathbf{p}_1')F_1(\mathbf{q}_1 + D\mathbf{e}, \mathbf{p}_2') - F_1(\mathbf{q}_1, \mathbf{p}_1)F_1(\mathbf{q}_1 - D\mathbf{e}, \mathbf{p}_2)\}.$$

Here the subscript + means that the integration extends over the half-sphere with $\mathbf{e} \cdot \boldsymbol{g} \geq 0$. To return to the notation used for the Boltzmann equation, we replace $nF_1 d^3p$ by fd^3v and obtain

$$\left(\frac{\partial}{\partial t} + \mathbf{v}_1 \cdot \nabla_1\right) f(\mathbf{q}, \mathbf{v}_1)$$
$$= D^2 \int d^3v_2 \int_+ d^2e \, \mathbf{e} \cdot \boldsymbol{g}[f(\mathbf{q}, \mathbf{v}_1')f(\mathbf{q} + D\mathbf{e}, \mathbf{v}_2') - f(\mathbf{q}, \mathbf{v}_1)f(\mathbf{q} - D\mathbf{e}, \mathbf{v}_2)].$$

The difference from the Boltzmann equation is that the two f's are evaluated at different positions.

Actually the two particles are not completely uncorrelated just before the collision. Correlations between them can be included by writing

$$F_2(1, 2) = F_1(1)F_1(2)\chi(1, 2).$$

In equilibrium χ is the pair correlation function, $\chi = g^{(2)}(r)$. For a gas of hard spheres, the density expansion of $g^{(2)}$ through linear terms is

$$\begin{aligned} g^{(2)}(r) &= 0, && r < D, \\ &= 1 + \frac{4}{3}\pi n D^3 \left[1 - \frac{3r}{4D} + \frac{1}{16}\left(\frac{r}{D}\right)^3\right], && D < r \leq 2D, \\ &= 1, && r > 2D. \end{aligned}$$

For χ, Enskog took the pair correlation function evaluated at contact

$$\chi = 1 + \frac{5}{12}\pi n D^3. \tag{7.31}$$

The density n is related to f by

$$n = \int d^3v \, f$$

and in general n, and consequently χ, is a function of position. Enskog used χ at the midpoint between the two particles, and the collision operator is then

$$J_E = D^2 \int d^3v_2 \int_+ d^2e \, \mathbf{e} \cdot \boldsymbol{g} \Big[f(\mathbf{q}_1, \mathbf{v}_1')f(\mathbf{q}_1 + D\mathbf{e}, \mathbf{v}_2')\chi(\mathbf{q}_1 + \tfrac{1}{2}D\mathbf{e})$$
$$- f(\mathbf{q}_1, \mathbf{v}_1)f(\mathbf{q}_1 - D\mathbf{e}, \mathbf{v}_2)\chi(\mathbf{q}_1 - \tfrac{1}{2}D\mathbf{e}) \Big].$$

The factor χ increases the rate of collisions by an amount which is proportional to the probability (in equilibrium) of finding the two particles in contact. One can think of χ as being due to two effects. One is that of excluded volume; the volume available to any particle is the total volume less that occupied by other particles. Hence, the density is effectively increased by the factor $[1 - \frac{4}{3}\pi n D^3]^{-1}$. The other effect is due to shielding; two particles which are about to collide may be prevented from doing so by collision with a third. Calculation of the shielding effect yields a reduction in the collision rate by a factor $1 - \frac{11}{12}\pi n D^3$. When these two factors are combined the result, through terms linear in the density, is Equation (7.31).

The transport coefficients on the basis of the Enskog theory will be worked out in Chapter 12. The difference in position of the two colliding particles contributes to collisional transfer of energy and momentum, and there is also a collisional-transfer contribution which is analogous to potential energy terms in the fluxes. The collisional-transfer contributions result in an increase in the thermal conductivity and viscosity. This is counterbalanced by the increase in collision frequency due to χ, which tends to decrease the transport coefficients. As will be seen, the former outweighs the latter so the net effect is an increase; however, coefficients of diffusion decrease with increasing density.

Both the excluded volume and the shielding effects are due to the presence of a third particle, and so the introduction of χ into the collision operator amounts to an approximate description of the triple collision operator. The approximation is actually quite good. The thermal conductivity and viscosity have been evaluated numerically by Sengers and colleagues on the basis of the kinetic equation with the triple collision operator, and the results for the density-dependent corrections differ from those obtained from the Enskog theory by only a few percent. However, the representation of triple collisions by χ does introduce some error and ambiguity which is more serious when diffusion in a multicomponent gas, or the Burnett coefficients, are treated.

7.9. SOLUTION TO THE HIERARCHY

The general term in the density expansion for the solution to the hierarchy has a fairly simple form, which we will exhibit and then verify its correctness. The series to all orders in n can be written as

$$\begin{aligned} F_s(1, \ldots, s; t) = &\sum_{k=0}^{\infty} \frac{n^k}{k!} \int dX_{s+1} \ldots dX_{s+k} \sum_{j=0}^{k} (-1)^j \binom{k}{j} \\ &\times S_{s+k-j}(1, \ldots, s+k-j) S_0(s+k-j+1, \ldots, s+k) \\ &\times F_{s+k}(1, \ldots, s+k; 0). \end{aligned} \tag{7.32}$$

Here $\binom{k}{j}$ denotes the binomial coefficient,

$$\binom{k}{j} = \frac{k!}{j!(k-j)!}.$$

For brevity the time argument $-t$ has been dropped in the streaming operators. (In the term with $j = 0$, $S_0(s + k + 1, \ldots, s + k)$ is to be interpreted as 1.)

To confirm that the preceding series is a solution, we first separate out the term with $k = 0$. Making the change of variables $k \to k + 1$ we get

$$\begin{aligned}
F_s(1, \ldots, s; t) &= S_s(1, \ldots, s)F_s(1, \ldots, s; 0) \\
&\quad + \sum_{k=0}^{\infty} \frac{n^{k+1}}{(k+1)!} \int dX_{s+1} \ldots dX_{s+k+1} \sum_{j=0}^{k+1} (-1)^j \binom{k+1}{j} \\
&\quad \times S_{s+k+1-j}(1, \ldots, s + k + 1 - j)S_0(s + k + 2 - j, \ldots, s + k + 1) \\
&\quad \times F_{s+k+1}(1, \ldots, s + k + 1; 0).
\end{aligned}$$

In the sum over j, we now separate out the term with $j = k + 1$ which yields

$$\begin{aligned}
F_s(1, \ldots, s; t) &= S_s(1, \ldots, s)\Bigg[F_s(1, \ldots, s; 0) \\
&\quad + \sum_{k=0}^{\infty} \frac{(-n)^{k+1}}{(k+1)!} \int dX_{s+1} \ldots dX_{s+1+k} S_0(s + 1, \ldots, s + 1 + k) \\
&\quad \times F_{s+1+k}(1, \ldots, s + 1 + k; 0)\Bigg] \\
&\quad + \sum_{k=0}^{\infty} \frac{n^{k+1}}{(k+1)!} \int dX_{s+1} \ldots dX_{s+1+k} \sum_{j=0}^{k} (-1)^j \binom{k+1}{j} \\
&\quad \times S_{s+1+k-j}(1, \ldots, s + 1 + k - j)S_0(s + 2 + k - j, \ldots, s + 1 + k) \\
&\quad \times F_{s+1+k}(1, \ldots, s + 1 + k; 0).
\end{aligned}$$

When this is substituted into the hierarchy the first term gives no contribution because of Equation (7.17) (with $-t$ in place of t). Using Equation (7.17) with s replaced by $s + 1 + k - j$ we obtain

$$\begin{aligned}
&\partial F_s(1, \ldots, s; t)/\partial t + \{F_s(1, \ldots, s; t), H_s\} \\
&\quad = n \sum_{k=0}^{\infty} \frac{n^k}{k!} \int dX_{s+1} \ldots dX_{s+1+k} \sum_{j=0}^{k} (-1)^j \binom{k}{j} (k + 1 - j)^{-1} \\
&\qquad \times \{Y, S_{s+1+k-j}(1, \ldots, s + 1 + k - j) \\
&\qquad \times S_0(s + 2 + k - j, \ldots, s + 1 + k)F_{s+1+k}(1, \ldots, s + 1 + k; 0)\}
\end{aligned}$$

where Y is the abbreviation

$$Y = H_{s+1+k-j}(1,\ldots,s + 1 + k - j)$$
$$+ K_{j-1}(s + 2 + k - j,\ldots,s + 1 + k) - H_s(1,\ldots,s).$$

The kinetic energy terms for particles $1,\ldots,s$ cancel while those from the set $s + 1,\ldots,s + k + 1$ give no contribution after the integration is carried out. All that remains in Y is the difference in potential energies, $U_{s+1+k-j}(1,\ldots,s + 1 + k - j) - U_s(1,\ldots,s)$. The pair potentials in $U_{s+1+k-j}$ which involve only particles in the set $s + 1,\ldots,s + 1 + k - j$ give no contribution after integration, and interactions between particles in the set $1,\ldots,s$ cancel. We are left with a sum of potentials with one particle in the set $1, \ldots, s$ and the other in $s + 1,\ldots,s + 1 + k - j$. The integral is symmetric in the $k + 1 - j$ particles in the latter set, and so there are $k + 1 - j$ identical contributions. For a typical such contribution, we choose the one with $j = k$ for which the net potential energy is just the quantity denoted previously by $U^{(s+1)}$. It is then straightforward to pick out the series (7.32) for F_{s+1} and the result is the hierarchy equation for F_s. Hence, it is confirmed that (7.32) is a solution to the hierachy.

The series (7.32) can be written in a different form which is symmetric in the particles whose phases are integrated. Consider the integral

$$\int dX_{s+1}\ldots dX_{s+k} \sum_{j=0}^{k} (-1)^j \binom{k}{j}$$
$$\times S_{s+k-j}(1,\ldots,s + k - j)S_0(s + k - j + 1,\ldots,s + k) F_{s+k}(1,\ldots,s + k;0).$$

Let the set of particles numbered $s + 1,\ldots,s + k$ be partitioned into disjoint subsets $\{j\}$ and $\{k - j\}$ containing j and $k - j$ members, respectively. The set $\{j\}$, which need not be those numbered $s + k - j + 1,\ldots,s + k$, may be taken as the particles acted on by S_0, and we indicate this by writing $S_0(\{j\})$. Then S_{s+k-j} acts on particles $1,\ldots,s$ plus those in the set $\{k - j\}$, which we indicate by $S_{s+k-j}(1,\ldots,s,\{k - j\})$. Thus, an equivalent form of the above expression is

$$\int dX_{s+1}\ldots dX_{s+k} \sum_{j=0}^{k} (-1)^j \binom{k}{j} S_{s+k-j}(1,\ldots,s,\{k - j\})\, S_0(\{j\})$$
$$\times F_{s+k}(1,\ldots,s + k;0).$$

The number of partitions into $\{j\}$ and $\{k - j\}$ is just the binomial coefficient, which we may leave out and instead sum over all partitions. We define an operator $T(1,\ldots,s;s + 1,s + k)$, which is symmetric in $1,\ldots,s$ and in $s + 1,\ldots,s + k$, by

$$T(1,\ldots,s;s + 1,\ldots,s + k) = \sum_{j=0}^{k} (-1)^j \sum_{P} S_{s+k-j}(1,\ldots,s,\{k - j\})\, S_0(\{j\})$$

where Σ_P denotes the sum over partitions into $\{j\}$ and $\{k - j\}$. Then the desired symmetric form for (7.32) is

$$F_s(1,\ldots,s;t) =$$

$$\sum_{k=0}^{\infty} \frac{n^k}{k!} \int dX_{s+1} \ldots dX_{s+k} T(1,\ldots,s;s+1,\ldots,s+k)\, F_{s+k}(1,\ldots,s+k;0).$$

The subscript on F_s is redundant when it can be determined from the arguments. Furthermore, the labeled particles can be replaced by any sets $\{s\},\{k\}$. This provides for a more abbreviated form of the series,

$$F(\{s\};t) = \sum_{k=0}^{\infty} \frac{n^k}{k!} \int dX_{\{k\}}\, T(\{s\};\{k\};-t)\, F(\{s\} + \{k\};0).$$

In this notation the formula for T is

$$T(\{s\};\{k\};t) = \sum_{j=0}^{k} (-1)^j \sum_P S(\{s\} + \{k - j\};t)\, S_0(\{j\};t).$$

Still another form can be obtained by using Liouville's theorem to replace S_0 by the interacting streaming operator. This yields

$$F(\{s\};t) = \sum_{k=0}^{\infty} \frac{n^k}{k!} \int dX_{\{k\}}\, T'(\{s\};\{k\};-t)\, F(\{s\} + \{k\};0)$$

where

$$T'(\{s\};\{k\};t) = \sum_{j=0}^{k} (-1)^j \sum_P S(\{s\} + \{k - j\};t) S(\{j\};t).$$

The coefficients in the series solution for $F_s(t)$ have close algebraic similarities with the cluster functions in the virial expansion for equilibrium distribution functions. Indeed the solution was constructed by Green and Cohen on the basis of this similarity. The solution in the form just given is due to Lewis.

7.10. LONG-TIME BEHAVIOR OF THE SOLUTION

Since a solution to the hierarchy is available, it is natural to ask why there is any need to develop a kinetic equation. The answer has to do with the behavior of the series (7.32) for large t. It turns out that the coefficient of n^k increases as t increases, and so the series is not directly useful for long-time effects such as occur in the hydrodynamic stage. The derivation of a kinetic equation is, in a sense, a procedure for resuming parts of the density expansion to all orders in the density to obtain a basis for describing the long-time behavior. However, the series solution may be used as it stands when only short-time effects are under discussion.

A general term is quite complex and only partial results have been obtained regarding its long-time behavior. Here only the linear (in n) term in F_1 will be analyzed in detail.

Consider the second term in the expansion for F_1, which we denote by F_1^1:

$$F_1^1(1;t) = n\int dX_2[S_2(1,2;-t) - S_0(1,2;-t)]\, F_2(1,2).$$

The argument $t = 0$ has been dropped in F_2. For simplicity we assume the particles to be hard spheres. If the two particles do not come into contact during the interval $(-t,0)$, then S_2 has the same effect as S_0 so the integrand as shown vanishes. In order for a collision to occur it is necessary first of all that $b \leq D$. It is also necessary that $\mathbf{r}\cdot\boldsymbol{g} \leq 0$; that is, the particles are receding so that in the backward motion they approach one another. To come into contact the particles must travel, relative to one another, a distance

$$s = \mathbf{r}\cdot\hat{\boldsymbol{g}} - \sqrt{D^2 - b^2}$$

and this distance cannot be greater than gt. The conditions

$$r \geq D, b \leq D, \quad \mathbf{r}\cdot\hat{\boldsymbol{g}} \geq 0, \quad s \leq gt \tag{7.33}$$

define a cylindrical region (in the space of relative coordinates) with axis along $\boldsymbol{g}$, circular cross section of radius D, length gt, and with hemispherical end-caps. The particles must lie within this collision cylinder if they are to collide within the assigned interval of time.

Assuming the conditions (7.33) to be satisfied, let $-t_c = -s/g$ be the time when a collision occurs. For times earlier than t_c, S_2 converts the momenta into initial values $\mathbf{p}_1', \mathbf{p}_2'$ for a collision such that the final momenta are $\mathbf{p}_1$, $\mathbf{p}_2$. At the point of collision, $\mathbf{r} = D\mathbf{e}$ where $\mathbf{e}$ is a unit vector. The net time traveled prior to the collision is $t - t_c$ so $S_2(-t)$ converts $\mathbf{r}$ into

$$\mathbf{r}(-t) = D\mathbf{e} - (t - t_c)\boldsymbol{g}', \quad \boldsymbol{g}' = \mathbf{v}_1' - \mathbf{v}_2'.$$

The $\mathbf{q}_\alpha$ at time $-t$ are

$$\mathbf{q}_1(-t) = \mathbf{q}_1 - \mathbf{v}_1 t_c - \mathbf{v}_1'(t - t_c),$$

$$\mathbf{q}_2(-t) = \mathbf{q}_2 - \mathbf{v}_2 t_c - \mathbf{v}_2'(t - t_c) = \mathbf{q}_1(-t) - \mathbf{r}(-t).$$

For simplicity we will assume the system to be spatially homogeneous. This means that F_1 is independent of position, while F_2 depends on $\mathbf{q}_1, \mathbf{q}_2$ only through $\mathbf{r}$. Then the effect of S_2 on F_2 is

$$S_2(-t)F_2(\mathbf{r},\mathbf{p}_1,\mathbf{p}_2) = F_2(\mathbf{r}(-t),\mathbf{p}_1',\mathbf{p}_2').$$

The motion under S_0 is noninteracting so the momenta are unchanged while $\mathbf{r}$ becomes

$$\mathbf{r}_0(-t) = \mathbf{r} - \boldsymbol{g}t.$$

Thus $S_0 F_2$ is

$$S_0(-t)F_2(\mathbf{r}, \mathbf{p}_1, \mathbf{p}_2) = F_2(\mathbf{r}_0(-t), \mathbf{p}_1, \mathbf{p}_2).$$

(This is to be set equal to zero if $r_0(-t)$ is less than D.)

To carry out the integration over $\mathbf{q}_2$ we use cylindrical polar coordinates as before. The minimum value of z is at the point of contact where it is equal to $z_0 = \sqrt{D^2 - b^2}$ and it ranges up to $z_0 + \mathcal{g}t$. The result for $F_1^1(t)$ is

$$F_1^1(t) =$$

$$n\int d^3p_2 \int_0^D bdb \int_0^{2\pi} d\varphi \int_{z_0}^{z_0+\mathcal{g}t} dz[F_2(\mathbf{r}(-t), \mathbf{p}_1', \mathbf{p}_2') - F_2(\mathbf{r}_0(-t), \mathbf{p}_1, \mathbf{p}_2)].$$

For fixed values of z the separation $r(-t)$ or $r_0(-t)$ increases as t increases. We will assume that when the particles are far apart they are uncorrelated so Equation (7.12) holds. Usually in gases, correlations extend only over distances which are about the same as the force range, so this assumption is valid if $r(-t), r_0(-t)$ are large in comparison to D. Since we have assumed F_1 to be independent of position, we get for large t

$$F_2(\mathbf{r}(-t), \mathbf{p}_1', \mathbf{p}_2') - F_2(\mathbf{r}_0(-t), \mathbf{p}_1, \mathbf{p}_2) \rightarrow F_1(\mathbf{p}_1')F_1(\mathbf{p}_2') - F_1(\mathbf{p}_1)F_1(\mathbf{p}_2).$$

The result for F_1^1 is

$$F_1^1(\mathbf{p}_1, t) = nt\int d^3p_2 \int \mathcal{g}\, bdb\, d\varphi\, [F_1(\mathbf{p}_1')F_1(\mathbf{p}_2') - F_1(\mathbf{p}_1)F_1(\mathbf{p}_2)].$$

This is just nt times the Boltzmann collision operator on F_1:

$$F_1^1 = ntJ_B[F_1]. \tag{7.34}$$

The region where the two particles are separated by a distance of the order of the force range has been neglected. However, this is a fixed region, and since the preceding result grows with t it is the dominant contribution for large t. Of course in equilibrium the expression (7.34) vanishes, and more generally the equilibrium expressions for the F_s are solutions to the hierarchy without the time-derivative term.

The long-time behavior of the third term in the expansion for F_1 has been analyzed by Green. The dominant contribution increases as t^2, and is the same as would be obtained by solving the Boltzmann equation with an expansion in powers of the time. In general the coefficient of n^k has a dominant part which grows with time as t^k. Thus, the density expansion is directly useful for solving the hierarchy only for short times.

Notes on Chapter 7

7.3 The BBGKY hierarchy was developed in Bogoliubov [1946], Born and Green [1949], Kirkwood [1946, 1947], and Yvon [1935]. The binary collision operator with allowance for the difference in position of the colliding particles and the triple collision operator were worked out by Bogoliubov. The discussion given here follows Cohen [1962a, 1962b, 1962c, 1963, 1973] who developed the generalized Boltzmann equation as far as the four-body collision operator. A useful survey covering the Boltzmann equation, the hierarchy, and time-dependent correlation functions is given in Dorfman and van Beijeren [1977].

For a one-dimensional system the kinetic equation can be worked out completely; see Lebowitz and Percus [1967], and Lebowitz, Percus, and Sykes [1968].

7.6 For a detailed discussion of the effects of the spatial extent of a collision, see Snider and Curtiss [1958, 1963].

7.7 The hydrodynamic equations and transport coefficients on the basis of Bogoliubov's kinetic equation were worked out by Choh [1958]; see also Garcia-Colin and Flores [1966a], and Garcia-Colin, Green, and Chaos [1966b]. Other references dealing with the properties of the triple collision operator are Resibois [1963, 1965, 1966], Weinstock [1963b], Green [1964], Sengers [1966, 1967, 1971], Hoegy and Sengers [1970], Sengers, Ernst, and Gillespie [1972], Sengers [1973], Sengers, Gillespie, and Peres-Esandi [1978], Boley and Desai [1973a, b], Brinser and Condiff [1973], and Eu [1974].

7.8 Enskog's factor χ can be related to the equation of state and then determined phenomenologically for real gases. The resulting Modified Enskog Theory has been successfully applied over a wide range of conditions; see Hanley, McCarty, and Cohen [1972], and van Beijeren and Ernst [1973a]. A generalization of Enskog's equation to deal with short times and displacements is discussed in Yip, Alley, and Alder [1982], and Leutheusser, Yip, Alder, and Alley [1983].

Enskog's equation for mixtures, and revisions of the older theory to avoid the difficulties mentioned in the last sentence of this section, are discussed in Garcia-Colin, Barajas, and Pina [1971], Barajas, Garcia-Colin, and Pina [1973], van Beijeren and Ernst [1973b], Kincaid [1978], Karkheck and Stell [1979a, b], and Grmela, Rosen, and Garcia-Colin [1981].

7.9 The solution to the hierarchy is obtained in Green [1956], Lewis [1961], and Cohen [1962a].

7.10 See Green [1956, 1958], Green and Piccirelli [1963], Cohen [1963], and Piccirelli [1966]. The behavior of the solution for short times is discussed in Sykes [1973].

8

THE CONSERVATION LAWS

8.1. CLASSICAL MECHANICS

The conservation laws were derived earlier from the Boltzmann equation, but actually the conservation laws hold at a more basic level than the Boltzmann equation. In this chapter the conservation laws will be derived from the microscopic equations of motion; in the process the conserved densities and their fluxes will be obtained as functions of the microscopic variables. The macroscopic conservation laws are obtained by taking an average of their microscopic counterparts. The conservation laws obtained in this way are not limited to a low-density gas but are general in their application. A general approach for the constitutive equations and transport coefficients will then be discussed in later chapters.

Initially we treat the particles as point masses whose motion is governed by classical mechanics. The density at $\mathbf{x}$ due to the α'th particle can be represented by a delta function $\delta(\mathbf{q}_\alpha - \mathbf{x})$ and the density due to all particles is

$$n = \sum_\alpha \delta(\mathbf{q}_\alpha - \mathbf{x}). \tag{8.1}$$

On taking the time derivative we obtain

$$\frac{\partial n}{\partial t} = -\sum_\alpha \frac{1}{m_\alpha}\mathbf{p}_\alpha \cdot \nabla\delta(\mathbf{q}_\alpha - \mathbf{x}) \tag{8.2}$$

where m_α is the mass of the α'th particle. It has been assumed that the number of particles is in fact conserved, as otherwise the number of terms in the sum would not be fixed. This result is just the equation of continuity

$$\frac{\partial n}{\partial t} + \nabla \cdot \mathbf{j} = 0$$

where the flux is given by

$$\mathbf{j} = \sum_\alpha \frac{\mathbf{p}_\alpha}{m_\alpha} \delta(\mathbf{q}_\alpha - \mathbf{x}).$$

If the particles are identical, then of course the masses m_α are all equal. If the system is a mixture of nonreacting species, then there is an equation of continuity for each species,

$$\frac{\partial n_s}{\partial t} + \nabla \cdot \mathbf{j}_s = 0.$$

The density n_s and flux $\mathbf{j}_s$ are given by formulas like (8.1) and (8.2) but with the sums extended only over the particles belonging to the s'th species.

The conservation law for mass is

$$\frac{\partial \rho}{\partial t} + \nabla \cdot \mathbf{g} = 0$$

where

$$\rho = \sum_\alpha m_\alpha \, \delta(\mathbf{q}_\alpha - \mathbf{x}), \ \mathbf{g} = \sum_\alpha \mathbf{p}_\alpha \, \delta(\mathbf{q}_\alpha - \mathbf{x}). \tag{8.3}$$

For a system consisting of more than one species, we have

$$\rho = \sum_s m_s n_s, \ \mathbf{g} = \sum_s m_s \mathbf{j}_s$$

where m_s is the mass of a particle of the s'th species.

From its form, $\mathbf{g}$ is seen to be the same as the momentum density. To obtain the conservation law for momentum, we calculate the time derivative of $\mathbf{g}$:

$$\frac{\partial g_i}{\partial t} = \sum_\alpha \left\{ \frac{p_{\alpha i} p_{\alpha j}}{m_\alpha} \frac{\partial \delta(\mathbf{q}_\alpha - \mathbf{x})}{\partial q_{\alpha j}} + \delta(\mathbf{q}_\alpha - \mathbf{x}) \frac{\partial p_{\alpha i}}{\partial t} \right\}.$$

This can be written as

$$\frac{\partial g_i}{\partial t} + \frac{\partial t_{ji}^K}{\partial x_j} = f_i$$

where

$$t_{ji}^K = \sum_\alpha \frac{p_{\alpha i} p_{\alpha j}}{m_\alpha} \delta(\mathbf{q}_\alpha - \mathbf{x}),\ f_i = \sum_\alpha \delta(\mathbf{q}_\alpha - \mathbf{x}) \frac{\partial p_{\alpha i}}{\partial t}.$$

The convective change in momentum due to particles moving in and out of a volume element, carrying momentum with them, is described by t_{ij}^K. The momentum in a volume element can also be changed by forces exerted from outside it, and this is accounted for by f_i.

Consider additive two-body forces with potential energy

$$U = \frac{1}{2} \sum_{\alpha\beta} U_{\alpha\beta},\ U_{\alpha\beta} = U(\mathbf{q}_\alpha - \mathbf{q}_\beta).$$

Then if the only forces are those due to U, we have

$$\begin{aligned} f_i &= -\sum_{\alpha\beta} \delta(\mathbf{q}_\alpha - \mathbf{x}) \frac{\partial U_{\alpha\beta}}{\partial q_{\alpha i}} \\ &= -\frac{1}{2} \sum_{\alpha\beta} [\delta(\mathbf{q}_\alpha - \mathbf{x}) - \delta(\mathbf{q}_\beta - \mathbf{x})] \frac{\partial U_{\alpha\beta}}{\partial q_{\alpha i}}. \end{aligned} \tag{8.4}$$

To express this as a divergence, use the identity

$$\begin{aligned} \delta(\mathbf{q}_\alpha - \mathbf{x}) - \delta(\mathbf{q}_\beta - \mathbf{x}) &= -\int_0^1 d\lambda \frac{d}{d\lambda} \delta(\mathbf{q}_\alpha - \mathbf{x} + \lambda(\mathbf{q}_\beta - \mathbf{q}_\alpha)) \\ &= -\frac{\partial}{\partial x_j}(q_{\alpha j} - q_{\beta j}) \int_0^1 d\lambda\, \delta(\mathbf{q}_\alpha - \mathbf{x} + \lambda(\mathbf{q}_\beta - \mathbf{q}_\alpha)). \end{aligned}$$

This enables us to write

$$f_i = -\frac{\partial t_{ji}^U}{\partial x_j}$$

where

$$t_{ji}^U = -\frac{1}{2} \sum_{\alpha\beta} (q_{\alpha j} - q_{\beta j}) \frac{\partial U_{\alpha\beta}}{\partial q_{\alpha i}} \int_0^1 d\lambda\, \delta(\mathbf{q}_\alpha - \mathbf{x} + \lambda(\mathbf{q}_\beta - \mathbf{q}_\alpha)).$$

Hence, we can express conservation of momentum by

$$\frac{\partial g_i}{\partial t} + \frac{\partial t_{ji}}{\partial x_j} = 0 \tag{8.5}$$

where

$$\begin{aligned} t_{ji} = t_{ji}^K + t_{ji}^U &= \sum_\alpha \frac{p_{\alpha i} p_{\alpha j}}{m_\alpha} \delta(\mathbf{q}_\alpha - \mathbf{x}) \\ &\quad - \frac{1}{2} \sum_{\alpha\beta} (q_{\alpha j} - q_{\beta j}) \frac{\partial U_{\alpha\beta}}{\partial q_{\alpha i}} \int_0^1 d\lambda\, \delta(\mathbf{q}_\alpha - \mathbf{x} + \lambda(\mathbf{q}_\beta - \mathbf{q}_\alpha)). \end{aligned} \tag{8.6}$$

The contribution of t_{ij}^U is referred to as collisional transfer of momentum.

An external force changes Equation (8.5) to

$$\frac{\partial g_i}{\partial t} + \frac{\partial t_{ji}}{\partial x_j} = F_i^{\text{ext}}$$

where

$$F_i^{\text{ext}} = \sum_\alpha F_{\alpha i}^{\text{ext}}\,\delta(\mathbf{q}_\alpha - \mathbf{x})$$

and $F_{\alpha i}^{\text{ext}}$ is the external force acting on the α'th particle.

Conservation of energy can be treated in a similar way. Including the contribution of additive two-body forces, we define the energy density by

$$\epsilon = \sum_\alpha \delta(\mathbf{q}_\alpha - \mathbf{x})\left[\frac{p_\alpha^2}{2m_\alpha} + \frac{1}{2}\sum_\beta U_{\alpha\beta}\right].$$

With each particle we have associated half of its potential energy of interaction with the other particles. On taking the time derivative, we get

$$\frac{\partial \epsilon}{\partial t} = -\nabla\cdot\mathbf{s}^K - \frac{1}{2}\sum_{\alpha\beta}\delta(\mathbf{q}_\alpha - \mathbf{x})\left[\frac{p_{\alpha j}}{m_\alpha} + \frac{p_{\beta j}}{m_\beta}\right]\frac{\partial U_{\alpha\beta}}{\partial q_{\alpha j}}. \tag{8.7}$$

Here $\mathbf{s}^K$ describes the convective flow of energy and is given by

$$\mathbf{s}^K = \sum_\alpha \delta(\mathbf{q}_\alpha - \mathbf{x})\frac{\mathbf{p}_\alpha}{m_\alpha}\left[\frac{p_\alpha^2}{2m_\alpha} + \frac{1}{2}\sum_{\alpha\beta} U_{\alpha\beta}\right].$$

The remainder of Equation (8.7) can be treated in the same way as (8.4). The result is the energy conservation law

$$\frac{\partial \epsilon}{\partial t} + \nabla\cdot\mathbf{s} = 0 \tag{8.8}$$

where the energy flux is $\mathbf{s} = \mathbf{s}^K + \mathbf{s}^U$ with

$$s_i^U = \frac{1}{2}\sum_{\alpha\beta}\int_0^1 d\lambda\,\delta(\mathbf{q}_\alpha - \mathbf{x} + \lambda(\mathbf{q}_\beta - \mathbf{q}_\alpha))\frac{p_{\alpha j}}{m_\alpha}(q_{\beta i} - q_{\alpha i})\frac{\partial U_{\alpha\beta}}{\partial q_{\alpha j}}.$$

In the presence of an external force, Equation (8.8) becomes

$$\frac{\partial \epsilon}{\partial t} + \nabla\cdot\mathbf{s} = w$$

where ϵ and $\mathbf{s}$ have the same form as previously and w is the rate at which work is done by the external force:

$$w = \sum_\alpha \delta(\mathbf{q}_\alpha - \mathbf{x})m_\alpha^{-1}\mathbf{p}_\alpha\cdot\mathbf{F}_\alpha^{\text{ext}}.$$

We will later make use of the volume-integrated fluxes and so list them here for reference. Let these be denoted by capital letters:

$$\mathbf{J}_s = \int d^3x\, \mathbf{j}_s, \quad T_{ij} = \int d^3x\, t_{ij}, \quad \mathbf{S} = \int d^3x\, \mathbf{s}.$$

Then we find

$$\mathbf{J}_s = \sum_\alpha \frac{\mathbf{p}_\alpha}{m_\alpha},$$

where the sum extends over particles of the s'th species. In addition

$$T_{ij} = \sum_\alpha \frac{p_{\alpha i} p_{\alpha j}}{m_\alpha} + \frac{1}{2} \sum_{\alpha\beta} (q_{\beta i} - q_{\alpha i}) \frac{\partial U_{\alpha\beta}}{\partial q_{\alpha j}},$$

$$S_i = \sum_\alpha \frac{p_{\alpha i}}{m_\alpha} \left[\frac{p_\alpha^2}{2m_\alpha} + \frac{1}{2} \sum_\beta U_{\alpha\beta} \right] + \frac{1}{2} \sum_{\alpha\beta} \frac{p_{\alpha j}}{m_\alpha} p_{\alpha j} (q_{\beta i} - q_{\alpha i}) \frac{\partial U_{\alpha\beta}}{\partial q_{\alpha j}}. \tag{8.9}$$

Of course the total momentum is given by $\mathbf{G} = \int d^3x\, \mathbf{g} = \Sigma \mathbf{p}_\alpha$.

In Chapter 1 a discussion was given of the behavior of the macroscopic conserved densities and fluxes under Galilean transformations. The transformation laws obtained there also apply to the microscopic functions. The argument can be carried through with little change and so will not be given.

8.2. QUANTUM MECHANICS

In quantum theory the q's and p's are operators satisfying the commutation relations

$$[q_{\alpha i}, p_{\beta j}] = i\hbar\, \delta_{ij}\, \delta_{\alpha\beta}, \quad [q_{\alpha i}, q_{\beta j}] = [p_{\alpha i}, p_{\beta j}] = 0.$$

Here the brackets denote the commutator, $[A, B] = AB - BA$, and $\hbar$ is Planck's constant divided by 2π. The time derivative of a function f of the q's and p's is determined by

$$i\hbar \frac{df}{dt} = [f, H].$$

We continue to use the Hamiltonian corresponding to additive two-body forces. The derivation of the conservation laws follows that for the classical case, the only difference being that one must pay attention to the order of noncommuting operators. If Equation (8.1) is used for n, we again obtain the equation of continuity but with $\mathbf{j}$ given by

$$\mathbf{j} = \frac{1}{2} \sum_\alpha m_\alpha^{-1} [\mathbf{p}_\alpha, \delta(\mathbf{q}_\alpha - \mathbf{x})]_+$$

where $[,]_+$ denotes the anticommutator, $[A, B]_+ = AB + BA$. Thus, a symmetrized product of noncommuting factors replaces the classical form. The remaining conservation laws are treated in a similar way. We define the momentum density by a symmetrized product,

$$\mathbf{g} = \frac{1}{2}\sum_\alpha [\mathbf{p}_\alpha, \delta(\mathbf{q}_\alpha - \mathbf{x})]_+.$$

The conservation law for momentum again has the form (8.5), but now one finds

$$\begin{aligned} t_{ji} &= \frac{1}{4\, m_\alpha}[p_{\alpha j}, [p_{\alpha i}, \delta(\mathbf{q}_\alpha - \mathbf{x})]_+]_+ \\ &\quad - \frac{1}{2}\sum_{\alpha\beta}(q_{\alpha j} - q_{\beta j})\frac{\partial U_{\alpha\beta}}{\partial q_{\alpha i}}\int_0^1 d\lambda\, \delta(\mathbf{q}_\alpha - \mathbf{x} + \lambda(\mathbf{q}_\beta - \mathbf{q}_\alpha)). \end{aligned} \tag{8.10}$$

The quantum mechanical expression for the energy density is

$$\epsilon = \frac{1}{2}\sum_\alpha \left[\frac{1}{2m_\alpha}[\delta(\mathbf{q}_\alpha - \mathbf{x}), p_\alpha^2]_+ + \delta(\mathbf{q}_\alpha - \mathbf{x})\sum_\beta U_{\alpha\beta}\right].$$

A straightforward but lengthy calculation yields the energy flux

$$\begin{aligned} s_i &= \sum_\alpha \frac{1}{8m_\alpha^2}[p_{\alpha i}, [p_\alpha^2, \delta(\mathbf{q}_\alpha - \mathbf{x})]_+]_+ + \sum_{\alpha\beta}\frac{1}{4m_\alpha}[p_{\alpha i}, \delta(\mathbf{q}_\alpha - \mathbf{x})U_{\alpha\beta}]_+ \\ &\quad + \sum_\alpha \frac{1}{4m_\alpha}\int_0^1 d\lambda \left[p_{\alpha j}, (q_{\beta i} - q_{\alpha i})\,\delta(\mathbf{q}_\alpha - \mathbf{x} + \lambda(\mathbf{q}_\beta - \mathbf{q}_\alpha))\frac{\partial U_{\alpha\beta}}{\partial q_{\alpha j}}\right]_+. \end{aligned}$$

The total fluxes are much simpler. Indeed $\mathbf{G}$ and T_{ij} have the same forms as classically, while

$$\begin{aligned} S_i &= \sum_\alpha \frac{1}{2m_\alpha^2}p_\alpha^2 p_{\alpha i} + \sum_\beta \frac{1}{4\, m_\alpha}[p_{\alpha i}, U_{\alpha\beta}]_+ \\ &\quad + \sum_{\alpha\beta}\frac{1}{4\, m_\alpha}\left[p_{\alpha j}, (q_{\beta i} - q_{\alpha i})\frac{\partial U_{\alpha\beta}}{\partial q_{\alpha j}}\right]_+. \end{aligned}$$

8.3. ANGULAR MOMENTUM

We next consider conservation of angular momentum and the related question of the symmetry of t_{ij}.

The angular momentum density due to translational motion is defined as the moment of the momentum density,

$$l_{ij} = -l_{ji} = x_i g_j - x_j g_i.$$

The corresponding flux of angular momentum is the moment of the flux of linear momentum,

$$m_{kij} = x_i t_{kj} - x_j t_{ki}.$$

The density is antisymmetric, and the flux antisymmetric in its last two subscripts:

$$l_{ij} = -l_{ji}, \ m_{kij} = -m_{kji}.$$

On taking the time derivative of l_{ij} and using the momentum conservation law, we obtain

$$\frac{\partial l_{ij}}{\partial t} + \frac{\partial m_{kij}}{\partial x_k} = t_{ij} - t_{ji}. \tag{8.11}$$

If the momentum flux is symmetric, the right-hand side vanishes and we have the conservation law for angular momentum.

Suppose the particles interact by central forces so the potential energy is a function only of the distance between them, $U_{\alpha\beta} = U(r_{\alpha\beta})$. Then

$$\frac{\partial U_{\alpha\beta}}{\partial q_{\alpha i}} = (q_{\alpha i} - q_{\beta i})\frac{1}{r_{\alpha\beta}}\frac{dU}{dr_{\alpha\beta}}$$

and it is seen immediately from Equation (8.6) that $t_{ij} = t_{ji}$. Generally $U_{\alpha\beta}$ will be a scalar under rotations, or otherwise the angular momentum will not be conserved. If $U_{\alpha\beta}$ is a function only of the coordinates, it will then depend only on $r_{\alpha\beta}$. However, a particle with internal structure can have an orientation, and the force between two such particles can depend on the angles between the radius vector and their orientations. One example is the dipole-dipole force between two particles with electric or magnetic dipole moments. In such cases t_{ij}, as previously defined, is not symmetric.

One nevertheless expects the angular momentum to satisfy a conservation law, at least if the potential energy is still a scalar under rotations, but it is necessary to include the internal angular momentum. Let $s_{\alpha ij} = -s_{\alpha ji}$ denote the internal angular momentum of the α'th particle, and define the density of internal angular momentum by

$$s_{ij} = \sum_{\alpha} s_{\alpha ij}\delta(\mathbf{q}_\alpha - \mathbf{x}).$$

We assume the total angular momentum $l_{ij} + s_{ij}$ to satisfy a conservation law

$$\frac{\partial(l_{ij} + s_{ij})}{\partial t} + \frac{\partial(m_{kij} + r_{kij})}{\partial x_k} = 0.$$

Here m_{kij} is as previously given, and $r_{kij} = -r_{kji}$ is the remainder needed to give the flux of total angular momentum. Note that because of Equation (8.11) we have

$$\frac{\partial s_{ij}}{\partial t} + \frac{\partial r_{kij}}{\partial x_k} = t_{ji} - t_{ij}. \tag{8.12}$$

Thus the antisymmetric part of t_{ij} describes a torque acting on the internal angular momentum.

It is a consequence of conservation of angular momentum that the momentum density can be defined in a different way such that its flux is symmetric. Clearly, the expression

$$t_{ij}^S = \frac{1}{2}(t_{ij} + t_{ji}) - \frac{1}{2}\frac{\partial}{\partial x_k}(r_{jki} + r_{ikj})$$

is symmetric in i,j. Using Equation (8.12) we get

$$t_{ji}^S = t_{ji} - \frac{1}{2}\frac{\partial}{\partial t}s_{ij} - \frac{1}{2}\frac{\partial}{\partial x_k}(r_{kij} + r_{jki} + r_{ikj}).$$

It follows from the antisymmetry of r_{kij} in i,j that $(r_{kij} + r_{jki} + r_{ikj})$ is antisymmetric in j,k so

$$\frac{\partial^2}{\partial x_j \partial x_k}(r_{kij} + r_{jki} + r_{ikj}) = 0.$$

Hence, we obtain

$$\frac{\partial}{\partial x_j} t_{ji}^S = -\frac{\partial}{\partial t}\left(g_i + \frac{1}{2}\frac{\partial}{\partial x_j}s_{ij}\right).$$

This yields the conservation law

$$\frac{\partial g_i'}{\partial t} + \frac{\partial t_{ji}^S}{\partial x_j} = 0$$

where

$$g_i' = g_i - \frac{1}{2}\frac{\partial s_{ji}}{\partial x_j}.$$

The second term in $\mathbf{g}'$ is the contribution to the linear momentum density of the internal angular momentum; a particle whose center of mass is stationary can contribute to linear momentum because of internal rotation. Since $\nabla \cdot \mathbf{g}' = \nabla \cdot \mathbf{g}$, either $\mathbf{g}'$ or $\mathbf{g}$ can be used as the mass flux.

A composite particle, such as a molecule, can have an internal angular momentum because of relative motion of its constituents, as opposed to electronic or nuclear spin. If the composite particle is treated as such, and if the forces on its elementary constituents are central, then the "canonical" pressure tensor (8.6) is symmetric. However a model is frequently useful in which the composite is treated as a particle with internal degrees of freedom, and with mass, energy, momentum, and internal angular momentum localized at the center of mass. The system of forces exerted on one molecule by another can be replaced by an effective force plus a torque, both depending on the internal states (including orientation) as well as on the separation of the mass centers. A symmetric pressure tensor can be constructed by the above procedure, but alternatively one can use the antisymmetric part of the canonical pressure tensor to describe the effective torques. The latter option appears to be a more practical approach to the behavior of large molecules in a fluid flow.

Constitutive equations have not been written for the flux of angular momentum because it is not a local quantity (but depends on the location of the axis). However, internal angular momentum is local, and it is sometimes useful to describe its motion by a "Fick's law" as in the theory of spin diffusion.

8.4. MACROSCOPIC CONSERVATION LAWS

The macroscopic conservation laws result when the microscopic ones are averaged over a statistical ensemble. The conserved densities and fluxes become, on averaging, functions (or functionals) of parameters (such as temperature and flow velocity) which occur in the ensemble, and the conservation laws then limit the dependence on space and time of these parameters.

The conserved densities and fluxes are sums of one- and two-particle quantities, and so their averages can be expressed as functionals of the one- and two-particle distribution functions. For example, the average mass density is

$$\langle \rho \rangle = \int dX_1 \,..\, dX_N \rho \, F_N$$

where ρ is given by Equation (8.3). Since F_N is symmetric in the N particles, ρ contributes N identical terms and we get

$$\langle \rho \rangle = (N/V)m \int dX_1 \delta(\mathbf{q}_1 - \mathbf{x}) F_1(X_1) = nm \int d^3p \, F_1(\mathbf{x}, \mathbf{p}).$$

Similarly, the average of the momentum density is

$$\langle \mathbf{g} \rangle = n \int d^3p \, \mathbf{p} \, F_1(\mathbf{x}, \mathbf{p}).$$

These expressions are the same as in the theory of the Boltzmann equation.

The pressure tensor contains both one- and two-particle terms, corresponding to the kinetic and potential parts. For the kinetic part we get

$$\langle t_{ij}^K \rangle = (n/m) \int d^3p \, p_i p_j F_1(\mathbf{x}, \mathbf{p})$$

which again is the same as in the theory of the Boltzmann equation. For the potential part, there are identical contributions from the $N(N-1)$ terms in the sum over α and β and we obtain

$$\langle t_{ji}^{U}\rangle = -\frac{1}{2}N(N-1)\int dX_1 \,..\, dX_N(q_{1j}-q_{2j})\frac{\partial U_{1,2}}{\partial q_{1i}} \times \int_0^1 d\lambda\, \delta(\mathbf{q}_1-\mathbf{x}+\lambda(\mathbf{q}_2-\mathbf{q}_1))F_N$$

$$= -\frac{1}{2}N(N-1)V^{-2}\int dX_1 dX_2(q_{1j}-q_{2j})\frac{\partial U_{1,2}}{\partial q_{1i}} \times \int_0^1 d\lambda\, \delta(\mathbf{q}_1-\mathbf{x}+\lambda(\mathbf{q}_2-\mathbf{q}_1))F_2(X_1,X_2).$$

To reduce this to a convenient form, change variables from $\mathbf{q}_1$, $\mathbf{q}_2$ to relative and mass-center coordinates $\mathbf{q} = \mathbf{q}_1 - \mathbf{q}_2$, $\mathbf{Q} = \frac{1}{2}(\mathbf{q}_1 + \mathbf{q}_2)$. The delta-function can be used to integrate over $\mathbf{Q}$. After taking the thermodynamic limit, we find

$$\langle t_{ji}^{U}\rangle = -\frac{1}{2}n^2\int d^3p_1 d^3p_2 d^3q\, q_j\frac{\partial U}{\partial q_i}\int_0^1 d\lambda\, F_2(\mathbf{x}-\lambda\mathbf{q},\mathbf{p}_1,\mathbf{x}+(1-\lambda)\mathbf{q},\mathbf{p}_2).$$

The energy density also has one-particle and two-particle contributions, which can be treated in the same way as in the preceding paragraph. In the thermodynamic limit, the total energy density is found to be

$$\langle\epsilon\rangle = n\int d^3p\,\frac{p^2}{2m}F_1(\mathbf{x},\mathbf{p}) - \frac{1}{2}n^2\int d^3p_1 d^3p_2 d^3q\, U(\mathbf{q})F_2(\mathbf{x},\mathbf{p}_1,\mathbf{x}-\mathbf{q},\mathbf{p}_2).$$

The energy flux is

$$\langle\mathbf{s}\rangle = n\int d^3p\frac{\mathbf{p}}{m}\frac{p^2}{2m}F_1(\mathbf{x},\mathbf{p}) + \frac{1}{2}n^2\int d^3p_1 d^3p_2 d^3q(1/m)\Big[U(\mathbf{q})\mathbf{p}_1F_2(\mathbf{x},\mathbf{p}_1,\mathbf{x}+\mathbf{q},\mathbf{p}_2) - \mathbf{q}(\mathbf{p}_1+\mathbf{p}_2)\cdot\frac{\partial U}{\partial\mathbf{q}}\int_0^1 d\lambda\, F_2(\mathbf{x}-(1-\lambda)\mathbf{q},\mathbf{p}_1,\mathbf{x}+\lambda\mathbf{q},\mathbf{p}_2)\Big].$$

Notes on Chapter 8

8.1 See Irving and Kirkwood [1950].

8.2 See Irving and Zwanzig [1951], and Zubarev [1970]. In solid crystals the most useful form for the conserved densities and fluxes is in terms of phonon variables rather than particle variables; the transformation is carried out in Hardy [1963], Choquard [1963], and DeVault [1966].

8.3 For the effects of internal molecular degrees of freedom on the pressure tensor see, for example, Sather and Dahler [1962], Condiff and

Dahler [1964], and Beenaker [1973]. The construction of a symmetric pressure tensor follows from the similar construction, due to Belinfante [1939], for the energy-momentum tensor in relativistic field theory. The treatment here follows McLennan [1966a]. A useful discussion of the differences between the two descriptions is given in Thomas and Snider [1970] and Olmsted and Snider [1976a, b].

9

FLUCTUATIONS AND DISSIPATION

9.1. INTRODUCTION

In this chapter we consider the connection between the hydrodynamic equations and the dynamics of spontaneous fluctuations. An example of this connection occurred in Chapter 2 where the diffusion constant for particles in Brownian motion was related to the velocity autocorrelation. There are similar relations, known as the Green-Kubo formulas, which express transport coefficients such as thermal conductivity and viscosity in terms of time-dependent correlation functions for fluctuations in the corresponding fluxes. The resulting theory is not limited to gases but applies to liquids and even solids. However, the generality is obtained at the price of a dependence on the solution to the equations of motion for the many-body system. In order to obtain specific numerical results, it is necessary to develop approximation methods for the many-body dynamics. A variety of methods is available for this purpose, but we will consider (in later chapters) only the application to gases. In this case the natural approach is based on the density, or cluster, expansion which leads back to the same analysis as occurs in kinetic theory.

The theory of fluctuations is interesting in itself. Fluctuations in density are responsible for scattering of radiation by a continuous medium; the scattering may be the phenomena of interest or it may be used to probe the material. Noise in electric circuits and devices is another example, where the goal is to describe the noise in terms of its intensity and spectral properties. It was shown by Nyquist that Johnson noise in circuits is related to resistance. The Nyquist theorem was generalized by Callen and Welton who discussed the connection between fluctuations and dissipation from a general point of view.

In equilibrium statistical mechanics, there are fluctuation formulas for the standard thermodynamic functions; the specific heat, for example, can be expressed in terms of the variance of fluctuations in energy. Roughly speaking, the intensity of fluctuations determines equilibrium thermodynamic functions while their time dependence determines transport coefficients.

The conserved densities and their fluxes will be denoted by ξ_α, $\boldsymbol{\psi}_\alpha$, $\alpha = 1,2,\ldots,$; this is the same notation as in Chapter 1 but now we mean the microscopic quantities obtained in Chapter 8. The conservation laws are

$$\frac{\partial \xi_\alpha}{\partial t} + \nabla \cdot \boldsymbol{\psi}_\alpha = 0.$$

Here the meaning of the time derivative is $\partial A/\partial t = \{A, H\}$, or in quantum theory $\partial A/\partial t = (1/i\hbar)[A, H]$, where H is the Hamiltonian. The extensive conserved variables will be denoted by $Y_\alpha = \int d^3x\, \xi_\alpha$.

9.2. EQUILIBRIUM ENSEMBLES

In this and the next section we collect together some of the basic results of equilibrium statistical mechanics with emphasis on the theory of fluctuations. For a system of identical particles, the grand canonical ensemble of Gibbs is

$$F = \exp(-Q + \nu N - \beta H) \tag{9.1}$$

where N is the number of particles and Q, ν, β are thermodynamic functions. In classical mechanics the normalization condition for F is

$$\sum_N \frac{1}{N!} \int d\Gamma_N F = 1. \tag{9.2}$$

Here $d\Gamma_N = dX_1 \cdots dX_N$ is the element of volume in the phase space for N particles and the factor $1/N!$ eliminates double counting of regions which differ only by permutations of the particles. For brevity we will use the abbreviation

$$\int d\Gamma\, F = \sum_N \frac{1}{N!} \int d\Gamma_N F.$$

In quantum theory the normalization condition is

$$\mathrm{Tr}\, F = 1$$

where Tr denotes the trace on the Hilbert space of functions which are symmetric or antisymmetric under an interchange of two particles, depending on their statistics, and includes a sum over the particle number. The average of a dynamical variable A is given by

$$\langle A \rangle = \int d\Gamma\, A\, F \tag{9.3}$$

classically, or by

$$\langle A \rangle = \mathrm{Tr}\, A\, F$$

in quantum theory.

The normalization condition determines Q as a function of ν, β, and the volume V, according to (in the classical form)

$$\exp Q = \int d\Gamma \exp(\nu N - \beta H). \tag{9.4}$$

It follows that

$$\langle N \rangle = (\partial Q/\partial \nu)_\beta, \quad \langle H \rangle = -(\partial Q/\partial \beta)_\nu \tag{9.5}$$

so Q is a potential for the thermodynamic particle number and internal energy. (In the differentiation it is understood that the volume is held constant in addition to the variables shown by the subscript.) The variation in Q is

$$\delta Q = \langle N \rangle\, \delta\nu - \langle H \rangle\, \delta\beta. \tag{9.6}$$

The entropy is given by

$$S = -\kappa\langle \ln F \rangle = \kappa(Q - \nu\langle N \rangle + \beta\langle H \rangle)$$

so S and Q are related by a Legendre transformation. Considering S to be a function of $\langle N \rangle$ and $\langle H \rangle$ and using (9.6) we get

$$\delta S = \kappa(-\nu\delta\langle N \rangle + \beta\delta\langle H \rangle).$$

Comparison with a standard relation of thermodynamics shows that $\beta = 1/\kappa T$, $\nu = \mu/\kappa T$ where T is the Kelvin temperature and μ the chemical potential. (Here we define ν in a slightly different way than in Chapter 1, where it was equal to μ/T.) Gibbs' formula for the chemical potential is

$$\mu\langle N \rangle = \langle H \rangle + pV - TS$$

where p is the pressure, and this shows $Q = pV/\kappa T$.

The ensemble (9.1) has been written for a system at rest, but the equilibrium ensemble for a moving system can be obtained by a Galilean transformation. If the velocity of the transformation is $\mathbf{w} = -\mathbf{u}$ then in the new frame of reference the system, assumed to be at rest in the original frame, moves with velocity $\mathbf{u}$. The transformation law for the energy is as given in Chapter 1, while N and the thermodynamic variables are unchanged. The ensemble for the moving system is obtained by the replacement

$$H \to H - \mathbf{u}\cdot\mathbf{G} + \frac{1}{2}Nmu^2$$

where **G** is the total momentum, which yields

$$F = \exp\left[-Q + \left(\nu - \frac{1}{2}\beta m u^2\right)N - \beta H + \beta \mathbf{u}\cdot\mathbf{G}\right]. \tag{9.7}$$

The internal energy $\langle H\rangle - \frac{1}{2}\langle N\rangle mu^2$ and average particle number $\langle N\rangle$ are unaffected by the Galilean transformation, and so are the same functions of β and ν as for a stationary system. Similarly Q is, for fixed β and ν, independent of **u**. However, Q can be converted into a potential for $\langle N\rangle$, $\langle H\rangle$, and $\langle \mathbf{G}\rangle$ by introducing

$$\zeta = \nu - \frac{1}{2}\beta\, mu^2, \quad \mathbf{y} = \beta\mathbf{u}. \tag{9.8}$$

Then Equations (9.5) become

$$\langle N\rangle = (\partial Q/\partial\zeta)_{\beta,\mathbf{y}}, \quad \langle H\rangle = -(\partial Q/\partial\beta)_{\zeta,\mathbf{y}}.$$

In addition the average momentum is given by

$$\langle \mathbf{G}\rangle = (\partial Q/\partial \mathbf{y})_{\zeta,\beta}.$$

Since Q is determined by ν and β, we have

$$(\partial Q/\partial \mathbf{y})_{\zeta,\beta} = (\partial Q/\partial\nu)_{\beta}\,(\partial\zeta/\partial\mathbf{y})_{\nu,\beta} = \langle N\rangle\, m\kappa T\mathbf{y}$$

which yields the obvious result

$$\langle \mathbf{G}\rangle = \langle N\rangle\, m\mathbf{u}. \tag{9.9}$$

The entropy obtained from the ensemble (9.7) by the rule $S = -\kappa\langle \ln F\rangle$ is

$$\begin{aligned} S &= \kappa[Q - \zeta\langle N\rangle + \beta\langle H\rangle - \beta\mathbf{u}\cdot\langle\mathbf{G}\rangle] \\ &= \kappa\left[Q - \nu\langle N\rangle + \beta\left(\langle H\rangle - \frac{1}{2}\langle N\rangle mu^2\right)\right]. \end{aligned}$$

Thus, the entropy is determined by $\langle N\rangle$ and the internal (rather than total) energy, and so its value is not changed by the Galilean transformation.

Denoting the conserved extensive variables such as N, H and **G** by Y_α, we may write the equilibrium ensemble as

$$F = \exp\left[-Q + \sum_\alpha \chi_\alpha Y_\alpha\right] \tag{9.10}$$

where χ_α denotes the intensive thermodynamic variable which is conjugate to Y_α. We also write $\chi(Y_\alpha) = \chi_\alpha$ and then

$$\chi(H) = -\beta, \quad \chi(N) = \zeta, \quad \chi(\mathbf{G}) = \mathbf{y}.$$

As before, we find that Q is a potential for the $\langle Y_\alpha\rangle$:

$$\partial Q/\partial\chi_\alpha = \langle Y_\alpha\rangle. \tag{9.11}$$

The entropy is

$$S = -\kappa\langle \ln F\rangle = \kappa\Big[Q - \sum_\alpha \chi_\alpha Y_\alpha\Big]$$

and χ_α can be obtained from S according to

$$\chi_\alpha = -\kappa^{-1}\,\partial S/\partial\langle Y_\alpha\rangle.$$

9.3. EQUILIBRIUM FLUCTUATIONS

We denote the deviation of a quantity from its average by $\Delta A = A - \langle A\rangle$. Taking second derivatives of (9.4) and using (9.5), we get

$$\langle \Delta N^2\rangle = \partial^2 Q/\partial\nu^2, \quad \langle \Delta H^2\rangle = \partial^2 Q/\partial\beta^2, \quad \langle \Delta N\,\Delta H\rangle = -\partial^2 Q/\partial\nu\,\partial\beta. \tag{9.12}$$

The first two of these are the variance, or autocorrelation, of N and H respectively, while the third is the covariance, or cross correlation, of N and H. The autocorrelation provides a measure of the amount by which a quantity fluctuates from its average value, and the cross correlation characterizes the simultaneous fluctuations of two different quantities. For brevity all such quantities will be referred to as correlations.

If the χ_α undergo small variations, then the corresponding variation in Q satisfies

$$\exp \delta Q = \Big\langle \exp\Big[\sum_\alpha \delta\chi_\alpha Y_\alpha\Big]\Big\rangle. \tag{9.13}$$

Except for the replacement of ik_α by $\delta\chi_\alpha$, the right-hand side is the same as the characteristic function (2.51) introduced in Chapter 2. The expansion of δQ in powers of $\delta\chi_\alpha$ is the same as the cumulant expansion. Hence, the derivatives of Q with respect to the χ_α are cumulants, as is illustrated by Equation (9.12).

The derivatives (9.12) can be expressed in terms of standard thermodynamic functions. For example the Gibbs-Duhem equation

$$\langle N\rangle d\mu = V dp - S dT$$

yields

$$\left(\frac{\partial\langle N\rangle}{\partial\nu}\right)_\beta = n\kappa T\left(\frac{\partial\langle N\rangle}{\partial p}\right)_\beta$$

where $n = \langle N\rangle/V$. This in turn gives

$$\langle \Delta N^2\rangle = n^2\kappa T V K_T$$

where $K_T = n^{-1}(\partial n/\partial p)_T$ is the isothermal compressibility. Similarly, we have

$$\langle \Delta H \, \Delta N \rangle = \left(\frac{\partial \langle H \rangle}{\partial \nu}\right)_\beta = \kappa T V \left(\frac{\partial \mathcal{U}}{\partial \mu}\right)_T = n\kappa T V \left(\frac{\partial \mathcal{U}}{\partial p}\right)_T$$

where $\mathcal{U} = \langle H \rangle / V$ and the last step follows from the Gibbs-Duhem equation. Some thermodynamic manipulations yield

$$\langle \Delta H \, \Delta N \rangle = n\kappa T V (h K_T - \alpha T)$$

where $\alpha = V^{-1}(\partial V/\partial T)_p$ is the coefficient of volume expansion and h is the enthalpy per unit volume. In a similar way, it can be shown that

$$\langle \Delta H^2 \rangle = \kappa T^2 V \rho C_V + (\kappa T V / K_T)(\alpha T - h K_T)^2,$$

where C_V is the specific heat (per unit mass) at constant volume. (If the energy fluctuations are calculated with the canonical ensemble, the first of these is replaced by $\langle \Delta H^2 \rangle = \kappa T^2 V \rho C_V$ which then yields a standard formula for the specific heat.) The preceding equations can be solved to express K_T, C_V, and α in terms of the correlations:

$$\begin{aligned} K_T &= (\kappa T n^2)^{-1} \frac{1}{V} \langle \Delta N^2 \rangle, \\ \rho C_V &= (\kappa^2 T^3 n^2 K_T)^{-1} \frac{1}{V^2} [\langle \Delta N^2 \rangle \langle \Delta H^2 \rangle - \langle \Delta N \, \Delta H \rangle^2], \qquad (9.14) \\ \alpha &= (\kappa T^2 n)^{-1} \frac{1}{V} [(h/n) \langle \Delta N^2 \rangle - \langle \Delta N \, \Delta H \rangle]. \end{aligned}$$

For finite volume the quantities $\langle \Delta N^2 \rangle / V$, $\langle \Delta H^2 \rangle / V$ and $\langle \Delta N \Delta H \rangle / V$ vary with the volume, but they have finite limits for $V \to \infty$. The thermodynamic limit should be taken in order that the left-hand sides of the preceding equations be the usual intensive thermodynamic variables. It would be too soon to take the limit before arriving at Equations (9.14), as the ensemble (9.1) is sensible only for a finite system. To save writing in the following, the convention will be adopted that once a quantity appears which has a well-defined thermodynamic limit, the limit is to be taken without an explicit statement to that effect.

Clearly the variances of N and H satisfy

$$\langle \Delta N^2 \rangle \geq 0, \quad \langle \Delta H^2 \rangle \geq 0.$$

In addition Schwarz's inequality yields

$$\langle \Delta N \, \Delta H \rangle^2 \leq \langle \Delta N^2 \rangle \langle \Delta H^2 \rangle.$$

From these inequalities we obtain the thermodynamic stability conditions $C_V \geq 0$, $K_T \geq 0$.

Equations (9.12) apply to correlations in the ensemble (9.1) for a system at rest. Correlations involving the momentum can be obtained from the ensemble (9.7) for a moving system. For example, we have

$$\langle \Delta G_i \, \Delta G_j \rangle = \partial^2 Q / \partial y_i \partial y_j = \partial \langle G_i \rangle / \partial y_j.$$

Equation (9.9) for the average momentum yields

$$\langle \Delta G_i \, \Delta G_j \rangle = \langle N \rangle m \kappa T \delta_{ij}. \tag{9.15}$$

The expression of correlations in terms of thermodynamic functions can to some extent be carried through for quantities other than the Y_α. Suppose A is any microscopic variable, independent of the thermodynamic variables χ_α. Then manipulations similar to those used previously show that

$$\partial \langle A \rangle / \partial \chi_\alpha = \langle \Delta A \, \Delta Y_\alpha \rangle, \quad \partial^2 \langle A \rangle / \partial \chi_\alpha \partial \chi_\beta = \langle \Delta A \, \Delta Y_\alpha \Delta Y_\beta \rangle, \quad \text{etc.} \tag{9.16}$$

Examples are

$$\partial \langle A \rangle / \partial \nu = \langle \Delta A \, \Delta N \rangle, \quad \partial \langle A \rangle / \partial \beta = -\langle \Delta A \, \Delta H \rangle.$$

The matrix g introduced in Chapter 1 is given by

$$\mathsf{g}_{\alpha\beta} = \frac{1}{V} \frac{\partial^2 Q}{\partial \chi_\alpha \partial \chi_\beta} = \frac{1}{V} \frac{\partial \langle Y_\alpha \rangle}{\partial \chi_\beta} = \frac{1}{V} \langle \Delta Y_\alpha \, \Delta Y_\beta \rangle. \tag{9.17}$$

An equivalent formula is

$$\mathsf{g}^{-1}_{\alpha\beta} = -\kappa^{-1} V \frac{\partial^2 S}{\partial \langle Y_\alpha \rangle \partial \langle Y_\beta \rangle} = V \frac{\partial \chi_\alpha}{\partial \langle Y_\beta \rangle}. \tag{9.18}$$

Stability conditions are contained in the statement that g is a non-negative matrix. This means that for arbitrary complex numbers a_α we have

$$\sum_{\alpha\beta} \bar{a}_\alpha \, a_\beta \mathsf{g}_{\alpha\beta} \geq 0,$$

as is evident from (9.17).

Fluctuations at different positions are characterized by correlation functions for local variables. An important example is the correlation function for density fluctuations:

$$C(\mathbf{x}_1 - \mathbf{x}_2) = \langle \Delta n(\mathbf{x}_1) \, \Delta n(\mathbf{x}_2) \rangle.$$

For a spatially homogeneous material, the average density $n = \langle n(\mathbf{x}) \rangle$ is independent of position and, as indicated, the correlation function depends only on the separation $\mathbf{x}_1 - \mathbf{x}_2$; for a fluid with spherical symmetry C depends only on the distance $r = |\mathbf{x}_1 - \mathbf{x}_2|$. The pair correlation function $g^{(2)}(r)$, also called the radial distribution function, is defined by the relation

$$C(\mathbf{x}_1 - \mathbf{x}_2) = n \, \delta(\mathbf{x}_1 - \mathbf{x}_2) + n^2 [g^{(2)}(r) - 1].$$

At low density the second term can be neglected so density fluctuations at different points are uncorrelated. The correlation functions for the full set of conserved densities will be denoted by

$$g_{\alpha\beta}(\mathbf{x}_1 - \mathbf{x}_2) = \langle \Delta\xi_\alpha(\mathbf{x}_1)\, \Delta\xi_\beta(\mathbf{x}_2)\rangle. \tag{9.19}$$

It follows from (9.17) that

$$\int d^3x\, g_{\alpha\beta}(\mathbf{x}) = \mathsf{g}_{\alpha\beta}. \tag{9.20}$$

In fluids $g_{\alpha\beta}(\mathbf{x})$ is ordinarily negligibly small unless $r = |\mathbf{x}|$ is about the same as the force range or less. However, the range of correlations is much larger at the gas-liquid critical point, and in systems such as solids with long-range order.

9.4. LOCAL EQUILIBRIUM

Suppose the thermodynamic variables β, ζ and $\mathbf{y}$ vary with position. The ensemble for local equilibrium is obtained by replacing the exponent of (9.7) by an integral as follows:

$$F_L = \exp\left\{-Q_L + \int d^3x\,[\zeta(\mathbf{x})n(\mathbf{x}) - \beta(\mathbf{x})\epsilon(\mathbf{x}) + \mathbf{y}(\mathbf{x})\cdot\mathbf{g}(\mathbf{x})]\right\} \tag{9.21}$$

or in the general form

$$F_L = \exp\left[-Q_L + \sum_\alpha \int d^3x\, \chi_\alpha(\mathbf{x})\xi_\alpha(\mathbf{x})\right]. \tag{9.22}$$

In this section we consider classical systems. The normalization condition for F_L is then stated as

$$\exp Q_L = \int d\Gamma \exp\left[\sum_\alpha \int d^3x\, \chi_\alpha(\mathbf{x})\xi_\alpha(\mathbf{x})\right],$$

which determines Q_L as a functional of the χ_α. If the thermodynamic variables are constant, then F_L reduces to the equilibrium ensemble.

The local equilibrium ensemble provides a mechanism for defining non-equilibrium thermodynamic variables. Let F be the ensemble which describes the actual state of the system. With F we associate an F_L with thermodynamic variables $\chi_\alpha(\mathbf{x})$ such that

$$\langle \xi_\alpha(\mathbf{x})\rangle_L = \langle \xi_\alpha(\mathbf{x})\rangle \tag{9.23}$$

where the brackets $\langle\ \rangle_L$ and $\langle\ \rangle$ denote averages over F_L and F respectively. This equation implicitly determines the thermodynamic variables $\chi_\alpha(\mathbf{x})$ as functionals of the averages $\langle \xi_\alpha(\mathbf{x})\rangle$, and will be adopted as the definition of the thermodynamic variables. The relation between this definition and the one introduced in Chapter 1 will be explored later.

For any ensemble F, let an entropy $S(F)$ be defined by

$$S(F) = -\kappa \langle \ln F \rangle.$$

This is Gibbs' definition of entropy for nonequilibrium states, which in general is not the same as the thermodynamic entropy defined in Chapter 1. Using an argument like that given in Chapter 3, we now show that the local-equilibrium ensemble has the property that it maximizes $S(F)$ subject to the constraints that the averages of the conserved densities take on given values. If F_L is associated with F according to Equation (9.23), we have

$$S(F_L) - S(F) = \kappa \langle \ln (F/F_L) \rangle.$$

The normalization conditions $\int d\Gamma\, F = \int d\Gamma\, F_L = 1$ allow us to write

$$S(F_L) - S(F) = \kappa \int d\Gamma\, [F \ln (F/F_L) - F + F_L]$$

and the inequality $x \ln x - x + 1 \geq 0$ yields the desired result

$$S(F_L) \geq S(F).$$

An argument of information theory shows that if the averages $\langle \xi_\alpha(\mathbf{x}) \rangle$, and nothing else, are known, then F_L is the best choice for the distribution function since it maximizes $S(F)$. Usually something else is known, for example, that the system has been freely evolving for a period of time, and the additional information is crucial in determining the distribution function F.

At this point we collect together some properties of the local equilibrium ensemble which will be needed later. In analogy to Equation (9.11) we obtain

$$\delta Q_L / \delta \chi_\alpha(\mathbf{x}) = \langle \xi_\alpha(\mathbf{x}) \rangle_L. \tag{9.24}$$

Particular examples are

$$\delta Q_L/\delta \zeta(\mathbf{x}) = \langle n(\mathbf{x}) \rangle_L, \quad \delta Q_L/\delta \beta(\mathbf{x}) = -\langle \epsilon(\mathbf{x}) \rangle_L, \quad \delta Q_L/\delta \mathbf{y}(\mathbf{x}) = \langle \mathbf{g}(\mathbf{x}) \rangle_L.$$

On taking second derivatives, we find

$$\delta^2 Q_L / \delta \chi_\alpha(\mathbf{x}_1) \delta \chi_\beta(\mathbf{x}_2) = \langle \Delta \xi_\alpha(\mathbf{x}_1)\, \Delta \xi_\beta(\mathbf{x}_2) \rangle_L. \tag{9.25}$$

Again ΔA denotes the deviation of a quantity A from its average, but now with the average calculated in local equilibrium

$$\Delta A = A - \langle A \rangle_L.$$

In particular fluctuations in density and energy density satisfy

$$\langle \Delta n(\mathbf{x}_1)\, \Delta n(\mathbf{x}_2) \rangle_L = \delta^2 Q_L / \delta \zeta(\mathbf{x}_1) \delta \zeta(\mathbf{x}_2),$$

$$\langle \Delta \epsilon(\mathbf{x}_1)\, \Delta \epsilon(\mathbf{x}_2) \rangle_L = \delta^2 Q_L / \delta \beta(\mathbf{x}_1) \delta \beta(\mathbf{x}_2).$$

Other second, and higher, derivatives of Q_L can be expressed in terms of similar correlation functions. (The discussion here is limited to classical mechanics. The modifications necessary in quantum theory will be described later.)

If A is a microscopic function, independent of the thermodynamic variables, then

$$\delta\langle A\rangle_L/\delta\chi_\alpha(\mathbf{x}) = \langle \Delta A\, \Delta\xi_\alpha(\mathbf{x})\rangle_L.$$

A conserved density $\xi_\alpha(\mathbf{x})$ depends on the field point $\mathbf{x}$ and the atomic positions in such a way that it is invariant to translations of the frame of reference; the same is true of a flux $\boldsymbol{\psi}_\alpha(\mathbf{x})$. If $A(\mathbf{x})$ is any function with this property, then

$$\partial\langle A(\mathbf{x})\rangle_L/\partial x_i = -\sum_\alpha \int d^3x'\, \langle \Delta A(\mathbf{x})\, \Delta\xi_\alpha(\mathbf{x}')\rangle_L\, \partial\chi_\alpha(\mathbf{x}')/\partial x_i'. \tag{9.26}$$

The identity

$$\langle\{A,H\}\rangle_L = \sum_\alpha \int d^3x\, \langle A\boldsymbol{\psi}_\alpha(\mathbf{x})\rangle_L \cdot \nabla\chi_\alpha(\mathbf{x}) \tag{9.27}$$

can be shown by shifting the Poisson bracket to F_L and using the conservation laws. Choosing $A = 1$ we get

$$\sum_\alpha \int d^3x\, \langle\boldsymbol{\psi}_\alpha(\mathbf{x})\rangle_L \cdot \nabla\chi_\alpha(\mathbf{x}) = 0 \tag{9.28}$$

so we can also write (9.27) as

$$\langle\{A,H\}\rangle_L = \sum_\alpha \int d^3x\, \langle \Delta A\, \Delta\boldsymbol{\psi}_\alpha(\mathbf{x})\rangle_L \cdot \nabla\chi_\alpha(\mathbf{x}).$$

The choice $A = \xi_\beta$ combined with the conservation laws and Equation (9.26) yield

$$\sum_\alpha \int d^3x'\, \langle[\Delta\xi_\beta(\mathbf{x})\Delta\boldsymbol{\psi}_\alpha(\mathbf{x}') - \Delta\boldsymbol{\psi}_\beta(\mathbf{x})\,\Delta\xi_\alpha(\mathbf{x}')]\rangle_L \cdot \nabla\chi_\beta(\mathbf{x}') = 0. \tag{9.29}$$

In general the thermodynamic variables are time dependent. If A is not an explicit function of the time, then its local-equilibrium average varies according to

$$\partial\langle A\rangle_L/\partial t = \sum_\alpha \int d^3x\, \langle \Delta A\, \Delta\xi_\alpha(\mathbf{x})\rangle_L\, \partial\chi_\alpha(\mathbf{x},t)/\partial t.$$

In the construction of the normal solution to the Boltzmann equation, the local Maxwellian was associated with the actual distribution function in such a way as to satisfy the conditions (9.23). In that case the averages of the conserved densities were related locally to the thermodynamic variables, in the same way as in equilibrium. The energy density, for example, satisfies

$$\langle \epsilon(\mathbf{x}) \rangle_L = \epsilon(\beta(\mathbf{x}), \nu(\mathbf{x}))$$

where $\epsilon(\beta, \nu)$ is the same function as in equilibrium. As a consequence of the local dependence, we have

$$\delta \langle \epsilon(\mathbf{x}_1) \rangle_L / \delta\beta(\mathbf{x}_2) = \delta(\mathbf{x}_1 - \mathbf{x}_2)\, \partial\epsilon/\partial\beta. \tag{9.30}$$

However, in the general case we get instead

$$\delta \langle \epsilon(\mathbf{x}_1) \rangle_L / \delta\beta(\mathbf{x}_2) = -\langle \Delta\epsilon(\mathbf{x}_1)\, \Delta\epsilon(\mathbf{x}_2) \rangle_L. \tag{9.31}$$

In a low-density gas correlation functions such as this vanish except when the points $\mathbf{x}_1$ and $\mathbf{x}_2$ coincide and so Equation (9.31) reduces to (9.30). In general, however, the range of correlations does not vanish. The local-equilibrium averages of the conserved densities then depend nonlocally on the thermodynamic variables; the same is true of the local-equilibrium averages of the fluxes. Thus, the definition of thermodynamic variables based on Equation (9.23) is in general not the same as the definition introduced in Chapter 1. We use the definition (9.23) because it is more convenient; of course one could maintain the definition of Chapter 1 but this would not eliminate the nonlocal dependence of the local-equilibrium average of the fluxes on the thermodynamic variables. However, our main concern is processes which vary slowly in comparison to the correlation length and for these the difference is unimportant.

Intuitively it is clear that the two definitions agree when the range of correlations is negligible, but this can be shown systematically as follows. Introduce an equilibrium ensemble F_0 which has the form (9.10) with parameters Q_0 and χ^0_α.

$$F_0 = \exp\left[-Q_0 + \sum_\alpha \chi^0_\alpha Y_\alpha\right].$$

Then F_L can be written as

$$F_L = F_0 \exp\left[-\delta Q + \sum_\alpha \int d^3x\, \xi_\alpha(\mathbf{x})\, \delta\chi_\alpha(\mathbf{x})\right] \tag{9.32}$$

where $\delta Q = Q_L - Q_0$, $\delta\chi_\alpha = \chi_\alpha - \chi^0_\alpha$. The normalization condition for F_L takes the form

$$\exp \delta Q = \left\langle \exp\left[\sum_\alpha \int d^3x\, \delta\chi_\alpha(\mathbf{x})\, \xi_\alpha(\mathbf{x})\right]\right\rangle_0 \tag{9.33}$$

where the brackets $\langle\ \rangle_0$ denote an average over F_0. Thus exp δQ is the characteristic function (or functional) for ξ_α and the cumulant expansion is

$$\delta Q = \sum_\alpha \int d^3x\, \delta\chi_\alpha(\mathbf{x})\, \langle \xi_\alpha(\mathbf{x})\rangle_0$$

$$+ \frac{1}{2}\sum_{\alpha,\beta} \int d^3x_1\, d^3x_2\, \delta\chi_\alpha(\mathbf{x}_1)\, \delta\chi_\beta(\mathbf{x}_2)\, \langle \Delta\xi_\alpha(\mathbf{x}_1)\Delta\xi_\beta(\mathbf{x}_2)\rangle_0 + \ldots$$

where now $\Delta\xi_\alpha = \xi_\alpha - \langle\xi_\alpha\rangle_0$. From this we obtain

$$\langle\xi_\alpha(\mathbf{x})\rangle_L = \langle\xi_\alpha(\mathbf{x})\rangle_0 + \sum_\beta \int d^3x'\, \delta\chi_\beta(\mathbf{x}')\, \langle \Delta\xi_\alpha(\mathbf{x})\Delta\xi_\beta(\mathbf{x}')\rangle_0$$

$$+ \frac{1}{2}\sum_{\beta,\gamma} \int d^3x_1\, d^3x_2\, \delta\chi_\beta(\mathbf{x}_1)\delta\chi_\gamma(\mathbf{x}_2)\, \langle \Delta\xi_\alpha(\mathbf{x})\, \Delta\xi_\beta(\mathbf{x}_1)\, \Delta\xi_\gamma(\mathbf{x}_2)\rangle_0 + \ldots$$

If the χ_α do not vary appreciably over distances for which correlations are important, then they can be pulled out from the integrals to give

$$\langle\xi_\alpha(\mathbf{x})\rangle_L = \langle\xi_\alpha(\mathbf{x})\rangle_0 + \sum_\beta \delta\chi_\beta(\mathbf{x}) \int d^3x'\, \langle \Delta\xi_\alpha(\mathbf{x})\, \Delta\xi_\beta(\mathbf{x}')\rangle_0$$

$$+ \frac{1}{2}\sum_{\beta,\gamma} \delta\chi_\beta(\mathbf{x})\, \delta\chi_\gamma(\mathbf{x}) \int d^3x_1\, d^3x_2\, \langle \Delta\xi_\alpha(\mathbf{x})\, \Delta\xi_\beta(\mathbf{x}_1)\, \Delta\xi_\gamma(\mathbf{x}_2)\rangle_0 + \ldots$$

This can be reduced to

$$\langle\xi_\alpha(\mathbf{x})\rangle_L = \langle\xi_\alpha\rangle_0 + \sum_\beta \delta\chi_\beta(\mathbf{x}) \frac{\partial\langle\xi_\alpha\rangle_0}{\partial\chi_\beta} + \frac{1}{2}\sum_{\beta,\gamma} \delta\chi_\beta(\mathbf{x})\delta\chi_\gamma(\mathbf{x}) \frac{\partial^2\langle\xi_\alpha\rangle_0}{\partial\chi_\beta\partial\chi_\gamma} + \ldots$$

which is the Taylor expansion for $\langle\xi_\alpha(\mathbf{x})\rangle_L$, expanded about the homogeneous equilibrium state. Thus, under the conditions assumed, $\langle\xi_\alpha(\mathbf{x})\rangle_L$ depends locally on $\chi_\alpha(\mathbf{x})$ with the same functional dependence as in equilibrium.

The same analysis applies to the fluxes: When the range of correlations is negligible, $\langle\boldsymbol{\psi}_\alpha(\mathbf{x})\rangle_L$ is a local function of $\chi_\alpha(\mathbf{x})$, with the same functional dependence as in equilibrium.

9.5. LOCAL EQUILIBRIUM IN QUANTUM THEORY

We briefly consider the algebraic modifications which are necessary in quantum theory because of the occurrence of noncommuting operators. The Y_α commute with one another so there is little change in the discussion of Sections 2 and 3, but local equilibrium must be treated differently. Let F_L be defined by Equation (9.21) as before. It is not possible to separate off a factor F_0

as in Equation (9.32) since the conserved densities do not commute with the Y_α. To generate the cumulant expansion, consider the function

$$f(\lambda) = \text{Tr} \exp(\eta + \lambda X) \tag{9.34}$$

where η is the exponent of the equilibrium ensemble,

$$\eta = -Q_0 + \sum_\alpha \chi_\alpha Y_\alpha$$

and X is for the moment any operator of interest. The expansion in powers of λ can be obtained from

$$e^{A+B} = e^A \left[1 + \int_0^1 d\mu\, B(\mu) + \int_0^1 d\mu_1 \int_0^{\mu_1} d\mu_2\, B(\mu_1)B(\mu_2) + \ldots \right]$$

where A, B are two operators which need not commute and $B(\mu) = \exp(-\mu A)B \exp(\mu A)$. An equivalent equation is

$$e^{A+B} = e^A T \exp\left[\int_0^1 d\mu\, B(\mu) \right]$$

where T is the "time-ordering" operator such that $T[B(\mu_1)\ldots R(\mu_p)]$ is a symmetric function of $\mu_1, \ldots, \mu_p$ which for $\mu_1 \geq \ldots \geq \mu_p$ has the value

$$T[B(\mu_1)\ldots B(\mu_p)] = B(\mu_1)\ldots B(\mu_p).$$

Equation (9.34) can be written

$$f(\lambda) = \left\langle T \exp\left[\lambda \int_0^1 d\mu\, X(\mu) \right] \right\rangle_0$$

where $X(\mu) = \exp(-\mu\eta) X \exp(\mu\eta)$. The cumulant expansion

$$f(\lambda) = \sum_{n=1}^{\infty} \frac{1}{n!} C_n \lambda^n$$

can be obtained from its classical counterpart by making the replacement

$$X \to \int_0^1 d\mu\, X(\mu)$$

and then properly ordering the products of $X(\mu)$'s. The first few cumulants are

$$C_1 = \langle X \rangle_0,$$

$$C_2 = \int_0^1 d\mu\, \langle \Delta X(\mu) \Delta X \rangle_0,$$

$$C_3 = \int_0^1 d\mu_1 \int_0^1 d\mu_2\, \langle T \Delta X(\mu_1) \Delta X(\mu_2) \Delta X \rangle_0.$$

By applying the preceding technique to the local-equilibrium ensemble we obtain, for example,

$$\delta Q_L/\delta\chi_\alpha(\mathbf{x})\,|_{\delta\chi=0} \;=\; \langle\xi_\alpha(\mathbf{x})\rangle_0,$$

$$\delta^2 Q_L/\delta\chi_\alpha(\mathbf{x}_1)\,\delta\chi_\beta(\mathbf{x}_2)\,|_{\delta\chi=0} \;=\; \int_0^1 d\mu\,\langle\Delta\xi_\alpha(\mathbf{x}_1;\mu)\,\Delta\xi_\beta(\mathbf{x}_2)\rangle_0.$$

The notation here is

$$\xi_\alpha(\mathbf{x};\mu) \;=\; e^{-\mu\eta}\,\xi_\alpha(\mathbf{x})e^{\mu\eta}.$$

Usually the conserved densities commute with those Y_α other than the Hamiltonian H, so this becomes

$$\xi_\alpha(\mathbf{x};\mu) \;=\; e^{\mu\beta H}\,\xi_\alpha(\mathbf{x})e^{-\mu\beta H}.$$

A dynamical variable depends on time (in the Heisenberg picture) according to

$$A(t) \;=\; S(t)A \;=\; \exp\,(itH/\hbar)A\,\exp\,(-itH/\hbar) \tag{9.35}$$

so we have $\xi_\alpha(\mathbf{x};\mu) = \xi_\alpha(\mathbf{x}, -i\hbar\mu\beta)$ where the argument following the comma refers to the "time" dependence.

The preceding cumulant expansion is actually not quite as general as its classical counterpart, since the statistical operator has been assumed to have the particular form (9.34).

9.6. LINEAR RESPONSE THEORY

To introduce the subject of time-dependent correlation functions and their relation to dissipation, we consider the response to an external conservative force. The linear response theory will be described in this section, but then we will use a different approach to develop the theory associated with the Green-Kubo formulas.

Let H denote the Hamiltonian for the system of interest, when in isolation. Suppose the system to be subject to an external force described by an additional term H' in the Hamiltonian, which we assume has the form $H' = A\,\alpha(t)$ where A depends on the microscopic state of the system and $\alpha(t)$ is a time-dependent parameter which measures the strength of the interaction. In classical mechanics the ensemble F satisfies the Liouville equation

$$\frac{\partial F}{\partial t} + \{F,H\} \;=\; -\{F,A\}\,\alpha(t). \tag{9.36}$$

As an initial condition we assume the system to be in equilibrium at time $t = -\infty$. Let F_0 denote the equilibrium ensemble, which we assume to have the form (9.1). The solution at time t can be expressed formally as

$$F(t) = F_0 - \int_{-\infty}^{t} ds\, S(s - t)\{F(s), A\}\, \alpha(s).$$

Here $S(t)$ is the streaming operator introduced in Chapter 7, which transforms a microscopic state into the state which occurs t seconds later. Let B denote a second dynamical variable, which may be the same as A but in any event does not depend explicitly on the time. We assume the equilibrium averages of both A and B to vanish. The average of B at time t, denoted by $\overline{B}(t)$, is

$$\overline{B}(t) = -\int_{-\infty}^{t} ds\, \alpha(s) \int d\Gamma\, B\, S(s - t)\{F(s), A\}.$$

This can be rearranged into

$$\overline{B}(t) = \int_{-\infty}^{t} ds\, \alpha(s) \int d\Gamma\, F(s)\{S(t - s)B, A\}.$$

We assume the external force, and so the deviation from equilibrium, to be small. Then $F(s)$ can be replaced by F_0 and the result written as

$$\overline{B}(t) = \int_{-\infty}^{t} ds\, \phi_{BA}(t - s)\, \alpha(s). \tag{9.37}$$

Here $\phi_{BA}(t)$ is given by

$$\phi_{BA}(t) = \langle\{B(t), A\}\rangle$$

where the brackets $\langle\ \ \rangle$ denote the average over F_0 and $B(t) = S(t)B$. Equation (9.37) is also valid in quantum theory with $\phi_{BA}(t)$ being

$$\phi_{BA}(t) = (1/i\hbar)\langle[B(t), A]\rangle \tag{9.38}$$

where the time dependence is given by Equation (9.35). Equation (9.37) gives the linear response of the quantity B to an external interaction proportional to A, and ϕ_{BA} is called the response function.

Of course ϕ_{BA} is a real quantity in classical mechanics if A and B are real; in quantum theory it is real if A and B are self-adjoint. Invariance of the equations of motion, and of the equilibrium ensemble, under a shift in the origin of time implies $\langle B(t)\, A\rangle = \langle B\, A(-t)\rangle$ and it follows that $\phi_{BA}(t) = -\phi_{AB}(-t)$.

A different expression for the response function can be obtained by using the canonical form (9.1) for the equilibrium ensemble. Then in classical mechanics $\{F_0, B\} = \beta F_0 \dot{B}$ where the dot denotes the time derivative, $\dot{B} = \{B, H\}$. This yields

$$\phi_{BA}(t) = \beta\langle\dot{B}(t)\, A\rangle.$$

To carry out a similar reduction in quantum theory, we use the operator identity

$$[B, e^{-\beta H}] = -\int_0^\beta d\lambda\, e^{-(\beta-\lambda)H}[B,H]\,e^{-\lambda H}$$

which yields

$$\phi_{BA}(t) = \int_0^\beta d\lambda\, \langle \dot{B}(t - i\hbar\lambda)\, A\rangle.$$

Because of invariance to a translation of time, this is equivalent to

$$\phi_{BA}(t) = \int_0^\beta d\lambda\, \langle \dot{B}\, A(-t + i\hbar\lambda)\rangle = -\int_0^\beta d\lambda\, \langle B\, \dot{A}(-t + i\hbar\lambda)\rangle.$$

As a specific example, consider the response to a uniform external electric field $\mathcal{E}$ which for simplicity we take to point in the x-direction. Then H' is given by

$$H' = -\sum_\alpha q_\alpha x_\alpha \mathcal{E}(t).$$

Here q_α, x_α are the charge and position of the α'th particle. The current density (in the x-direction) averaged over the volume of the system is $j = J/V$ where $J = \Sigma_\alpha q_\alpha \dot{x}_\alpha$ and its ensemble average is

$$\bar{j}(t) = \int_{-\infty}^t ds\, \phi(t - s)\mathcal{E}(s).$$

In this case the response function corresponds to the variables $\dot{A} = -J$, $B = J/V$ and is given by

$$\phi(t) = \int_0^\beta d\lambda \frac{1}{V}\langle J(t - i\hbar\lambda)\, J\rangle. \tag{9.39}$$

For a sinusoidal field, $\mathcal{E} \propto \exp(i\omega t)$, we get

$$\bar{j}(t) = \sigma(\omega)\,\mathcal{E}(t)$$

which yields Kubo's formula for the frequency-dependent conductivity:

$$\sigma(\omega) = \int_0^\infty dt \exp(-i\omega t) \int_0^\beta d\lambda \frac{1}{V}\langle J(-t - i\hbar\lambda) J\rangle. \tag{9.40}$$

By using the cyclic property of the trace one can show that

$$\langle B(t)\, A\rangle = \langle A\, B(t + i\hbar\beta)\rangle. \tag{9.41}$$

It follows that the response function (9.39) is an even function of the time. Consequently, the real part of the conductivity (which describes the dissipation) can be written as the Fourier transform

$$\mathrm{Re}\,\sigma(\omega) = \frac{1}{2}\int_{-\infty}^{\infty} dt \exp(i\omega t)\,\phi(t)$$

which with the aid of (9.41) can be converted to

$$\mathrm{Re}\,\sigma(\omega) = \frac{1}{2\hbar\omega}[1 - \exp(-\beta\hbar\omega)]\int_{-\infty}^{\infty} dt \exp(i\omega t)\frac{1}{V}\langle J(t)\,J\rangle.$$

In classical mechanics the time-dependent autocorrelation function for a variable A is defined by

$$C(t) = \langle A(t)\,A\rangle. \tag{9.42}$$

Let $\hat{A}(\omega, T)$ denote the Fourier component of $A(t)$, calculated over a finite interval:

$$\hat{A}(\omega, T) = \int_{-T}^{T} dt \exp(i\omega t)A(t).$$

The quantity

$$W(\omega) = \lim_{T\to\infty}\frac{1}{2T}|\hat{A}(\omega, T)|^2$$

is called the power spectrum. According to the Weiner-Khinchin theorem, the power spectrum is the Fourier transform of the correlation function:

$$W(\omega) = \int_{-\infty}^{\infty} \exp(i\omega t)\,C(t).$$

(W is not power in the physical sense unless A is chosen appropriately.)

In quantum theory $A(t)$ and A need not commute and as a consequence $C(t)$ (as previously defined) need not be real even if A is self-adjoint. Instead we have (for self-adjoint A)

$$\overline{C}(t) = C(-t). \tag{9.43}$$

The real and imaginary parts, $C(t) = R(t) + iI(t)$, are given by

$$R(t) = \frac{1}{2}\langle[A(t), A]_+\rangle, \quad I(t) = \frac{1}{2}\hbar\,\phi_{AA}(t)$$

where $[,]_+$ denotes the anticommutator, $[A,B]_+ = AB + BA$. It follows from Equation (9.41) that $C(t) = C(-t - i\hbar\beta)$. As a result $R(t)$ and $\phi_{AA}(t)$ can be expressed in terms of $C(t)$,

$$R(t) = \frac{1}{2}[C(t) + C(t - i\hbar\beta)], \quad \phi_{AA}(t) = \frac{1}{i\hbar}[C(t) - C(t - i\hbar\beta)].$$

In quantum theory the power spectrum is defined in terms of R:

$$W(\omega) = \int_{-\infty}^{\infty} dt \exp(i\omega t)\, R(t).$$

and can be written

$$W(\omega) = \frac{1}{2}[1 + \exp(-\beta\hbar\omega)] \int_{-\infty}^{\infty} dt \exp(i\omega t)\langle A(t)\, A\rangle.$$

For the case of electrical conduction, we define $W(\omega)$ by

$$W(\omega) = \int_{-\infty}^{\infty} dt \exp(i\omega t) \frac{1}{2V}\langle [J(t), J]_{+}\rangle.$$

In view of the identities obtained in the foregoing, this is related to the conductivity by

$$W(\omega) = 2\, E(\omega)\, \mathrm{Re}\, \sigma(\omega) \tag{9.44}$$

where $E(\omega) = \frac{1}{2}\hbar\omega \coth(\frac{1}{2}\beta\hbar\omega)$ is the average energy of a Planck oscillator. Equation (9.44) is the Nyquist theorem. At the conditions normally of interest, $\beta\hbar\omega << 1$ so it reduces to $W(\omega) = 2\kappa T\, \mathrm{Re}\, \sigma(\omega)$.

Another illustration of linear response theory is provided by Brownian motion under gravity. In this case the external force is $F = -mgz$ so we take $A = z$, $B = v_z$. The average velocity satisfies $\overline{v}_z = \mu F$ where the mobility μ is given by

$$\mu = \frac{1}{\kappa T} \int_{0}^{\infty} dt \langle v_z(t) v_z \rangle.$$

Comparison with Equation (2.56) yields Einstein's relation between diffusion and mobility, $D = \kappa T \mu$.

9.7. INITIAL VALUE PROBLEM

Processes such as thermal conduction and viscosity occur as "responses" to internal gradients rather than to external conservative fields, and so an approach different from that of the previous section is needed. Indeed even for electrical conduction, the conductivity is defined in terms of the total field rather than the external field, so linear response theory is valid only to the extent that the internal field can be neglected.

To describe the effects of internal gradients, we first construct a nonequilibrium ensemble corresponding to the initial value problem for an isolated system. Suppose the conserved densities are given definite values at an initial time τ, and it is desired to obtain their values at a later time t. We describe the initial state by a local equilibrium ensemble

$$F(\tau) = F_L(\tau).$$

The initial thermodynamic variables $\chi_\alpha(\tau)$ are determined by the requirement that the averages of the conserved densities take on the assigned values. The distribution function at later times is then determined by the Liouville equation, which has the formal solution

$$\begin{aligned} F(t) &= S(\tau - t)F_L(\tau) \\ &= \exp\left[-Q_L(\tau) + \sum_\alpha d^3x\, \chi_\alpha(\mathbf{x},\tau)\xi_\alpha(\mathbf{x},\tau - t)\right] \end{aligned} \tag{9.45}$$

where $\xi_\alpha(\mathbf{x},t) = S(t)\xi_\alpha(\mathbf{x})$. Thus $F(t)$ is determined as a functional of the initial values $\chi_\alpha(\mathbf{x},\tau)$.

We consider states near equilibrium, for which the hydrodynamic equations are linear. Thus, we suppose the χ_α to be close to constant values χ_α^0 which define an equilibrium ensemble F_0. Separating out F_0 as a factor, we write $F(t)$ in analogy to (9.32) as

$$F(t) = F_0 \exp\left[-\delta Q(\tau) + \sum_\alpha \int d^3x\, \delta\chi_\alpha(\mathbf{x},\tau)\, \xi_\alpha(\mathbf{x},\tau - t)\right]$$

where $\delta Q(\tau) = Q_L(\tau) - Q_0$, $\delta\chi_\alpha(\mathbf{x},\tau) = \chi_\alpha(\mathbf{x},\tau) - \chi_\alpha^0$, and Q_0 is the normalizing constant for F_0. The normalization condition for $F(t)$ can be written in a form similar to (9.33):

$$\exp \delta Q(\tau) = \left\langle \exp\left[\sum_\alpha \int d^3x\, \delta\chi_\alpha(\mathbf{x},\tau)\xi_\alpha(\mathbf{x},\tau - t)\right]\right\rangle_0. \tag{9.46}$$

Denoting the average over $F(t)$ by a bar, we get

$$\bar{\xi}_\alpha(\mathbf{x},t) = \left\langle \xi_\alpha(\mathbf{x}) \exp\left[-\delta Q(\tau) + \sum_\beta \int d^3x'\, \delta\chi_\beta(\mathbf{x}',\tau)\xi_\beta(\mathbf{x}',\tau - t)\right]\right\rangle_0. \tag{9.47}$$

Through linear terms in the deviation from equilibrium, Equation (9.46) yields

$$\delta Q(\tau) = \sum_\alpha \int d^3x\, \delta\chi_\alpha(\mathbf{x},\tau)\, \langle \xi_\alpha(\mathbf{x},\tau - t)\rangle_0$$

and (9.47) then reduces to

$$\delta\bar{\xi}_\alpha(\mathbf{x},t) = \sum_\beta \int d^3x'\, G_{\alpha\beta}(\mathbf{x} - \mathbf{x}',t - \tau)\, \delta\chi_\beta(\mathbf{x}',\tau). \tag{9.48}$$

Here $\delta\bar{\xi}_\alpha = \bar{\xi}_\alpha - \langle\xi_\alpha\rangle_0$ and $G_{\alpha\beta}$ denotes the space- and time-dependent correlation function

$$G_{\alpha\beta}(\mathbf{x} - \mathbf{x}',t) = \langle \Delta\xi_\alpha(\mathbf{x})\Delta\xi_\beta(\mathbf{x}', -t)\rangle_0 = \langle \Delta\xi_\alpha(\mathbf{x},t)\Delta\xi_\beta(\mathbf{x}')\rangle_0. \tag{9.49}$$

The equilibrium reference state has been assumed to be spatially homogeneous so the correlation function depends on the two positions only through their difference. Equation (9.48) can be written more briefly in matrix form as

$$\delta\bar{\xi}(\mathbf{x},t) = \int d^3x' \, G(\mathbf{x} - \mathbf{x}', t - \tau)\, \delta\chi(\mathbf{x}',\tau). \tag{9.50}$$

We see that G is a Green's function for hydrodynamics in the sense that it determines the averaged conserved densities at time t in terms of the initial values of the thermodynamic variables. The initial value problem is thereby "solved," but the solution in this form is too implicit to be directly useful. Our goal instead is to derive the hydrodynamic equations, from which the solution to the initial value problem can be constructed if that is the ultimate goal. The relation between Equation (9.50) and the hydrodynamic equations will be discussed in the next section, with the particular aim of extracting the transport coefficients.

9.8. GREEN-KUBO FORMULAS

The ensemble average of the conservation laws can be written

$$\frac{\partial}{\partial t}\delta\bar{\xi}(\mathbf{x},t) = -\nabla \cdot \delta\bar{\boldsymbol{\psi}}(\mathbf{x},t)$$

where $\delta\bar{\boldsymbol{\psi}}$ denotes the deviation of the average fluxes from their (constant) equilibrium averages. The hydrodynamic equations are obtained by using the constitutive relations to express $\delta\bar{\boldsymbol{\psi}}$ as a functional of the thermodynamic variables χ. We write the resulting equations schematically as

$$\frac{\partial}{\partial t}\delta\bar{\xi}(\mathbf{x},t) = -B\,\delta\chi(\mathbf{x},t) \tag{9.51}$$

where for linear processes B is a linear operator on functions of position. The relation between $\delta\bar{\xi}$ and $\delta\chi$ is determined by Equation (9.23), which in the case at hand is equivalent to

$$\delta\bar{\xi}_\alpha(\mathbf{x},t) = \sum_\beta \int d^3x' \, g_{\alpha\beta}(\mathbf{x} - \mathbf{x}')\, \delta\chi_\beta(\mathbf{x}',t) \tag{9.52}$$

where $g_{\alpha\beta}(\mathbf{x} - \mathbf{x}') = G_{\alpha\beta}(\mathbf{x} - \mathbf{x}', 0)$ is the same function as defined in Equation (9.19). With the aid of this relation, we can eliminate $\delta\chi$ and write Equation (9.51) entirely in terms of $\delta\bar{\xi}$.

The analysis is somewhat simpler in the Fourier representation. The Fourier transform will be denoted by a hat as in

$$\delta\hat{\xi}(\mathbf{k},t) = \int d^3x \exp(i\mathbf{k}\cdot\mathbf{x})\, \delta\bar{\xi}(\mathbf{x},t), \quad \hat{G}(\mathbf{k},t) = \int d^3x \exp(i\mathbf{k}\cdot\mathbf{x}) G(\mathbf{x},t).$$

Equation (9.52) transforms to

$$\delta\hat{\xi}(\mathbf{k},t) = \hat{g}(\mathbf{k})\,\delta\hat{\chi}(\mathbf{k},t)$$

and the hydrodynamic equation (9.51) becomes

$$\frac{\partial}{\partial t}\delta\hat{\xi}(\mathbf{k},t) = -\hat{B}(\mathbf{k})\,\delta\hat{\chi}(\mathbf{k},t) = -\hat{A}(\mathbf{k})\delta\hat{\xi}(\mathbf{k},t) \tag{9.53}$$

where

$$\hat{A}(\mathbf{k}) = \hat{B}(\mathbf{k})\hat{g}^{-1}(\mathbf{k}).$$

In the long wavelength limit, we have $\hat{g}(0) = \mathsf{g}$ where g is the matrix (9.17) of thermodynamic derivatives.

After the Fourier transformation, Equation (9.50) becomes

$$\delta\hat{\xi}(\mathbf{k},t) = \hat{G}(\mathbf{k},t-\tau)\delta\hat{\chi}(\mathbf{k},\tau). \tag{9.54}$$

The problem now is to obtain the connection between this equation and the hydrodynamic equation (9.53). To eliminate the dependence on the initial data in the preceding equation, take the time derivative and use the equation itself to obtain

$$\frac{\partial}{\partial t}\delta\hat{\xi}(\mathbf{k},t) = \frac{\partial\hat{G}(\mathbf{k},t-\tau)}{\partial t}\hat{G}^{-1}(\mathbf{k},t-\tau)\,\delta\hat{\xi}(\mathbf{k},t). \tag{9.55}$$

This equation does not have the hydrodynamic form (9.53) since the "coefficient" on the right is an explicit function of time; the hydrodynamic equation holds only after the ageing period. To compare the two equations we move the initial time τ back into the remote past and obtain the identification

$$\hat{A}(\mathbf{k}) = -\lim_{t\to\infty}\frac{\partial\hat{G}(\mathbf{k},t)}{\partial t}\hat{G}^{-1}(\mathbf{k},t). \tag{9.56}$$

This result provides the connection between linearized hydrodynamics and time-dependent correlation functions. It is important that the thermodynamic limit of the correlation functions be taken prior to the large-t limit as otherwise Poincare recurrences cause the preceding expression to take on meaningless values.

The Chapman-Enskog expansion amounts to an expansion in powers of $\mathbf{k}$, and the transport coefficients are obtained as coefficients in this expansion. To illustrate the procedure, we first consider the case of a single independent mode, say the diffusion of a B-particle. Then ξ is the density n, and

$$G(\mathbf{x},t) = \langle\Delta n(\mathbf{x},t)\,\Delta n\rangle_0.$$

The expansion in powers of **k** is most easily worked out by using the conservation law for particle number, which yields

$$\frac{\partial}{\partial t}G(\mathbf{x},t) + \frac{\partial}{\partial x_i}H_i(\mathbf{x},t) = 0, \quad \frac{\partial}{\partial t}H_i(\mathbf{x},t) + \frac{\partial}{\partial x_j}K_{ij}(\mathbf{x},t) = 0. \tag{9.57}$$

Here H_i and K_{ij} are correlation functions involving the particle flux **j**:

$$H_i(\mathbf{x},t) = \langle j_i(\mathbf{x},t)n\rangle_0, \quad K_{ij}(\mathbf{x},t) = \langle j_i(\mathbf{x},t)j_j\rangle_0.$$

(It is not necessary to write $\Delta\mathbf{j}$ since the equilibrium average of the flux vanishes.) On taking the Fourier transform of these equations we get

$$\frac{\partial}{\partial t}\hat{G}(\mathbf{k},t) - ik_i\hat{H}_i(\mathbf{k},t) = 0, \quad \frac{\partial}{\partial t}\hat{H}_i(\mathbf{k},t) - ik_j\hat{K}_{ij}(\mathbf{k},t) = 0. \tag{9.58}$$

In the case of Brownian motion, the pair correlation function is

$$G(\mathbf{x},0) = g(\mathbf{x}) = \bar{n}\delta(\mathbf{x})$$

so $\hat{G}(\mathbf{k},0) = \bar{n}$ where $\bar{n}$ is the average density. In addition it follows from invariance under time reversal that $\hat{H}_i(\mathbf{k},0) = 0$. Equations (9.58) can, therefore, be written in integral form as

$$\hat{G}(\mathbf{k},t) = \bar{n} + ik_i\int_0^t ds\,\hat{H}_i(\mathbf{k},s), \quad \hat{H}_i(\mathbf{k},t) = ik_j\int_0^t ds\,\hat{K}_{ij}(\mathbf{k},s).$$

This yields for Equation (9.56)

$$\hat{A}(\mathbf{k}) = k_ik_j\lim_{t\to\infty}\int_0^t ds\,\hat{K}_{ij}(\mathbf{k},s)\,\hat{G}^{-1}(\mathbf{k},t).$$

To lowest nonvanishing order in **k**, this is

$$\hat{A}(\mathbf{k}) = Dk^2 + \ldots$$

where the diffusion constant D is given by

$$D = (3\bar{n})^{-1}\int_0^\infty dt\,\hat{K}_{ii}(0,t)$$

$$= (3\bar{n})^{-1}\int_0^\infty dt\int d^3x\langle\mathbf{j}(\mathbf{x},t)\cdot\mathbf{j}\rangle_0.$$

In practice the density of B-particles is usually so small that correlations between different particles are negligible. The diffusion constant can then be written in terms of the velocity autocorrelation for a single particle, as in Equation (2.56).

The procedure in the general case is similar, although more complicated algebraically. Equations (9.57) are replaced by

$$\frac{\partial}{\partial t}G_{\alpha\beta}(\mathbf{x},t) + \frac{\partial}{\partial x_i}H_{\alpha i,\beta}(\mathbf{x},t) = 0, \quad \frac{\partial}{\partial t}H_{\alpha i,\beta}(\mathbf{x},t) + \frac{\partial}{\partial x_j}K_{\alpha i,\beta j}(\mathbf{x},t) = 0 \tag{9.59}$$

where

$$H_{\alpha i,\beta}(\mathbf{x},t) = \langle \Delta\psi_{\alpha i}(\mathbf{x},t)\Delta\xi_\beta\rangle_0, \quad K_{\alpha i,\beta j}(\mathbf{x},t) = \langle \Delta\psi_{\alpha i}(\mathbf{x},t)\,\Delta\psi_{\beta j}\rangle_0.$$

After the Fourier transformation the equations in matrix form appear the same as (9.58), and integration on time yields

$$\begin{aligned}\hat{G}(\mathbf{k},t) &= \hat{g}(\mathbf{k}) + ik_i\int_0^t ds\,\hat{H}_i(\mathbf{k},s),\\ \hat{H}_i(\mathbf{k},t) &= \hat{h}_i(\mathbf{k}) + ik_j\int_0^t ds\,\hat{K}_{ij}(\mathbf{k},s).\end{aligned} \tag{9.60}$$

Here, as before, $\hat{g}$ is the initial value of $\hat{G}$, while $\hat{h}_i$ is defined by

$$\hat{h}_i(\mathbf{k}) = \hat{H}_i(\mathbf{k},0).$$

In the case just treated of Brownian motion the corresponding quantity $\hat{h}$ vanished, but this need not be the case if the set ξ_α includes variables with different parities under time reversal.

The quantity $\hat{G}_{\alpha\beta}(0,t) = \langle \Delta Y_\alpha\, \Delta\xi_\beta\rangle_0$ is time independent, since Y_α is a constant of the motion. Setting $t = 0$ we get $\hat{G}(0,t) = \hat{g}(0) = \mathsf{g}$ where g is the matrix (9.17). A similar argument shows that $\hat{H}_i(0,t)$ is also time independent, with value which we denote by h_i

$$\mathsf{h}_i = \hat{H}_i(0,t) = \hat{h}_i(0).$$

Like g, h_i can be expressed as a thermodynamic derivative:

$$\mathsf{h}_{\alpha i,\beta} = \int d^3x\,\langle \Delta\psi_{\alpha i}(\mathbf{x})\,\Delta\xi_\beta\rangle_0 = \partial\psi^0_{\alpha i}/\partial\chi_\beta$$

where the superscript 0 denotes the equilibrium average. It follows from Equation (9.29) that h_i is symmetric,

$$\mathsf{h}_{\alpha i,\beta} = \mathsf{h}_{\beta i,\alpha}.$$

The elements of h_i are the coefficients of $\partial\delta\chi/\partial x_i$ in the Euler equations.

It follows from the time independence of $\hat{G}(0,t)$ that $\hat{A}(0) = 0$. We write the expansion for $\hat{A}$ as

$$\hat{A}(\mathbf{k}) = -ik_iA_i^{(1)} + k_ik_jA_{ij}^{(2)} + \ldots \tag{9.61}$$

and find for $A_i^{(1)}$

$$\begin{aligned}A_i^{(1)} &= i\,\partial\hat{A}/\partial k_i|_{\mathbf{k}=0}\\ &= -i\lim_{t\to\infty}\left[\frac{\partial^2\hat{G}}{\partial k_i\partial t}\hat{G}^{-1} - \frac{\partial\hat{G}}{\partial t}\hat{G}^{-1}\frac{\partial\hat{G}}{\partial k_i}\hat{G}^{-1}\right]_{\mathbf{k}=0}.\end{aligned}$$

The second term drops out because of the time independence of $\hat{G}(0,t)$. The first term is evaluated with the aid of Equation (9.58) and we get

$$A_i^{(1)} = \lim_{t\to\infty}\hat{H}_i(0,t)\hat{g}^{-1}(0) = \mathsf{h}_i\mathsf{g}^{-1}.$$

The corresponding value for $\hat{B}$ is $\hat{B}_i^{(1)} = \mathsf{h}_i$, which is the same result as obtained from the Euler equations. The limit $t \to \infty$ is not necessary; at the Euler approximation ageing to hydrodynamics takes place instantaneously.

The coefficient of the second term in (9.61) is

$$A_{ij}^{(2)} = -\frac{1}{2}\frac{\partial^2}{\partial k_i \partial k_j} \lim_{t\to\infty} \frac{\partial \hat{G}(\mathbf{k},t)}{\partial t} \hat{G}^{-1}(\mathbf{k},t)\bigg|_{\mathbf{k}=0}.$$

It can be shown that

$$\partial \hat{g}(\mathbf{k})/\partial k_i|_{\mathbf{k}=0} = 0, \quad \partial \hat{h}_j(\mathbf{k})/\partial k_i|_{\mathbf{k}=0} = 0$$

so some of the terms drop out and the result can be reduced to

$$A_{ij}^{(2)} = \int_0^\infty dt\, [\hat{K}_{ij}(0,t) - \mathsf{h}_i \mathsf{g}^{-1} \mathsf{h}_j]\mathsf{g}^{-1}.$$

Using the definition of h_i, we combine the two bracketed terms into the form

$$\hat{K}_{\alpha i,\beta j}(0,t) - [\mathsf{h}_i \mathsf{g}^{-1} \mathsf{h}_j]_{\alpha\beta} = \int d^3x\, \langle \psi_{\alpha i}^{-}(\mathbf{x})\, \psi_{\beta j}(\mathbf{x}, -t)\rangle_0$$

where

$$\psi_{\alpha i}^{-} = \Delta\psi_{\alpha i} - \sum_{\beta} (\mathsf{h}_i \mathsf{g}^{-1})_{\alpha\beta} \Delta\xi_\beta.$$

The $A^{(2)}$-coefficients relate the fluxes to the gradients of the conserved densities. When the constitutive relations are written in the form (1.59), the gradients of the thermodynamic variables have coefficients $L = A^{(2)}\mathsf{g}$ given by

$$L_{\alpha i,\beta j} = \int_0^\infty dt \int d^3x\, \langle \psi_{\alpha i}^{-}(\mathbf{x},t)\, \psi_{\beta j}\rangle_0. \tag{9.62}$$

This is a general expression of the Green-Kubo formulas.

The time-dependent correlation function in (9.62) occurs in several different contexts and will be abbreviated as M:

$$M_{\alpha i,\beta j}(t) = \int d^3x\, \langle \psi_{\alpha i}^{-}(\mathbf{x},t)\, \psi_{\beta j}\rangle_0. \tag{9.63}$$

Taking spatial homogeneity into account, we replace $\psi_{\beta j}$ by its volume average and write

$$M_{\alpha i,\beta j} = \frac{1}{V}\langle \Psi_{\alpha i}^{-}(t)\, \Psi_{\beta j}\rangle_0 \tag{9.64}$$

where the thermodynamic limit is understood. Here $\Psi_{\alpha i} = \int d^3x\, \psi_{\alpha i}(\mathbf{x})$ and $\Psi_{\alpha i}^{-}$ has a similar definition in terms of $\psi_{\alpha i}^{-}(\mathbf{x})$. Strictly speaking, the thermodynamic limit of the correlation function should be taken first, then the volume integral, as in Equation (9.63). We use the form (9.64) for convenience, but if a difference should occur then (9.63) is to be considered as primary.

It can be shown from the definitions of g and h_i that M_{ij} has the symmetric form

$$M_{\alpha i,\beta j} = \frac{1}{V}\langle\Psi^-_{\alpha i}(t)\Psi^-_{\beta j}\rangle_0. \tag{9.65}$$

9.9. SUBTRACTED FLUXES

The $\Psi^-_{\alpha i}$'s are "subtracted" fluxes, analogous to the quantities **S**, **T**, defined in Equation (4.13), which occur in the normal solution to the Boltzmann equation. The extra terms in $\Psi^-_{\alpha i}$ are linear combinations of the Y_α whose presence, as we now show, is needed for convergence of the time integral.

Let the subtracted fluxes be written as

$$\Psi^-_{\alpha i} = \Delta\Psi_{\alpha i} - \sum_\beta E_{\alpha i,\beta}\,\Delta Y_\beta \tag{9.66}$$

where the coefficients $E_{\alpha i,\beta}$ are given by

$$E_{\alpha i,\beta} = [\mathsf{h}_i\mathsf{g}^{-1}]_{\alpha\beta}.$$

Using the definitions of h_i and g, we get

$$E_{\alpha i,\beta} = \sum_\mu (\partial\psi^0_{\alpha i}/\partial\chi_\mu)(\partial\chi_\mu/\partial\xi^0_\beta) = \partial\psi^0_{\alpha i}/\partial\xi^0_\beta. \tag{9.67}$$

The derivatives can also be written in terms of the extensive variables:

$$E_{\alpha i,\beta} = \partial\Psi^0_{\alpha i}/\partial Y_\beta.$$

Let A be any function of the microscopic state, and denote its time average by $\langle A\rangle_i$:

$$\langle A\rangle_t = \lim_{T\to\infty}\frac{1}{T}\int_0^T dt\,A(t).$$

The time average is a constant of the motion, which for an ergodic system can be expressed as a function of the Y_α:

$$\langle A\rangle_t = \mathcal{A}(Y_\alpha).$$

Let this be expanded about the averages Y^0_α:

$$\langle A\rangle_t = \mathcal{A}(Y^0_\alpha) + \sum_\beta(\partial\mathcal{A}/\partial Y^0_\beta)[Y_\beta - Y^0_\beta] + \ldots \tag{9.68}$$

In the equilibrium ensemble fluctuations in Y_α are small, and the terms not written out in this expansion can be shown to give contributions to the correlation functions which vanish in the thermodynamic limit. Comparing Equation (9.68) with (9.66), we see

$$\Psi^{-} = \Psi - \langle\Psi\rangle_t. \tag{9.69}$$

Thus Ψ^{-} has a vanishing time average, which is necessary in order for the time integral in the Green-Kubo formula to exist. This is analogous to the situation in kinetic theory where the subtracted fluxes **S**, **T** contain linear combinations of the summational invariants which are needed for solvability of the integral equations (4.16).

Needless to say, the preceding argument is not sufficient to prove convergence of the time integral. Some effects related to the question of convergence will be discussed briefly in Section 14.

9.10. VISCOSITY, THERMAL CONDUCTION, DIFFUSION

For a specific application, consider a simple fluid with n, ϵ and $\mathbf{g}$ being the conserved densities. As noted previously, the conjugate thermodynamic variables are $\chi(n) = \zeta$, $\chi(\epsilon) = -\beta$, and $\chi(\mathbf{g}) = \mathbf{y}$, respectively.

To calculate the subtracted fluxes we need the equilibrium averages of the fluxes, which are

$$\mathbf{s}^0 = \left(h + \frac{1}{2}\rho u^2\right)\mathbf{u}, \quad t^0_{ij} = p\delta_{ij} + \rho u_i u_j.$$

In the derivatives (9.67) the conserved densities are the independent variables so $\mathbf{u}$ is to be replaced by $\mathbf{g}/\rho$. The subtracted mass flux $\mathbf{g}^{-}$ vanishes since $\mathbf{g}$ is a conserved density as well as a flux. It follows that none of the ψ^* depend on $\nabla\chi(n)$, as was pointed out in Chapter 1. The transport coefficients are Galilean invariants so we may take the equilibrium reference state to be at rest. With $\mathbf{u} = 0$ the derivatives of the energy flux are

$$\partial s^0_i/\partial g^0_j = (h/\rho)\delta_{ij}, \quad \partial s^0_i/\partial\rho = \partial s^0_i/\partial\epsilon = 0.$$

Thus, the subtracted energy flux is

$$\mathbf{s}^{-} = \mathbf{s} - (h/\rho)\mathbf{g}.$$

In a similar way we find

$$t^{-}_{ij} = t_{ij} - \delta_{ij}[p + (\partial p/\partial n)\Delta n + (\partial p/\partial\epsilon)\Delta\epsilon].$$

Because of spherical symmetry, the energy flux has no contributions depending on the velocity gradients, and the momentum flux none depending on the temperature gradient. Thus, in a notation $L_{ij}(\xi_\alpha, \xi_\beta)$ instead of $L_{\alpha i,\beta j}$

the coefficients $L_{ij}(\epsilon, \mathbf{g})$ and $L_{ij}(\mathbf{g}, \epsilon)$ vanish. The equation for the energy flux reduces to

$$s_i^* = -L_{ij}(\epsilon,\epsilon)\frac{\partial \chi(\epsilon)}{\partial x_j} = -(1/3\kappa T^2)\,L_{jj}(\epsilon,\epsilon)\frac{\partial T}{\partial x_i}.$$

The thermal conductivity can be identified as

$$\lambda = (1/3\kappa T^2)L_{jj}(\epsilon,\epsilon) = (1/3\kappa T^2)\int_0^\infty dt\int d^3x\langle \mathbf{s}^-(\mathbf{x},t)\cdot\mathbf{s}^-\rangle_0.$$

Equivalent expressions are

$$\lambda = \frac{1}{3\kappa T^2}\int_0^\infty dt\frac{1}{V}\langle \mathbf{S}^-(t)\cdot\mathbf{S}^-\rangle_0 = \frac{1}{\kappa T^2}\int_0^\infty dt\frac{1}{V}\langle S_x^-(t)S_x^-\rangle_0. \tag{9.70}$$

Here $\mathbf{S}^- = \mathbf{S} - (h/\rho)\mathbf{G}$ where $\mathbf{S}$ is the total energy flux given by Equation (8.9) and $\mathbf{G}$ is the total momentum.

In a similar way we find for the momentum flux

$$t_{ij}^* = -L_{jk}(g_i,g_l)\frac{\partial \chi(g_l)}{\partial x_k} = \frac{1}{\kappa T}\int_0^\infty dt\int d^3x\langle t_{ij}^-(\mathbf{x},t)\,t_{kl}^-\rangle_0\,\frac{\partial u_l}{\partial x_k}.$$

The spherical symmetry enables one to reduce this to the form (1.35) with the shear and bulk viscosities given by

$$\eta = \frac{1}{10\kappa T}\int_0^\infty dt\,\frac{1}{V}\left[\langle T_{ij}^-(t)T_{ij}^-\rangle_0 - \frac{1}{3}\langle T_{ii}^-(t)T_{jj}^-\rangle_0\right],$$

$$\eta' = \frac{1}{9\kappa T}\int_0^\infty dt\,\frac{1}{V}\langle T_{ii}^-(t)T_{jj}^-\rangle_0.$$

Here T_{ij}^- is given by

$$T_{ij}^- = T_{ij} - \delta_{ij}[pV + (\partial p/\partial n)\Delta N + (\partial p/\partial \epsilon)\,\Delta H]$$

where T_{ij} is the volume-integrated momentum flux given in Equation (8.9). Spherical symmetry implies

$$\frac{1}{\kappa T}\int_0^\infty dt\,\frac{1}{V}\langle T_{ij}^-(t)T_{kl}^-\rangle_0 = \eta\left(\delta_{ik}\delta_{jl} + \delta_{il}\delta_{jk} - \frac{2}{3}\delta_{ij}\delta_{kl}\right) + \eta'\delta_{ij}\delta_{kl}$$

from which alternative forms for η and η' can be obtained, such as

$$\eta = \frac{1}{\kappa T}\int_0^\infty dt\,\frac{1}{V}\langle T_{xy}(t)T_{xy}\rangle_0,$$

$$\eta' - \frac{2}{3}\eta = \frac{1}{\kappa T}\int_0^\infty dt\,\frac{1}{V}\langle T_{xx}^-(t)T_{yy}^-\rangle_0. \tag{9.71}$$

It was seen in Chapter 4 that the bulk viscosity vanishes for an ideal monatomic gas, and this is readily confirmed from the Green-Kubo formula for η'. In this case $(\partial p/\partial\epsilon) = \frac{2}{3}$ so $T_{ij}^{-} = T_{ij} - \frac{2}{3}\delta_{ij}H$. If the potential energy is neglected (as it can be at low density) in H and T_{ij} then $T_{ii} = 2\ H$ so $T_{ii}^{-} = 0$ and consequently $\eta' = 0$. It can be shown that the bulk viscosity vanishes through linear terms of the density expansion, unless the particles have internal structure when it may be nonvanishing at the lowest order in density.

In a nonreacting mixture the particle number is conserved for each species. Let n_r, $\mathbf{j}_r$ denote the density and flux for the r'th species; the conjugate thermodynamic variable is $\chi(n_r) = \nu_r - \frac{1}{2}\,\beta m_r u^2$. The equilibrium particle flux is

$$\mathbf{j}_r^0 = (n_r^0/\rho)\mathbf{g}^0$$

which yields

$$\mathbf{j}_r^{-} = \mathbf{j}_r - (n_r^0/\rho)\mathbf{g}.$$

It is straightforward to show that the coefficients occurring in Equation (1.45) are given by

$$L_r = \frac{1}{3}\int_0^\infty dt\,\frac{1}{V}\,\langle \mathbf{S}^{-}(t)\cdot\mathbf{J}_r^{-}\rangle_0,$$

$$L_r' = \frac{1}{3\kappa T^2}\int_0^\infty dt\,\frac{1}{V}\,\langle \mathbf{J}_r^{-}(t)\cdot\mathbf{S}^{-}\rangle_0,$$

$$L_{rp} = \frac{1}{3}\int_0^\infty dt\,\frac{1}{V}\,\langle \mathbf{J}_r^{-}(t)\cdot\mathbf{J}_p^{-}\rangle_0.$$

9.11. AGEING PERIOD AND HYDRODYNAMIC STAGE

In order for Equation (9.55) to take on the hydrodynamic form (9.53) for large times, it is necessary that $G(\mathbf{x},t)$ asymptotically obey the linearized hydrodynamic equations. The assumption that this is in fact the case is called Onsager's assumption on the regression of fluctuations. The time dependence of $G(\mathbf{x},t)$ is determined by the microscopic equations of motion, and so the validity of Onsager's assumption is subject to verification by analysis based on the microscopic equations. For a gas described by the Boltzmann equation, the verification is straightforward and will be given in Chapter 11, but in any case it is clear that Onsager's assumption must hold if the hydrodynamic equations are valid.

To state it somewhat differently, if for large times $G(\mathbf{x},t)$ satisfies a set of linear, first order (in the time), partial differential equations, then these are, by definition, the hydrodynamic equations. If that is not the case, then the system cannot be described by hydrodynamic equations, except perhaps in some generalized sense. (Here we are concerned only with positive values of the time. Correlation functions with negative t are related to those with positive t by time reversal, as discussed in Chapter 10.)

If it were the case that $G(\mathbf{x},t)$ satisfied the hydrodynamic equations for all (positive) t, then we could express it in the form

$$\hat{G}(\mathbf{k},t) = \exp[-t\hat{A}(\mathbf{k})]\hat{g}(\mathbf{k}).$$

However, this requires correction, even for large t, because the hydrodynamic equations do not hold during the ageing period. Let $G^h(\mathbf{x},t)$ denote a function which satisfies the hydrodynamic equations for all t and which is asymptotically equal to $G(\mathbf{x},t)$ for large t. Then we can express $\hat{G}^h$ by

$$\hat{G}^h(\mathbf{k},t) = \exp[-t\hat{A}(\mathbf{k})]\,X(\mathbf{k}) \tag{9.72}$$

where $X(\mathbf{k})$ is determined by the requirement that G^h approach G for large t:

$$X(\mathbf{k}) = \lim_{t\to\infty} \exp[t\hat{A}(\mathbf{k})]\hat{G}(\mathbf{k},t). \tag{9.73}$$

The remainder of G will be denoted by G^m:

$$G(\mathbf{x},t) = G^h(\mathbf{x},t) + G^m(\mathbf{x},t).$$

Then

$$\lim_{t\to\infty} \exp[t\hat{A}(\mathbf{k})]\hat{G}^m(\mathbf{k},t) = 0$$

so G^m approaches zero more rapidly than G^h for large t.

Because of the propagation of hydrodynamic disturbances, $G(\mathbf{x},t)$ extends over macroscopic distances for $t \neq 0$. In this respect it is quite different from its initial value, which is the short-ranged function $g(\mathbf{x})$. In the hydrodynamic stage G^h is the dominant part of G, while G^m describes the microscopic relaxation which is important during the ageing period. In the following we will show that $X(\mathbf{k})$ differs, for small $\mathbf{k}$, from $\hat{g}(\mathbf{k})$ by terms of order k^2 and higher, and it follows that $\hat{G}^m$ is smaller (for small $\mathbf{k}$) than $\hat{G}^h$ by the same order. The hydrodynamic part of G is more dominant for processes which vary more slowly.

The explicit form of the hydrodynamic part G^h can be obtained from Equation (9.72). We will only treat the lowest order, when $X(\mathbf{k})$ can be replaced by g. Consider a fluid satisfying the Navier-Stokes equations, and let ξ_α denote the variables

$$\xi_1 = (\rho\sqrt{TK_T})^{-1}\, m\,\Delta n,$$

$$\xi_2 = (T\sqrt{\rho C_V})^{-1}\,[\Delta\epsilon - (\partial\epsilon^0/\partial n^0)_T\,\Delta n],$$

$$\xi_3 = \Delta g_x/\sqrt{\rho T},\ \xi_4 = \Delta g_y/\sqrt{\rho T},\ \xi_5 = \Delta g_z/\sqrt{\rho T}$$

where Δ denotes the deviation from the equilibrium average. With these variables as a basis, the matrix g reduces to the unit matrix, and the matrix A is the same as given in Equation (6.1). We can evaluate $\hat{G}^h$ by using the eigenvalues and eigenvectors of A obtained in Chapter 6. In particular

$$\hat{G}^h_{\alpha\beta}(\mathbf{k},t) = \sum_{\mu} R_{\alpha\mu} R_{\beta\mu} \exp(-\zeta_\mu t)$$

where ζ_μ are the eigenvalues (6.2) and R is the matrix (6.3). In the lowest approximation R_0 for R the 11 element, for example, is

$$\hat{G}^h_{11}(\mathbf{k},t) = \frac{\gamma - 1}{\gamma} \exp(-D_T k^2 t) + \frac{1}{\gamma} \exp(-\Gamma k^2 t) \{\cos(ckt) + [\Gamma + (\gamma - 1)D_T](k/c) \sin(ckt)\},$$

while the 44 element is

$$\hat{G}^h_{44}(\mathbf{k},t) = \exp(-\nu k^2 t). \tag{9.74}$$

These approach zero on a time scale set by the hydrodynamics; the decay rate is smaller for smaller $\mathbf{k}$.

A corresponding representation for the correlation functions H and K can be obtained from Equations (9.59). For example, $\hat{K}_{44}$is determined by

$$\frac{\partial^2}{\partial t^2} \hat{G}_{44}(\mathbf{k},t) = -k^2 \hat{K}_{44}(\mathbf{k},t).$$

We use the hydrodynamic approximation (9.74) for $\hat{G}_{44}$; this equation has been written for positive times but in this case negative times can be included by writing $|t|$ in place of t. The result is

$$\hat{K}_{44}(\mathbf{k},t) = -\nu^2 k^2 \exp(-\nu k^2 |t|) + 2\nu\delta(t).$$

The first term decays with time in the same way as a hydrodynamic mode while the second term represents the ageing period. This level of description is the Markovian limit, in which ageing takes place instantaneously and the correlation function satisfies the hydrodynamic equation down to $t = 0$. The Green-Kubo relation

$$\nu = \int_0^\infty dt\, \hat{K}_{44}(0,t)$$

is satisfied but only as an identity. Evaluation of the transport coefficients requires a more realistic description of the ageing period but nevertheless this argument illustrates a general feature: In the limit $\mathbf{k} \to 0$ the hydrodynamic part of a correlation function for subtracted fluxes vanishes while its microscopic part decays rapidly with time during the ageing period. This corresponds in kinetic theory to the determination of the transport coefficients through integral equations containing the linearized collision operator but not the convective term. In a sense, the purpose of the Green-Kubo formulas is to reduce a problem involving internal gradients to one which is spatially homogeneous.

The difference between $X(\mathbf{k})$ and $\hat{g}(\mathbf{k})$ arises because of initial slip; the initial conditions imposed on a solution to the hydrodynamic equations must be adjusted from the actual initial conditions in order to properly extrapolate through the ageing period. Let $\delta\hat{\xi}^h(\mathbf{k}, 0)$ denote the adjusted initial values; $\delta\hat{\xi}(\mathbf{k}, t)$ (for large t) is then obtained from Equation (9.53):

$$\delta\hat{\xi}(\mathbf{k},t) = \exp[-t\hat{A}(\mathbf{k})]\,\delta\hat{\xi}^h(\mathbf{k},0).$$

This agrees with (9.54) for large t if

$$\delta\hat{\xi}^h(\mathbf{k},0) = X(\mathbf{k})\,\delta\hat{\chi}(\mathbf{k},0) \tag{9.75}$$

To evaluate $X(\mathbf{k})$, we put

$$\Delta X(\mathbf{k}) = X(\mathbf{k}) - \hat{g}(\mathbf{k})$$

and expand in powers of $\mathbf{k}$. It can be shown from Equation (9.73) that

$$\partial\Delta X(\mathbf{k})/\partial k_i\big|_{\mathbf{k}=0} = 0.$$

In addition after some manipulation it is found that

$$\partial^2\Delta X(\mathbf{k})/\partial k_i\,\partial k_j\big|_{\mathbf{k}=0} = 2\lim_{t\to\infty}\left[\int_0^t ds\,s\,M_{ij}(s) + t\int_t^\infty ds\,M_{ij}(s)\right]. \tag{9.76}$$

The expansion for $X(\mathbf{k})$ can be written

$$X(\mathbf{k}) = \hat{g}(\mathbf{k}) + \frac{1}{2}X_{ij}^{(2)}k_i k_j + \ldots$$

where $X_{ij}^{(2)}$ denotes the quantity (9.76).

We will work out the adjusted initial conditions for a fluid on the basis of (9.75), assuming the range of equilibrium correlations to be small enough that $\hat{g}(\mathbf{k})$ can be replaced by $\hat{g}(0) = \mathsf{g}$. It can be shown that there is no correction to the initial value for the density through terms of second order in $\mathbf{k}$. After inverting the Fourier transform, we find for the momentum density

$$\mathbf{g}^h(\mathbf{x}, 0) = \mathbf{g}(\mathbf{x}, 0) - \alpha_1\nabla^2\mathbf{u}(\mathbf{x}, 0) - (\tfrac{1}{3}\alpha_1 + \alpha_2)\nabla\nabla\cdot\mathbf{u}(\mathbf{x}, 0)$$

where

$$\alpha_1 = \frac{1}{10\kappa T}\lim_{t\to\infty}\frac{1}{V}\left[\int_0^t ds\,s\,f_1(s) + t\int_t^\infty ds\,f_1(s)\right],$$

$$\alpha_2 = \frac{1}{9\kappa T}\lim_{t\to\infty}\frac{1}{V}\left[\int_0^t ds\,s\,f_2(s) + t\int_0^\infty ds\,f_2(s)\right]$$

Here f_1,f_2 are abbreviations for the functions

$$f_1(s) = \langle T_{ij}^-(s)T_{ij}^-\rangle_0 - \frac{1}{3}\langle T_{ii}^-(s)T_{jj}^-\rangle_0,$$

$$f_2(s) = \langle T_{ii}^-(s)T_{jj}^-\rangle_0.$$

Similarly, the adjusted initial value for the energy density is found to be

$$\epsilon^h(\mathbf{x}, 0) = \epsilon(\mathbf{x}, 0) - \mu\nabla^2 T(\mathbf{x}, 0)$$

where

$$\mu = \frac{1}{3\kappa T^2}\lim_{t\to\infty}\frac{1}{V}\left[\int_0^t ds\, s\langle \mathbf{S}^-(s)\cdot\mathbf{S}^-\rangle_0 + t\int_t^\infty ds\,\langle \mathbf{S}^-(s)\cdot\mathbf{S}^-\rangle_0\right].$$

In terms of the temperature, the initial condition is

$$T^h(\mathbf{x}, 0) = T(\mathbf{x}, 0) - (\mu/\rho C_V)\nabla^2 T(\mathbf{x}, 0).$$

For a gas described by the Boltzmann equation, the expansion in powers of $\mathbf{k}$ is equivalent to the perturbation expansion discussed in Chapter 6. The proof of analyticity given there implies that $\hat{A}(\mathbf{k})$ and $X(\mathbf{k})$ are analytic in $\mathbf{k}$ (for hard-core interactions). However long-time tails, which are similar to those discussed in Chapter 2 for Brownian motion and will be discussed further in Chapter 12, destroy the analytic dependence on $\mathbf{k}$. The expansion is legitimate to the order worked out for $\hat{A}(\mathbf{k})$ but breaks down in the next order, while the second-order approximation obtained for $X(\mathbf{k})$ is actually nonexistent when the dynamical processes responsible for long-time tails are taken into account.

9.12. NONEQUILIBRIUM ENSEMBLE

Emphasis in the preceding discussion has been placed on the time-dependent correlation functions, but a treatment is also possible in which the nonequilibrium ensemble is a central object of interest. In this section we put the ensemble into a form similar to the distribution function which occurs in the normal solution to the Boltzmann equation, which is convenient for the derivation of the hydrodynamic equations. This can be done on the basis of the hydrodynamic initial value problem, but for generality we also allow for the specification of boundary values. In the next section the Green-Kubo formulas will be rederived and a brief discussion given of the Burnett approximation.

When boundary values are specified, it is necessary to consider an open system which interacts with its surroundings. The surroundings may be treated as reservoirs which are maintained at definite values of the thermodynamic variables and are distributed in a continuous manner around the boundary of the system. The force exerted on the system by the reservoirs is in general nonconservative: A system acting under the influence of a conservative force has an equilibrium thermodynamic state described by an ensemble such as (9.1) with the external potential included in H, but thermodynamic equilibrium is not possible if the system interacts with two or more reservoirs at different conditions.

To describe the reservoir forces, let the system and reservoirs be combined into a larger system, called the universe. The universe is isolated, and its Hamiltonian can be written $H_u = H + H_r + U$ where H is the Hamiltonian for the system of interest, H_r is that for the reservoirs and U is the interaction between them. The distribution function F_u for the universe satisfies the Liouville equation

$$\frac{\partial F_u}{\partial t} + \{F_u, H_u\} = 0. \tag{9.77}$$

An ensemble for the universe entails one for the system and surroundings separately, given by

$$F = \int d\Gamma_r F_u, \; F_r = \int d\Gamma F_u.$$

Here $d\Gamma_r$ and $d\Gamma$ denote elements of volume in the phase space for the reservoirs and system respectively. The normalization condition for F_u is

$$\int d\Gamma_r d\Gamma \, F_u = 1$$

and this implies that F and F_r are separately normalized.

The time dependence of F is governed by an equation which can be obtained by integrating the Liouville equation over the degrees of freedom for the reservoirs. The procedure is similar to the one used to derive the BBGKY hierarchy, but now the interest is in a system containing many particles rather than in clusters of only a few particles. On integrating (9.77) over the phase space for the surroundings we get

$$\frac{\partial F}{\partial t} + \{F,H\} + \sum_\alpha \frac{\partial}{\partial \mathbf{p}_\alpha} \cdot \mathbf{f}_\alpha F = 0. \tag{9.78}$$

Here the sum extends over the variables for the system alone, and the $\mathbf{f}_\alpha$ are effective forces determined by

$$\mathbf{f}_\alpha F = - \int d\Gamma_r F_u \frac{\partial U}{\partial \mathbf{q}_\alpha}.$$

We have assumed the potential energy U to be momentum independent, but $\mathbf{f}_\alpha$ in general depends on both coordinates and momenta.

If U is the potential for a long-range force, such as gravitational or electromagnetic forces, the "reservoirs" can be located far from the system, and the system and surroundings can be assumed to be statistically uncorrelated. Then $F_u = FF_r$ and the equation for $\mathbf{f}_\alpha$ reduces to

$$\mathbf{f}_\alpha = -\frac{\partial}{\partial \mathbf{q}_\alpha}\int d\Gamma_r F_r U.$$

In this case the effective forces are conservative, with a potential equal to the average of U over the distribution for the reservoirs. However a reservoir which maintains, say, a given temperature must be in contact with the system; the short-range correlations cause $\mathbf{f}_\alpha$ to be nonconservative and momentum dependent.

A function A of the system's coordinates and momenta depends on time according to

$$\frac{dA}{dt} = \{A,H\} + \sum_\alpha \mathbf{f}_\alpha \cdot \frac{\partial A}{\partial \mathbf{p}_\alpha}. \tag{9.79}$$

If A has in addition an explicit dependence on time, then

$$\frac{dA}{dt} = \frac{\partial A}{\partial t} + \{A,H\} + \sum_\alpha \mathbf{f}_\alpha \cdot \frac{\partial A}{\partial \mathbf{p}_\alpha}. \tag{9.80}$$

Hence Equation (9.78) for F can be written

$$\frac{dF}{dt} + F\sum_\alpha \frac{\partial}{\partial \mathbf{p}_\alpha}\cdot \mathbf{f}_\alpha = 0. \tag{9.81}$$

Let $d\Gamma_s$ denote an element of volume in the phase space for the system, and $d\Gamma_s(t)$ the element into which it transforms after t seconds. The relation between the two is

$$d\Gamma_s(t) = J(t)d\Gamma_s$$

where J is the Jacobian

$$J(t) = \frac{\partial(q(t),p(t))}{\partial(q,p)}.$$

For conservative systems $J = 1$, but in the present case explicit calculation gives

$$\frac{dJ}{dt} = J\sum_\alpha \frac{\partial}{\partial \mathbf{p}_\alpha}\cdot \mathbf{f}_\alpha.$$

Thus, a region of phase space does not maintain a fixed volume as it moves. Equation (9.81) conserves probability since F changes with time in a way which compensates for the changing volume in phase space.

Let Equation (9.81) be written as

$$\frac{d\eta}{dt} = q \tag{9.82}$$

where $\eta = \ln F$ and

$$q = -\sum_\alpha \frac{\partial}{\partial \mathbf{p}_\alpha} \cdot \mathbf{f}_\alpha. \tag{9.83}$$

We suppose different reservoirs to be statistically independent in which case their forces are additive, so q can be written as a sum of contributions from each reservoir:

$$q = \sum_r q_r.$$

Each q_r has a form like (9.83) but with only the forces from that reservoir included. To determine the dependence on the thermodynamic state of a reservoir, suppose the system to interact with only a single reservoir whose thermodynamic state is maintained at χ^r_α. Equation (9.82) becomes

$$\frac{d\eta}{dt} = q_r.$$

In this situation the system will come to equilibrium with the reservoir, so this equation should be satisfied by the equilibrium ensemble with parameters χ^r_α. This yields

$$q_r = \sum_\alpha \chi^r_\alpha \left(\frac{dY_\alpha}{dt}\right)_r.$$

The subscript r on the right-hand side means that in calculating the time derivative according to Equation (9.79), only the forces due to the r'th reservoir are to be included. Thus we obtain for q

$$q = \sum_r \sum_\alpha \chi^r_\alpha \left(\frac{dY_\alpha}{dt}\right)_r. \tag{9.84}$$

The reservoirs are distributed along the boundary of the system and so can be labeled by their position. Hence, we write $\chi_\alpha(\mathbf{x})$ rather than χ^r_α. The time derivative of Y_α due to interaction with the reservior at $\mathbf{x}$ can be expressed as $-\boldsymbol{\psi}_\alpha \cdot d\mathbf{A}$ where $d\mathbf{A}$ is the element of area at $\mathbf{x}$. Writing the sum over r in Equation (9.84) as an integral over $\mathbf{x}$, we get

$$q = -\sum_\alpha \int \chi_\alpha(\mathbf{x}) \boldsymbol{\psi}_\alpha(\mathbf{x}) \cdot d\mathbf{A}.$$

This is to be used in conjunction with (9.82) to determine η.

If the external forces are not explicitly time dependent, the solution to Equation (9.82) can be written

$$\eta(t) = S(-t)\eta(0) + \int_0^t ds\, S(s - t) q$$

where $S(t)$ is the solution operator to the equations of motion (9.80). Using local equilibrium as the initial condition at time $t = 0$, we get

$$\eta(t) = -Q_L + \sum_\alpha \int d^3x\, \chi_\alpha(\mathbf{x})\, \xi_\alpha(\mathbf{x}, -t) - \sum_\alpha \int_0^t ds \int \chi_\alpha(\mathbf{x}) \boldsymbol{\psi}_\alpha(\mathbf{x}, s - t) \cdot d\mathbf{A}.$$

Here Q_L and χ_α without time arguments are the initial values. This equation determines the nonequilibrium ensemble as a functional of the initial and boundary values of the thermodynamic variables. The surface integral can be converted to a volume integral, which then depends on the thermodynamic variables $\chi_\alpha(\mathbf{x}, t)$ defined throughout the system and as functions of time. With the aid of the conservation laws, we obtain

$$\eta(t) = -Q_L + \sum_\alpha \int d^3x\, \chi_\alpha(\mathbf{x}, t)\, \xi_\alpha(\mathbf{x})$$
$$- \sum_\alpha \int_0^t ds \int d^3x \left[\frac{\partial \chi_\alpha(\mathbf{x}, s)}{\partial s} \xi_\alpha(\mathbf{x}, s - t) + \nabla \chi_\alpha(\mathbf{x}, s) \cdot \boldsymbol{\psi}_\alpha(\mathbf{x}, s - t) \right].$$

At each time t we associate with $F(t) = \exp \eta$ a local equilibrium ensemble

$$F_L(t) = \exp\left[-Q_L(t) + \int d^3x \sum_\alpha \chi_\alpha(\mathbf{x}, t) \xi_\alpha(\mathbf{x}) \right]. \tag{9.85}$$

The parameters $\chi_\alpha(\mathbf{x}, t)$ are fixed by the condition (9.23) which we rewrite as

$$\langle \xi_\alpha(\mathbf{x}); t \rangle_L = \langle \xi_\alpha(\mathbf{x}); t \rangle \tag{9.86}$$

where $\langle \;\; ; t \rangle_L$ and $\langle \;\; ; t \rangle$ denote averages over $F_L(t)$ and $F(t)$ respectively. To put $F(t)$ into a form which is adapted to the Chapman-Enskog expansion, let a factor $F_L(t)$ be extracted:

$$F(t) = F_L(t) e^D. \tag{9.87}$$

Then D is given by

$$D = Q_L(t) - Q_L$$
$$- \int_0^t ds \sum_\alpha \int d^3x \left[\frac{\partial \chi_\alpha(\mathbf{x}, s)}{\partial s} \xi_\alpha(\mathbf{x}, s - t) + \nabla \chi_\alpha(\mathbf{x}, s) \cdot \boldsymbol{\psi}_\alpha(\mathbf{x}, s - t) \right].$$

The difference in Q's can be calculated by

$$Q_L(t) - Q_L = \int_0^t ds \frac{d}{ds} Q_L(s) = \sum_\alpha \int d^3x \frac{\delta Q_L(s)}{\delta \chi_\alpha(\mathbf{x}, s)} \frac{\partial \chi_\alpha(\mathbf{x}, s)}{\partial s}$$
$$= \sum_\alpha \int d^3x \langle \xi_\alpha(\mathbf{x}); s \rangle_L \frac{\partial \chi_\alpha(\mathbf{x}, s)}{\partial s}$$

where we have used (9.24). Hence, D can be written as

$$D = -\int_0^t ds \sum_\alpha \int d^3x$$

$$\times \left\{ [\xi_\alpha(\mathbf{x}, s - t) - \langle \xi_\alpha(\mathbf{x}); s \rangle_L] \frac{\partial \chi_\alpha(\mathbf{x}, s)}{\partial s} + \boldsymbol{\psi}_\alpha(\mathbf{x}, s - t) \cdot \nabla \chi_\alpha(\mathbf{x}, s) \right\}$$

which in view of (9.28) can also be written in the form

$$D = -\int_0^t ds \sum_\alpha \int d^3x \cdot \left\{ [\xi_\alpha(\mathbf{x}, s - t) - \langle \xi_\alpha(\mathbf{x}); s \rangle_L] \frac{\partial \chi_\alpha(\mathbf{x}, s)}{\partial s} \right.$$

$$\left. + [\boldsymbol{\psi}_\alpha(\mathbf{x}, s - t) - \langle \boldsymbol{\psi}_\alpha(\mathbf{x}); s \rangle_L] \cdot \nabla \chi_\alpha(\mathbf{x}, s) \right\}.$$

9.13. HYDRODYNAMIC EQUATIONS

In this section we consider the derivation of the hydrodynamic equations from the ensemble (9.87). The Green-Kubo formulas will be obtained again and a brief discussion given of the Burnett approximation.

In lowest order D is set equal to zero, so F reduces to F_L. The hydrodynamic equations are then the Euler equations, modulo the nonlocality mentioned in Section 4. In proceeding to the next, or Navier-Stokes, order, the time derivatives in D are eliminated by using the Euler equations, which gives

$$D = -\int_0^t ds \sum_\alpha \int d^3x$$

$$\times \left\{ \boldsymbol{\psi}_\alpha(\mathbf{x}, s - t) - \langle \boldsymbol{\psi}_\alpha(\mathbf{x}); s \rangle_L - \sum_\beta E_{\beta\alpha} [\xi_\beta(\mathbf{x}, s - t) - \langle \xi_\beta(\mathbf{x}); s \rangle_L] \right\} \cdot \nabla \chi_\alpha(\mathbf{x}, s).$$

Here $E_{\beta\alpha}$ denotes the coefficients (9.67). Through terms linear in gradients, F is given by

$$F = F_L(1 + D) \tag{9.88}$$

which yields for the irreversible fluxes

$$\boldsymbol{\psi}_\alpha^*(\mathbf{x}, t) = \langle \boldsymbol{\psi}_\alpha(\mathbf{x}); t \rangle - \langle \boldsymbol{\psi}_\alpha(\mathbf{x}); t \rangle_L = \langle \boldsymbol{\psi}_\alpha(\mathbf{x}) D; t \rangle_L. \tag{9.89}$$

In the approximation being considered it is consistent to replace the local-equilibrium averages by averages over an equilibrium reference state with thermodynamic variables equal to those at the point $\mathbf{x}$. Since the equilibrium ensemble is time independent, we obtain

$$\boldsymbol{\psi}_\alpha^*(\mathbf{x}, t) = -\sum_\beta \int_0^t ds \int d^3x' \, \langle \boldsymbol{\psi}_\alpha(\mathbf{x}) \boldsymbol{\psi}_\beta^-(\mathbf{x}', -s) \rangle_0 \, \nabla \chi_\beta(\mathbf{x}', t - s) \tag{9.90}$$

where ψ_α^- are the subtracted fluxes obtained previously. In the approximation at hand we may treat the gradients of the thermodynamic variables as constants. In addition we take the limit $t \to \infty$ in order to get past the ageing period and reach a steady state. The result is Equation (1.59) with the coefficients $L_{\alpha\beta}$ being given by the Green-Kubo formulae as obtained previously.

In Equation (9.48) it would not be correct to extract the thermodynamic variables, assumed to be slowly varying, from the integral since $G(\mathbf{x} - \mathbf{x}', t)$ is not a short ranged function. In the case of Equation (9.90), this procedure is valid because the correlation function for the subtracted fluxes is dominated by its short-ranged microscopic part, but it is necessary to proceed more carefully when deriving the Burnett or higher-order equations.

The ensemble for slowly varying processes near equilibrium can be obtained by applying the previous method to Equation (9.88), which yields

$$F = F_L\left[1 - \sum_\alpha \int_{-\infty}^{0} ds\, \mathbf{\Psi}_\alpha^-(s) \cdot \nabla\chi_\alpha\right].$$

This result is only formal since the infinite time limit should be preceded by the thermodynamic limit, and F does not exist in the thermodynamic limit. Nevertheless, it can be used to calculate averages if the limits are then taken in the proper order. For a simple fluid the ensemble is

$$F = F_L\left[1 - \frac{1}{\kappa T^2}\int_{-\infty}^{0} dt\, S_i^-(t)\frac{\partial T}{\partial x_i} - \frac{1}{\kappa T}\int_{-\infty}^{0} dt\, T_{ij}^-(t)\frac{\partial u_i}{\partial x_j}\right]$$

which has an evident similarity with the distribution function (4.18) obtained as the normal solution to the Boltzmann equation. Fluctuations in a non-equilibrium steady state can be described on the basis of this result. For example, the correlation function for fluctuations in density is

$$\begin{aligned}\langle \Delta n(\mathbf{x})\, \Delta n(\mathbf{x}', t)\rangle &= \langle \Delta n(\mathbf{x})\, \Delta n(\mathbf{x}', t)\rangle_L - \frac{1}{\kappa T^2}\int_{-\infty}^{0} ds\, \langle \Delta n(\mathbf{x})\, \Delta n(\mathbf{x}', t) S_i^-(s)\rangle_L \frac{\partial T}{\partial x_i} \\ &\quad - \frac{1}{\kappa T}\int_{-\infty}^{0} ds \langle \Delta n(\mathbf{x}) \Delta n(\mathbf{x}', t)\, T_{ij}^-(s)\rangle_L \frac{\partial u_i}{\partial x_j}.\end{aligned}$$

The hydrodynamic equations at the Burnett order are obtained by extending the analysis to the next order of the Chapman-Enskog expansion. The calculations are rather lengthy and so only a brief discussion will be given. We note the particular points in the preceding analysis where the approximation must be improved; the linearization (9.88), the replacement of local equilibrium by strict equilibrium in going from Equation (9.89) to (9.90), and the elimination of the nonlocality in space and time contained in (9.90). As might be expected, the linear Burnett coefficients are the easiest to obtain;

one method is to continue the expansion in powers of **k** developed in Section 8.

To illustrate the form of the Burnett coefficients, we list a few of the contributions to the momentum flux:

$$t_{ij}^* = -(2\eta_1/\rho)\left[\frac{\partial^2 p}{\partial x_i \partial x_j} - K_T \frac{\partial p}{\partial x_i}\frac{\partial p}{\partial x_j}\right] - (\eta_2/\rho)\delta_{ij}[\nabla^2 p - K_T(\nabla p)^2]$$

$$+ \eta_3 \frac{\partial^2 T}{\partial x_i \partial x_j} + \eta_4 \delta_{ij} \nabla^2 T + \eta_5 \frac{\partial T}{\partial x_i}\frac{\partial T}{\partial x_j} + \delta_{ij}\eta_6(\nabla T)^2 + \ldots$$

Here K_T is the isothermal compressibility and $\eta_1, .., \eta_6$ are Burnett coefficients. The form of the terms in both t_{ij}^* and s_i^* could (for a simple fluid) be written down on the basis of spherical symmetry, but because of time reversal invariance the number of independent coefficients is less than that expected on the basis of spherical symmetry alone; furthermore, symmetry would allow independent coefficients where the compressibility appears. For a gas the number of terms is reduced by the condition $t_{ii}^* = 0$ which gives

$$\eta_1 = -\frac{3}{2}\eta_2 = -\frac{1}{2}\rho\gamma_8, \quad \eta_3 = -3\eta_4 = \gamma_7, \quad \eta_5 = -3\eta_6 = \gamma_9$$

where γ_7, γ_8, γ_9 are coefficients introduced in Equation (4.52).

As examples, we list the correlation function formulas for η_1, η_3, and η_5:

$$\eta_1 = \frac{1}{\kappa T}\int_0^\infty dt\, t \frac{1}{V}\langle T_{xy}(t)\, T_{xy}\rangle_0,$$

$$\eta_3 = -\frac{2}{\kappa T^2}\int_0^\infty dt \int d^3xy\, \langle t_{xy}(0,t) s_x^-(\mathbf{x})\rangle_0,$$

$$\eta_5 = -\frac{2}{T}\eta_3$$

$$-\frac{2}{\kappa^2 T^4}\int_0^\infty dt \int d^3xy \left\langle t_{xy}(0,t)\left\{S_x^-\left[\Delta\epsilon(\mathbf{x}) - \frac{h}{\rho_0}\Delta\rho(\mathbf{x})\right] - \kappa T^2 \frac{\partial(h/\rho_0)}{\partial T} g_x(\mathbf{x})\right\}\right\rangle_0.$$

The formula for η_1 depends on the same correlation function as appears in the Green-Kubo formula for the viscosity, but the integrand contains an extra factor of the time; η_3 involves a spatial moment. (Consequently, these coefficients vanish in the Markovian limit.) Other linear coefficients also depend on space and time moments of two-point correlation functions, while nonlinear coefficients, such as η_5, involve correlations for three or more variables.

For a gas described by the Boltzmann equation, the correlation function formulas for the Burnett coefficients reduce to the same results as obtained in Chapter 4. However long-time tails cause the Burnett (and higher-order) coefficients to diverge, which means that the expansion used is not valid. In the linear case the fluxes are not analytic functions of the wavenumber, and nonlinear contributions are not analytic in the amplitude of the disturbance.

9.14. SLOW INTERNAL RELAXATION

For systems contemplated in the foregoing, correlation functions for the conserved densities describe slowly varying hydrodynamic processes while those for the fluxes carry the fast relaxation which determines the transport coefficients. However, there is a variety of dynamical effects which destroy the sharp separation between the hydrodynamic stage and the ageing period. Persistent correlations have already been mentioned; the long-time tails decay to zero fast enough for the Green-Kubo formulas to converge in three dimensions but for two-dimensional systems these integrals diverge, as do the corresponding formulas for Burnett coefficients in three dimensions. This complication is present even for monatomic gases.

Polyatomic gases exhibit slow internal relaxation, due for example to vibrational transitions. The rate of relaxation may be comparable to hydrodynamic rates such as the frequency of a sound wave. This slow relaxation can be incorporated by proceeding with the expansion in spatial derivatives as before, but keeping the time integral as it is in Equation (9.90). If the initial time is chosen at $t = 0$ we get

$$\boldsymbol{\psi}^*(t) = -\int_0^t ds\, M(t-s)\nabla\chi(s). \tag{9.91}$$

The system "remembers" its history, and M is the memory, or relaxation, function. If for simplicity we suppose χ to be a purely diffusive mode satisfying

$$\frac{\partial\chi}{\partial t} + \nabla\cdot\boldsymbol{\psi}^* = 0$$

then its equation of motion is

$$\frac{\partial\chi(t)}{\partial t} = \int_0^t ds\, M(t-s)\nabla^2\chi(s).$$

In the extreme case of a very long relaxation time, the correlation function has effectively a nonvanishing long-time limit:

$$\lim_{t\to\infty} M_{\alpha\beta}(t) = M^{\infty}_{\alpha\beta}.$$

We then rewrite Equation (9.91) as

$$\boldsymbol{\psi}^*(t) = -\int_0^t ds[M(t-s) - M^{\infty}]\nabla\chi(s) - M^{\infty}\nabla\int_0^t ds\,\chi(s). \tag{9.92}$$

If M approaches its limit rapidly, the first term can be described by a transport coefficient

$$L = \int_0^\infty ds[M(s) - M^\infty].$$

To interpret the second term in (9.92), suppose the system to be a solid and let χ be velocity; then the time integral of χ is the displacement X and ∇X is the strain. Thus, in a general sense M^∞ is an elastic coefficient. The equation of motion for χ is now

$$\frac{\partial^2 \chi}{\partial t^2} - L\nabla^2 \frac{\partial \chi}{\partial t} - M^\infty \nabla^2 \chi = 0.$$

Depending on the values of the parameters, χ may have an oscillating, rather than diffusive, behavior.

In the case of sinusoidal variations, $\chi \sim \exp i\omega t$, the initial condition is imposed in the remote past:

$$\boldsymbol{\psi}^*(t) = -\int_{-\infty}^{t} ds\, M(t - s)\nabla\chi(s).$$

This can be written in terms of a frequency-dependent transport coefficient

$$\boldsymbol{\psi}^* = -L(\omega)\nabla\chi, \quad L(\omega) = \int_{-\infty}^{0} dt M(t) \exp(-i\omega t).$$

A different kind of modification is necessary if the set of conservation laws is not entirely fixed. A given Hamiltonian has definite conservation laws which are satisfied by virtue of the equations of motion. However in addition to the quantities which are universally (for the given Hamiltonian) conserved, there can be others which are time independent (or nearly so) only under certain conditions. To fix ideas, consider again a polyatomic gas. At high temperature the conservation laws are those appropriate to dissociated atoms. If the atoms are treated as elementary (that is, if the possibilities of ionization or nuclear processes are ignored), then these are the universal conservation laws which hold under all conditions. At low temperatures when reactions are negligible, there are in addition conservation laws for the numbers of molecules of the various kinds which can be formed from the given atoms. There is also a change in the set of thermodynamic variables; a chemical potential can be defined and assigned a value for each of the molecular species. The derivation of Green-Kubo formulas on the basis of the universal conservation laws does not lead to subtracted fluxes which account for the extra conserved quantities, and the resulting time integrals may diverge. Of course in this case the correct procedure at low temperature is clear; one should base the derivation on the conservation laws for the molecular system. At intermediate temperatures reactions may proceed at a rate comparable to the hydrodynamic rates of interest. In this case, fluctuations in concentration of the different species do not decay at a rate which can be treated as rapid, and the hydrodynamic equations must be modified to account for the coupling between the conserved densities and the slow fluctuations.

Slow fluctuations are common at thermodynamic phase transitions. Localized fluctuations to structures resembling the new phase occur with increasing probability as the transition is approached, and their lifetime increases corresponding to the stability of the new phase. Depending on the nature of the transition, the form of the conservation laws and hydrodynamic equations may be different in the two phases. In any case, slow modes near the transition modify the hydrodynamics by their coupling to the conserved densities.

Notes on Chapter 9

9.3 Our treatment of the static properties of fluctuations follows that of Callen [1985].

9.4 For the information theory argument referred to, see Jaynes [1963, 1979], and other references given therein.

9.6 See Nyquist [1928], Callen and Welton [1951], Kubo [1957a, 1966], and van Kampen [1971]. For nonlinear effects in the response to an external field, see Bernard and Callen [1959], Lax [1960], Peterson [1967], Tanaka, Mourjani, and Morita [1967], and Ronis [1984b].

9.7 The use of the grand-canonical ensemble in this context, with local equilibrium as an initial condition, is due to Mori [1958a, b].

9.8 See Zwanzig [1964], Mori [1965a], McLennan [1967], Dufty [1968a], and Dufty and McLennan [1968b].

9.10 The Green-Kubo formulas were derived by M. S. Green [1954] using a generalization of the Fokker-Planck equation (Green [1952]), and by Kubo, Yokota, and Nakajima [1957b] using Onsager's assumption on the regression of fluctuations. As noted previously, Kubo's formula for the conductivity was derived by linear response theory in Kubo [1957a]. M. S. Green used a microcanonical ensemble so his results do not contain the fluctuations of conserved variables which appear in the subtracted fluxes. The Green-Kubo formulas with the subtracted fluxes as described here were obtained by Mori [1958a, b, 1962]; M. S. Green [1960] showed that his formulas go over into Mori's in the thermodynamic limit. It was shown by H. S. Green [1961] that application of the Chapman-Enskog method leads in a straightforward way to formulas with the subtracted fluxes. Derivations of the Green-Kubo formulas are reviewed in Zwanzig [1965c] and Brush [1972].

Beyond their role in the Green-Kubo formulas, time-dependent correlation functions are widely used for modeling phenomena in gases and dense systems; for reviews see Curtiss [1967], Egelstaff [1967], Berne and Forster [1971], Hynes [1977], and Boon and Yip [1980].

The proof that for monatomic gases the bulk viscosity vanishes through terms linear in the density is given in Ernst [1966a].

9.11 The method used to obtain correlation functions in the Markovian limit is developed in Landau and Lifshitz [1959] who work out the detailed form for fluctuations of pressure and of the energy and momentum fluxes. Other calculations using the Markovian limit are given in Rytov [1958a, b, c], Kadanoff and Martin [1963], Mountain [1966], and Bixon and Zwanzig [1969].

9.12 Open systems are treated in Bergmann and Lebowitz [1955], Lebowitz and Bergmann [1957], and Lebowitz and Shimony [1962]. The method discussed here was developed in McLennan [1959, 1960, 1963a]; the nonequilibrium ensemble was also constructed by Zubarev [1962, 1965, 1966, 1970, 1974].

Initial conditions modified for initial slip are worked out on the basis of the Boltzmann equation in Grad [1963a].

9.13 Correlation function expressions for the linear Burnett coefficients are obtained in McLennan [1973], and for the nonlinear coefficients in Dufty [1976a], Wong, McLennan, Lindenfeld, and Dufty [1978], de Schepper and Cohen [1978], Ronis [1979], Brey, Zwanzig, and Dorfman [1981], Dufty and Marchetti [1981], and Brey [1983]. For the reduction of these expressions to the corresponding results obtained from the Boltzmann equation, see Dufty [1976a, b], and Dufty and Lindenfeld [1979]. The theory of fluctuations in a nonequilibrium state is reviewed in Tremblay [1984].

9.14 For discussions of internal relaxation, see Meixner and Reik [1959], Zwanzig [1965a], and Zwanzig and Mountain [1965b]. There is an extensive literature on slow modes at phase transitions; see for example the review by Gunton [1979].

10

SYMMETRIES OF CORRELATION FUNCTIONS

10.1. INVARIANCE UNDER TRANSLATIONS OF TIME

The space- and time-dependent correlation functions have certain general properties which follow from underlying symmetries. One of these is Onsager's theorem, which will be proven in the next section. In this section we consider the consequences of invariance to translations of the time axis. Among other things, this invariance is responsible for the positivity of transport coefficients such as the thermal conductivity and viscosity.

Consider a system governed by classical mechanics, and define $C_{AB} = \langle \overline{A} B \rangle$ where the average is taken over an equilibrium ensemble. For generality we have allowed for the possibility that A or B be complex, and the bar denotes the complex conjugate. We note first of all that C_{AB} has the algebraic properties of the scalar product in a Hilbert space. In particular $\|A\| = \sqrt{C_{AA}}$ defines a norm, and C_{AB} satisfies the triangle inequality:

$$C_{AB} \leq \sqrt{C_{AA} C_{BB}}.$$

Let $C(t) = \langle \overline{A}(t) A \rangle$ denote the time-dependent correlation function for a single variable A. Invariance of the microscopic equations of motion, and of the equilibrium ensemble, under translations of the time implies $\langle \overline{A}(t) A \rangle = \langle \overline{A}(\tau + t) A(\tau) \rangle$. Setting $\tau = -t$ we find

$$\overline{C}(t) = C(-t). \tag{10.1}$$

In addition for arbitrary complex numbers a_p we have

$$\sum_{p,q=1}^{n} \bar{a}_p a_q C(t_p - t_q) = \left\langle \left| \sum_{p=1}^{n} a_p A(t_p) \right|^2 \right\rangle$$

for any positive integer n. It follows that

$$\sum_{p,q=1}^{n} \bar{a}_p a_q C(t_p - t_q) \geq 0. \tag{10.2}$$

In a similar way it can be shown that

$$\int dt \int du\, \bar{a}(t) a(u) C(t - u) \geq 0 \tag{10.3}$$

where $a(t)$ is any function such that the integral exists.

A function satisfying the conditions (10.1) and (10.2) is said to be of positive type, or a positive definite function. A function of positive type need not be positive for all values of its argument; the cosine function, for example, is readily shown to be positive type. Another way of stating the conditions is that $C(t_p - t_q)$ are the elements of a Hermitian non-negative matrix.

In quantum theory the definition of $C(t)$ is replaced by $C(t) = \langle A^\dagger(t) A \rangle$ where the dagger denotes the adjoint. It is readily confirmed that this quantity is a function of positive type. The transport coefficients depend on response functions with Kubo's extra temperature integral as in Equation (9.39),

$$\phi(t) = \beta^{-1} \int_0^\beta d\lambda \langle e^{-\lambda H} A^\dagger(t)\, e^{\lambda H} A \rangle.$$

This function also satisfies the conditions (10.1) and (10.2).

The inequality (10.2) is equivalent to the requirement that the determinant of the quadratic form be non-negative, for all n. Taking $n = 1$ we get the obvious result, $C(0) \geq 0$. With $n = 2$ we find a form of the triangle inequality,

$$|C(t)| \leq C(0). \tag{10.4}$$

Thus, a correlation function is bounded by its initial value. With $n = 3$, we get

$$C(0)[|C(t_1 - t_2)|^2 + |C(t_2 - t_3)|^2 + |C(t_3 - t_1)|^2] \leq C^3(0) + 2\,\mathrm{Re}\, C(t_1 - t_2) C(t_2 - t_3) C(t_3 - t_1).$$

We set $t = t_1 - t_2$, $\tau = t_2 - t_3$ and rearrange this inequality into

$$C(0)\,|C(t) - C(t + \tau)|^2 \leq C(0)[C^2(0) - |C(\tau)|^2] - 2\,\mathrm{Re}\{[C(0) - C(\tau)]\, C(t) \bar{C}(t + \tau)\}.$$

Letting τ approach 0, we see that C is continuous for all t if it is continuous at $t = 0$. It also follows that if $C(\tau)$ is equal to its initial value for some τ, then it is periodic with period τ. If for some τ, $C(\tau) = -C(0)$ then C is periodic with period 2τ.

A general characterization of functions of positive type is provided by a theorem due to Bochner: If a continuous function $C(t)$ has a Fourier transform

$$\hat{C}(\omega) = \int_{-\infty}^{\infty} dt\, e^{i\omega t}\, C(t)$$

then $C(t)$ is positive type if and only if $\hat{C}(\omega)$ is non-negative for all (real) ω.

It follows from (10.1) that $\hat{C}(\omega)$ is real. In the case of main interest $C(t)$ is real and we have

$$\hat{C}(\omega) = \hat{C}(-\omega). \tag{10.5}$$

A straightforward calculation shows that $\exp(-\mathsf{s}|t|)$ is a function of positive type if s is real and positive. More generally, the function

$$\begin{aligned} C(t) &= \exp(-\mathsf{s}t), \quad t \geq 0, \\ &= \exp(\bar{\mathsf{s}}t), \quad t \leq 0 \end{aligned} \tag{10.6}$$

is positive type if $\mathrm{Re}\,\mathsf{s} \geq 0$.

Let $C(t)$ and $C_1(t)$ be two functions of positive type. Bochner's theorem and the convolution theorem for Fourier transforms show that

$$C_2(t) = \int_{-\infty}^{\infty} ds\, C_1(s) C(t - s)$$

is also positive type. Choosing C_1 to be the function (10.6) and taking $t = 0$ we find an inequality for the Laplace transform of C:

$$\mathrm{Re}\int_0^{\infty} dt\, e^{-\mathsf{s}t}\, C(t) \geq 0 \quad \text{for } \mathrm{Re}\,\mathsf{s} \geq 0. \tag{10.7}$$

Because of (10.4) the Laplace transform exists and defines an analytic function of s throughout the half plane $\mathrm{Re}\,\mathsf{s} > 0$. An alternative derivation can be obtained from the inequality (10.3) with $a(t) = \exp(-\mathsf{s}t)$, $t \geq 0$, $a(t) = 0$, $t < 0$. This gives

$$\int_0^{\infty} dt \exp[-(\mathsf{s} + \bar{\mathsf{s}})t] \int_{-\infty}^{t} d\tau \exp(\bar{\mathsf{s}}\tau)\, C(\tau) \geq 0$$

and integrating by parts we obtain

$$(\mathsf{s} + \bar{\mathsf{s}})^{-1}\, \mathrm{Re}\int_0^{\infty} dt\, e^{-\mathsf{s}t}\, C(t) \geq 0.$$

In the integration by parts s has been assumed to lie in the right half-plane so this again yields (10.7).

The preceding inequalities have a direct application to transport coefficients. Taking the limit $s \to 0$ in Equation (10.7), we get for real C

$$L = \int_0^\infty dt\, C(t) \geq 0 \tag{10.8}$$

provided that the integral exists. Transport coefficients such as thermal conductivity and viscosity have this form and so must be non-negative. An additional consequence is

$$\lim_{t\to\infty} C(t) = \lim_{s\to 0} s \int_0^\infty dt e^{-st} C(t) \geq 0 \tag{10.9}$$

provided that the first limit exists. If it does not vanish this limit is a coefficient of elasticity, and so we see that such coefficients are also non-negative.

The frequency moments are defined by

$$\overline{\omega^p} = [2\pi C(0)]^{-1} \int_{-\infty}^{\infty} d\omega\, \omega^p\, \hat{C}(\omega)$$

where the factor in front is introduced so that for $p = 0$ the moment is equal to 1. The frequency moments are related to the short-time behavior of $C(t)$ since formally

$$\overline{\omega^p} = [2\pi C(0)]^{-1} (-i)^p [d^p C(t)/dt^p]|_{t=0}.$$

It follows from Equation (10.5) that the odd order moments vanish. Not all moments need exist; their existence can be related to differentiability of the potential energy between particles. The positivity of $\hat{C}(\omega)$ implies an inequality for the frequency moments

$$\overline{(\omega^{p+q})^2} \leq \overline{\omega^{2p}}\; \overline{\omega^{2q}}$$

whose proof follows that for the triangle inequality.

The discussion is readily extended to a matrix of correlation functions,

$$C_{\alpha\beta}(t) = \langle \overline{A}_\alpha(t)\, A_\beta \rangle. \tag{10.10}$$

This satisfies

$$\overline{C}_{\alpha\beta}(t) = C_{\beta\alpha}(-t) \tag{10.11}$$

and it is readily verified that for arbitrary $a_{p\alpha}$

$$\sum_{p\alpha} \sum_{q\beta} \overline{a}_{p\alpha}\, a_{q\beta}\, C_{\alpha\beta}(t_p - t_q) \geq 0. \tag{10.12}$$

A matrix satisfying Equations (10.11) and (10.12) is called a matrix of positive type. Each diagonal element $C_{\alpha\alpha}(t)$ is a function of positive type; a nondiagonal element need not be positive type but is restricted by

$$|C_{\alpha\beta}(t)| \leq \sqrt{C_{\alpha\alpha}(t)C_{\beta\beta}(t)}.$$

A function

$$C_y(t) = \sum_{\alpha\beta} \bar{y}_\alpha y_\beta C_{\alpha\beta}(t)$$

is positive type, and Bochner's theorem shows that the Fourier transform

$$\hat{C}_{\alpha\beta}(\omega) = \int_{-\infty}^{\infty} dt\, e^{i\omega t}\, C_{\alpha\beta}(t)$$

is a non-negative matrix. This implies, for example, that

$$\left|\hat{C}_{\alpha\beta}(\omega)\right| \leq \sqrt{\hat{C}_{\alpha\alpha}(\omega)\,\hat{C}_{\beta\beta}(\omega)}.$$

In addition there are inequalities analogous to (10.8) and (10.9); the matrices

$$L_{\alpha\beta} = \int_0^{\infty} dt\, C_{\alpha\beta}(t), \quad M^{\infty}_{\alpha\beta} = \lim_{t\to\infty} C_{\alpha\beta}(t)$$

are non-negative.

When the constitutive equations are limited to the Navier-Stokes approximation, the entropy production has the form

$$\sigma = \sum_{\alpha\beta} \bar{y}_\alpha y_\beta L_{\alpha\beta}$$

where the y_α are gradients of the thermodynamic functions. From what has just been shown, we see that σ is non-negative:

$$\sigma \geq 0. \tag{10.13}$$

If instead of the Navier-Stokes equations we use relaxation equations like (9.91), the entropy production is

$$\sigma = \sum_{\alpha\beta} \bar{y}_\alpha(t) \int_0^t ds\, C_{\alpha\beta}(t - s) y_\beta(s)$$

which does not have a definite sign. However, its time integral can be expressed as

$$\int_0^t dt\, \sigma(t) = \frac{1}{2}\sum_{\alpha\beta} \int_0^t dt_1 \int_0^t dt_2\, \bar{y}_\alpha(t_1) y_\beta(t_2)\, C_{\alpha\beta}(t_1 - t_2)$$

which is non-negative.

Inequalities such as (10.8) and (10.13) underlay the entropy principle. It is remarkable that they are consequences of no more than invariance under time translations. Indeed it is not necessary that the averages be calculated with an equilibrium ensemble, so long as it is stationary in time. Thus, fluctuations in a nonequilibrium steady state have correlations of positive type.

In the case of turbulence, a stationary ensemble evidently does not exist since the hydrodynamic variables do not approach definite limiting values with the passage of time. Consequently arguments such as given do not apply. However, if the boundary conditions are time independent, the time average is stationary and fluctuations calculated on the basis of the time average are positive type.

10.2. ONSAGER'S THEOREM

We next consider the consequences of invariance to a time reversal, including Onsager's theorem.

By time reversal is meant the operation

$$\mathbf{q}_\alpha \to \mathbf{q}'_\alpha = \mathbf{q}_\alpha, \quad \mathbf{p}_\alpha \to \mathbf{p}'_\alpha = -\mathbf{p}_\alpha, \quad t \to t' = -t.$$

It is readily confirmed that the microscopic equations of motion are invariant under this operation, provided that there are no velocity-dependent or time-dependent forces.

For a function $A(\mathbf{q},\mathbf{p})$ of the coordinates and momenta, let τ denote the operation

$$\tau A(\mathbf{q},\mathbf{p}) = A(\mathbf{q}',\mathbf{p}') = A(\mathbf{q}, -\mathbf{p}).$$

The Liouville operator is odd in the momenta so $\tau \mathcal{L} A = -\mathcal{L}\tau A$; the time evolution operator $S(t)$ satisfies $\tau S(t)A = S(-t)\tau A$. Clearly $\tau^2 = 1$ so τ has eigenvalues ± 1. Functions which are even or odd in the momenta are eigenfunctions of τ with eigenvalues $+1$ or -1 respectively. Regarding the conserved densities and fluxes, it is easily seen that n, ϵ and $\mathbf{t}$ are even, while $\mathbf{g}$ and $\mathbf{s}$ are odd.

Consider the time-dependent correlation functions (10.10) for a set of variables A_α. The average entails an integral over the $\mathbf{q}_\alpha, \mathbf{p}_\alpha$ in which we change variables to $\mathbf{q}'_\alpha, \mathbf{p}'_\alpha$. Assuming the equilibrium ensemble to be invariant under τ we get

$$C_{\alpha\beta}(t) = \langle \tau\{[S(t)\overline{A}_\alpha]\, A_\beta\}\rangle = \langle [S(-t)\tau \overline{A}_\alpha]\,\tau A_\beta\rangle = \langle [S(t)\tau A_\beta]\,\tau \overline{A}_\alpha\rangle$$

where the last step follows from invariance to a translation of time. The volume integrated fluxes which occur in the Green-Kubo formulas have a definite parity under time reversal,

$$\tau A_\alpha = \tau_\alpha A_\alpha$$

where $\tau_\alpha = \pm 1$. Hence in this case

$$C_{\alpha\beta}(t) = \tau_\alpha \tau_\beta C_{\beta\alpha}(t) = \tau_\alpha \tau_\beta C_{\alpha\beta}(-t). \qquad (10.14)$$

It follows that the transport coefficients

$$L_{\alpha\beta} = \int_0^\infty dt\, C_{\alpha\beta}(t)$$

satisfy

$$L_{\alpha\beta} = \tau_\alpha \tau_\beta L_{\beta\alpha} \tag{10.15}$$

which is Onsager's theorem. A density and its flux have opposite parities under τ so in this equation the τ_α can be taken to be the parities of either the densities or the fluxes.

In contrast to the argument of the previous section, it is essential here that the average be taken over an equilibrium ensemble. A nonequilibrium ensemble, even for a steady state, is not in general invariant under τ, so the implications of invariance under time reversal must be treated differently.

With magnetic interactions the equations of motion are invariant to an operation of time reversal which includes a reversal of the direction of the magnetic field. This reversal is automatic if the source of the field is itself part of the system. However with a specified external field **B** the correlation functions and transport coefficients are functions of **B** and the reversal must be stated explicitly. If the system supports a spontaneous magnetization, the internal field must be reversed since it is an order parameter whose specification destroys time reversal invariance of the ensemble. In either case Equation (10.15) becomes

$$L_{\alpha\beta}(\mathbf{B}) = \tau_\alpha \tau_\beta \mathrm{L}_{\beta\alpha}(-\mathbf{B}).$$

10.3. SPACE-TIME SYMMETRIES

Consider next a space- and time-dependent correlation function, say the function (9.49):

$$G_{\alpha\beta}(\mathbf{x} - \mathbf{x}', t) = \langle \Delta\xi_\alpha(\mathbf{x}, t)\Delta\xi_\beta(\mathbf{x}')\rangle.$$

Invariance to space-time translations implies

$$G_{\alpha\beta}(\mathbf{x}, t) = G_{\beta\alpha}(-\mathbf{x}, -t) \tag{10.16}$$

and non-negativity is expressed by a generalization of Equation (10.12):

$$\sum_{pr\alpha}\sum_{qs\beta} \bar{a}_{pr\alpha}\, a_{qs\beta}\, G_{\alpha\beta}(\mathbf{x}_r - \mathbf{x}_s, t_p - t_q) \geq 0.$$

Time reversal invariance yields

$$G_{\alpha\beta}(\mathbf{x}, t) = \tau_\alpha \tau_\beta G_{\alpha\beta}(\mathbf{x}, -t). \tag{10.17}$$

Let the Fourier transform on position and time be denoted by $S(\mathbf{k},\omega)$:

$$S_{\alpha\beta}(\mathbf{k},\omega) = \int_{-\infty}^{\infty} dt \int d^3x \exp[i(\omega t - \mathbf{k}\cdot\mathbf{x})] G_{\alpha\beta}(\mathbf{x},t).$$

The element with $\xi_\alpha = \xi_\beta = n$ occurs in the theory of scattering of radiation and is called the scattering function. It is readily shown that

$$S_{\alpha\beta}(\mathbf{k},\omega) = \overline{S}_{\beta\alpha}(\mathbf{k},\omega) = \overline{S}_{\alpha\beta}(-\mathbf{k},-\omega) = \tau_\alpha\tau_\beta S_{\alpha\beta}(\mathbf{k},-\omega).$$

In particular a diagonal element $S_{\alpha\alpha}$ is real and even in $\mathbf{k}$ and ω. An application of Bochner's theorem shows S to be a non-negative matrix; the scattering function in particular is non-negative (for all $\mathbf{k}$ and ω).

The preceding results imply corresponding restrictions on the hydrodynamic representation (9.72). Let $\hat{G}_{\alpha\beta}(\mathbf{k},t)$ denote the Fourier transform of $G_{\alpha\beta}(\mathbf{x},t)$ on position. Transforming Equations (10.16) and (10.17), we write the result in matrix form as

$$\hat{G}(\mathbf{k},t) = \hat{G}^\dagger(\mathbf{k},-t) = \tau\hat{G}^\dagger(\mathbf{k},t)\,\tau.$$

It follows that

$$X^\dagger(\mathbf{k})\exp[-t\hat{A}^\dagger(\mathbf{k})] = \tau\exp[-t\hat{A}(\mathbf{k})]\,X(\mathbf{k})\,\tau$$

from which we get

$$X^\dagger(\mathbf{k}) = \tau X(\mathbf{k})\,\tau, \quad [\hat{A}(\mathbf{k})\,X(\mathbf{k})]^\dagger = \tau\hat{A}(\mathbf{k})X(\mathbf{k})\,\tau. \qquad (10.18)$$

If expanded to second order in wave number, the last equation yields Onsager's theorem.

Equation (9.72) provides the asymptotic representation of G for t large and positive. The form for t large and negative can be obtained from time reversal

$$\hat{G}(\mathbf{k},t) = \exp[t\hat{A}^\tau(\mathbf{k})]\,X^\tau(\mathbf{k}), \quad t < 0$$

where

$$\hat{A}^\tau(\mathbf{k}) = \tau\hat{A}(\mathbf{k})\tau, \quad X^\tau(\mathbf{k}) = \tau X(\mathbf{k})\tau.$$

The quantity A^τ determines the "antihydrodynamic" equations.

Complex eigenvalues of $\hat{A}(\mathbf{k})$ describe propagating hydrodynamic modes such as sound. Equation (10.18) shows $\hat{A}(\mathbf{k})$ to be equivalent to its Hermitian conjugate, so any complex eigenvalues must occur in complex-conjugate pairs. In addition the wave vectors $\mathbf{k}$ and $-\mathbf{k}$ have sets of eigenvalues which are the same (that is, if ζ is an eigenvalue of $\hat{A}(\mathbf{k})$, it is also an eigenvalue of $\hat{A}(-\mathbf{k})$). The following can also be shown: Let n be the number of conserved densities ξ_α with even parity ($\tau_\alpha = 1$), and m the number with odd parity. Then the number of complex eigenvalues of $\hat{A}(\mathbf{k})$ does not exceed $2n$ or $2m$, whichever is smaller. Thus a propagating mode arises from the coupling (in lowest order of the wave vector) of an odd variable with an even variable, as illustrated by the modes $g_{2,3}$ of Equation (6.14) for sound in a gas. (For a simple fluid $n = 2$ and $m = 3$ which would allow four propagating modes but spherical symmetry reduces this to two.)

Additional restrictions follow from spatial symmetries. Consider in particular a system with a center of symmetry, that is, one which is invariant under the spatial inversion $\mathbf{x} \to -\mathbf{x}$. For functions $A(\mathbf{q},\mathbf{p})$ we define the operator σ by

$$\sigma A(\mathbf{q},\mathbf{p}) = A(-\mathbf{q},-\mathbf{p}).$$

Clearly $\sigma^2 = 1$ and σ has eigenvalues ± 1. Functions which are even or odd in the q's and p's are eigenfunctions, with eigenvalue equal to their spatial parity. Let $\sigma_\alpha = \pm 1$ denote the spatial parities of the conserved densities,

$$\sigma\, \xi_\alpha(\mathbf{x}) = \sigma_\alpha\, \xi_\alpha(-\mathbf{x}).$$

The correlation function $G_{\alpha\beta}$ satisfies

$$G_{\alpha\beta}(\mathbf{x},t) = \sigma_\alpha \sigma_\beta\, G_{\alpha\beta}(-\mathbf{x},t).$$

In fact the densities of particle number, momentum, and energy (as well as their fluxes) have the same parity under σ as under τ, $\sigma_\alpha = \tau_\alpha$. Combining the preceding equation with (10.16) and (10.17), we get

$$G_{\alpha\beta}(\mathbf{x},t) = G_{\beta\alpha}(\mathbf{x},t) = G_{\alpha\beta}(-\mathbf{x},-t).$$

This implies, for example, that

$$\hat{A}^{\tau}(\mathbf{k}) = \hat{A}(-\mathbf{k}), \quad X^{\tau}(\mathbf{k}) = X(-\mathbf{k}).$$

We review the symmetry arguments which restrict the form of the constitutive equations in the Navier-Stokes approximation. It has been noted that the gradient of the pressure does not occur so the most general linear relations between the irreversible fluxes and the gradients of the thermodynamic variables are

$$s_i^* = -\lambda_{ij}\frac{\partial T}{\partial x_j} - a_{ijk}\frac{\partial u_j}{\partial x_j},$$

$$t_{ij}^* = -b_{ijk}\frac{\partial T}{\partial x_k} - \eta_{ijkl}\frac{\partial u_k}{\partial x_l}.$$

The Green-Kubo formulas for λ_{ij} and η_{ijkl} are like those given before, while

$$a_{ijk} = \frac{1}{\kappa T}\int_0^\infty dt \frac{1}{V}\langle S_i^-(t)\, T_{jk}^-\rangle,$$

$$b_{ijk} = \frac{1}{\kappa T^2}\int_0^\infty dt \frac{1}{V}\langle T_{ij}^-(t) S_k^-\rangle.$$

It follows from Onsager's theorem that $a_{ijk} = -Tb_{jki}$. Invariance under the rotation group implies the form used in Chapter 1 for λ_{ij} and η_{ijkl}, and in addition $a_{ijk} = a\epsilon_{ijk}$, $b_{ijk} = b\epsilon_{ijk}$ where ϵ_{ijk} is the Levi-Civita alternating symbol. If the symmetric pressure tensor is used, a and b must vanish. Alternatively, a and b vanish if the system is invariant under spatial inversions. In the absence of inversion symmetry, the energy flux can contain a term proportional to $\nabla \times \mathbf{u}$ and the asymmetric pressure tensor describes a torque along ∇T. For linear processes the divergence of such terms in the fluxes vanish, but if a, b depend on pressure or density there can be contributions to the rates of change of energy and momentum.

Notes on Chapter 10

10.1 Functions of positive type occur in probability theory and measure theory. See for example Cooke [1953], Feller [1957], and Donoghue [1969].

10.2 References to Onsager's theorem are cited under Section 1.9. The implications of time-reversal invariance for a nonequilibrium steady state are worked out in Dufty and Rubi [1987].

10.3 The number of propagating modes is discussed in Lekkerkerker and Laidlaw [1974], McLennan [1974], van Oost and Lekkerker [1978], Grmela and Iscoe [1978], and Goldhirsch and Procaccia [1981].

Parity nonconserving terms in the constitutive equations for a material lacking inversion symmetry were introduced by Pomeau [1971a].

11

DENSITY EXPANSION FOR CORRELATION FUNCTIONS

11.1. INTRODUCTION

In this chapter correlation functions for a monatomic gas will be discussed on the basis of an expansion in powers of the density. We consider a system satisfying classical mechanics, but the modifications necessary in quantum theory are straightforward. The results will be used in the next chapter to discuss density-dependent corrections to the transport coefficients.

The density expansion for a time-dependent correlation function can be obtained directly, but the resulting series is not uniformly convergent for all values of the time. A rearrangement of the series is necessary to provide a description of the long-time behavior and a basis for the determination of the transport coefficients. Correlation functions and the generalized Boltzmann equation are complementary approaches to the same problem, and so the analysis to follow has close similarities with the discussion in Chapter 7.

11.2. TIME-INDEPENDENT CORRELATIONS

We first discuss time-independent correlation functions and their density expansion. Consider a classical gas of identical particles, and let A, B be sums of single-particle functions:

$$A = \sum_{\alpha} a(X_{\alpha}), \quad B = \sum_{\alpha} b(X_{\alpha}). \tag{11.1}$$

Here X_α denotes the q's and p's for the α'th particle. Let C_{AB} denote the correlation of A and B in the thermodynamic limit,

$$C_{AB} = \lim_{V\to\infty} \langle \Delta\bar{A}\, \Delta B\rangle \tag{11.2}$$

where the average is taken over an equilibrium ensemble, and Δ denotes the deviation from the average value. For generality we have allowed for the possibility that A, B be complex, and the complex conjugate is shown by the bar.

Considered as a functional of a and b, C_{AB} has the properties of a Hermitian scalar product which we emphasize by writing

$$(a,b) = C_{AB}. \tag{11.3}$$

It is readily verified that (a,b) is linear in b and antilinear in a, has the Hermitian symmetry $\overline{(a,b)} = (b,a)$, defines a norm $\|a\|$ by $\|a\|^2 = (a,a) \geq 0$, and satisfies the Cauchy inequality $|(a,b)| \leq \|a\|\,\|b\|$. The set of functions with finite norm forms a Hilbert space, which we denote by $\mathscr{C}$.

A Hilbert space of functions A, B can be defined in a similar way if the volume is kept finite, but these functions do not have thermodynamic limits.

The definition (11.3) of the scalar product is not appropriate if A, B are global functions, depending on the $\mathbf{q}_\alpha$ only through the differences $\mathbf{q}_\alpha - \mathbf{q}_\beta$. Then the thermodynamic limit in (11.2) does not exist and the interesting quantity is instead

$$(a,b) = \lim_{V\to\infty} \frac{1}{V}\langle \Delta\bar{A}\, \Delta B\rangle.$$

We describe equilibrium by the grand canonical ensemble (9.1). The reduced distribution function for a cluster of s particles is defined by

$$F^{(s)}(X_1,\ldots,X_s) = \lim_{V\to\infty} e^{-Q} \sum_{N=s}^{\infty} \frac{1}{(N-s)!} e^{\nu N} \int dX_{s+1}\ldots dX_N e^{-\beta H_N}. \tag{11.4}$$

(These are related to the functions of Chapter 7 by $F^{(s)} = n^s F_s$.) The momentum dependence can be separated out to give

$$F^{(s)}(X_1,\ldots,X_s) = n^s \phi(p_1)\ldots\phi(p_s) g^{(s)}(\mathbf{q}_1,\ldots,\mathbf{q}_s)$$

where n is the equilibrium density and ϕ is the Maxwellian

$$\phi(p) = (2\pi m\kappa T)^{-3/2} \exp\left(-p^2/2m\kappa T\right).$$

The reduced distribution function $g^{(s)}$ is given by

$$n^s g^{(s)}(\mathbf{q}_1,\ldots,\mathbf{q}_s) = \lim_{V\to\infty} e^{-Q} \sum_{N=s}^{\infty} \frac{1}{(N-s)!} z^N \int d^3q_{s+1}\ldots d^3q_N e^{-\beta U_N} \tag{11.5}$$

where U_N is the potential energy for the N-particle system and z is the activity

$$z = (2\pi m\kappa T)^{3/2} e^{\nu}.$$

An equivalent formula for $g^{(s)}$ is

$$n^s g^{(s)}(\mathbf{x}_1,\ldots,\mathbf{x}_s) = \lim_{V\to\infty} \langle N(N-1)\cdots(N-s+1)\delta(\mathbf{x}_1-\mathbf{q}_1)\cdots\delta(\mathbf{x}_s-\mathbf{q}_s)\rangle.$$

(Note that $g^{(1)} = 1$.) The density expansion for $g^{(s)}$ can be obtained from (11.5). Writing the expansion as

$$g^{(s)} = \sum_{r=0}^{\infty} g_r^{(s)} n^s \tag{11.6}$$

we list the first two coefficients

$$g_0^{(s)} = \exp(-\beta U_s),$$

$$g_1^{(s)} = \int d^3q_{s+1}\{\exp[-\beta U_{s+1}(\mathbf{q}_1,\ldots\mathbf{q}_{s+1})] - \exp[-\beta U_s(\mathbf{q}_1,\ldots,\mathbf{q}_s)]$$
$$- s\exp[-\beta U_s(\mathbf{q}_1,\ldots,\mathbf{q}_s)](\exp[-\beta U_2(\mathbf{q}_1,\mathbf{q}_{s+1})] - 1)\}.$$

Spatial homogeneity implies that $g^{(s)}(\mathbf{q}_1,\ldots,\mathbf{q}_s)$ is determined by $s - 1$ displacements, say $\mathbf{q}_2 - \mathbf{q}_1,\ldots,\mathbf{q}_s - \mathbf{q}_1$. For large separations $g^{(s)}$ approaches 1,

$$g^{(s)}(\mathbf{q}_1,\ldots,\mathbf{q}_s) \to 1, \quad \text{for} \quad |\mathbf{q}_\alpha - \mathbf{q}_\beta| \to \infty, \quad \alpha,\beta = 1,\ldots,s.$$

In gases $g^{(s)}$ differs significantly from 1 only when the separations between particles are about the same as the force range or less.

After substituting (11.1) into (11.3) we get

$$(a,b) = \lim_{V\to\infty}[\langle N\bar{a}(X_1)b(X_1)\rangle + \langle N(N-1)\bar{a}(X_1)b(X_2)\rangle - \langle N\bar{a}(X_1)\rangle\langle Nb(X_2)\rangle].$$

This can be expressed in terms of reduced distribution functions with the result

$$(a,b) = n\int dX_1\phi(p_1)\bar{a}(X_1)b(X_1)$$
$$+ n^2\int dX_1dX_2\phi(p_1)\phi(p_2)[g^{(2)}(\mathbf{q}_1 - \mathbf{q}_2) - 1]\bar{a}(X_1)b(X_2). \tag{11.7}$$

If a and b are local functions with the form

$$a = a(\mathbf{x}) = \delta(\mathbf{q} - \mathbf{x})\alpha(\mathbf{p}), \quad b = b(\mathbf{x}) = \delta(\mathbf{q} - \mathbf{x})\beta(\mathbf{p}) \tag{11.8}$$

then the correlation function is

$$C_{AB}(\mathbf{x},\mathbf{x}') = (a(\mathbf{x}),b(\mathbf{x}'))$$
$$= n\delta(\mathbf{x} - \mathbf{x}')\langle\bar{\alpha}\beta\rangle_M + n^2[g^{(2)}(\mathbf{x} - \mathbf{x}') - 1]\langle\bar{\alpha}\rangle_M\langle\beta\rangle_M$$

where $\langle\ \rangle_M$ denotes the Maxwellian average

$$\langle\alpha\rangle_M = \int d^3p\,\phi(p)\,\alpha(\mathbf{p}).$$

For example, the density-density correlation function is obtained by the choice $\alpha = \beta = 1$:

$$(n(\mathbf{x}),n(\mathbf{x}')) = n\delta(\mathbf{x} - \mathbf{x}') + n^2[g^{(2)}(\mathbf{x} - \mathbf{x}') - 1].$$

If the Maxwellian average of a or b vanishes, then only the first term in (11.7) remains. Thus, the correlation function for the momentum density, obtained with $\alpha = p_i$, $\beta = p_j$, is

$$(g_i(\mathbf{x}),g_j(\mathbf{x}')) = nm\kappa T\delta_{ij}\delta(\mathbf{x} - \mathbf{x}').$$

At low density (or, as just noted, when a,b have vanishing Maxwellian averages) (,) can be replaced by the leading term in (11.7) which will be denoted by $(,)_M$,

$$(a,b)_M = n\int dX\,\phi(p)\bar{a}(X)b(X)$$

In general we have

$$(a,b) = (a,Qb)_M = (Qa,b)_M$$

where Q is given by

$$Qb(X_1) = b(X_1) + n\int dX_2\phi(p_2)[g^{(2)}(\mathbf{q}_1 - \mathbf{q}_2) - 1]b(X_2).$$

The inverse of Q is

$$Q^{-1}b(X_1) = b(X_1) - n\int dX_2\phi(p_2)D(\mathbf{q}_1 - \mathbf{q}_2)b(X_2)$$

where D, which is called the direct correlation function, satisfies the integral equation

$$D(\mathbf{q}_1 - \mathbf{q}_2) = g^{(2)}(\mathbf{q}_1 - \mathbf{q}_2) - 1$$
$$- n\int d^3q_3D(\mathbf{q}_1 - \mathbf{q}_3)[g^{(2)}(\mathbf{q}_3 - \mathbf{q}_2) - 1].$$

Let $R^\dagger$ denote the adjoint of an operator R with respect to (,), and R^+ its adjoint with respect to $(,)_M$:

$$(a,Rb) = (R^\dagger a,b), \quad (a,Rb)_M = (R^+a,b)_M. \tag{11.9}$$

Then the two are related

$$R^\dagger = Q^{-1}R^+Q.$$

The advantage of this formula is that R^+ may be evident. Consider, for example, the operation of multiplication by the momentum $\mathbf{p}$. Then $\mathbf{p}^+ = \mathbf{p}$ and we get

$$(\mathbf{p}^\dagger b)(X_1) =$$

$$\mathbf{p}_1b(X_1) + n\int dX_2\phi(p_2)\{\mathbf{p}_1[g^{(2)}(\mathbf{q}_1 - \mathbf{q}_2) - 1] - \mathbf{p}_2D(\mathbf{q}_1 - \mathbf{q}_2)\}b(X_2).$$

Equation (11.1) does not include all cases of interest; the energy density and energy and momentum fluxes, for example, contain sums of two-particle functions. The space $\mathscr{C}$ can be generalized to include two-particle functions as follows. Suppose A and B have the forms

$$\begin{aligned} A &= \sum_{\alpha} a^{(1)}(X_\alpha) + \frac{1}{2}\sum_{\alpha\beta} a^{(2)}(X_\alpha, X_\beta), \\ B &= \sum_{\alpha} b^{(1)}(X_\alpha) + \frac{1}{2}\sum_{\alpha\beta} b^{(2)}(X_\alpha, X_\beta). \end{aligned} \tag{11.10}$$

Here the sums run from 1 to N but in the double sums terms with $\alpha = \beta$ are to be excluded since they could be counted in the first sum. We assume $a^{(2)}$ and $b^{(2)}$ to be symmetric functions of their arguments, and the factor $\frac{1}{2}$ has been included so that each term is counted only once. The space $\mathscr{C}$ now has elements which are pairs of functions, $a = \{a^{(1)}, a^{(2)}\}$ with the sum $a + b$ defined by $a + b = \{a^{(1)} + b^{(1)}, a^{(2)} + b^{(2)}\}$. The scalar product is again defined by (11.3) and a straightforward but tedious calculation gives

$$\begin{aligned} (a,b) = {} & (a^{(1)}, b^{(1)}) \\ & + n^2 \int dX_1 dX_2 \phi(1)\phi(2) g^{(2)}(1,2) \\ & \times \left\{ \frac{1}{2}\bar{a}^{(2)}(1,2) b^{(2)}(1,2) + \bar{a}^{(1)}(1) b^{(2)}(1,2) + \bar{a}^{(2)}(1,2) b^{(1)}(1) \right\} \\ & + n^3 \int dX_1 dX_2 dX_3 \phi(1)\phi(2)\phi(3) \Big\{ g^{(3)}(1,2,3) \bar{a}^{(2)}(1,2) b^{(2)}(1,3) \\ & + \frac{1}{2}[g^{(3)}(1,2,3) - g^{(2)}(2,3)][\bar{a}^{(1)}(1) b^{(2)}(2,3) + \bar{a}^{(2)}(2,3) b^{(1)}(1)] \Big\} \\ & + \frac{1}{4} n^4 \int dX_1 \cdots dX_4 \phi(1) \cdots \phi(4) \\ & \times [g^{(4)}(1,2,3,4) - g^{(2)}(1,2) g^{(2)}(3,4)] \bar{a}^{(2)}(1,2) b^{(2)}(3,4). \end{aligned}$$

For brevity the coordinates and momenta have been replaced by the particle labels.

This procedure is readily generalized to include functions which depend on the variables for three, four, and so on, particles. Actually the extra generality will not be needed in the following, and henceforth, we will work only with quantities A, B which are sums of single-particle functions.

In quantum theory Equation (11.1) is written with sums of single-particle operators. By the same procedure, C_{AB} converts the algebra of single-particle operators into a Hilbert space.

11.3. TIME-DEPENDENT CORRELATIONS

Let A, B be functions with the form (11.1), and consider the time-dependent correlation function

$$C_{AB}(t) = \lim_{V\to\infty} \langle \Delta A \Delta B(t) \rangle.$$

Because of interactions $B(t)$ is not a sum of one-particle functions; the time dependence of, say, $b(X_1)$ involves the phases X_α of all particles. Nevertheless, $C_{AB}(t)$ is a bilinear functional of a, b and it follows that there exists a linear operator $\Lambda(t)$ on $\mathscr{C}$ such that

$$C_{AB}(t) = (a, \Lambda(t) b).$$

An operator $\Lambda(t)$ can be defined in the same way in quantum theory.

Clearly we have $\Lambda(0) = 1$. That $\Lambda(t)$ is bounded, indeed a contraction, follows from invariance to translations of the time as expressed by Equation (10.4). Equation (10.11) shows the adjoint of $\Lambda(t)$ to be

$$\Lambda^\dagger(t) = \Lambda(-t).$$

Other symmetries of the correlation functions are reflected in $\Lambda(t)$; for example, if the system is invariant to time reversal, then

$$\Lambda(t) = \tau\Lambda(-t)\tau = \tau\Lambda^\dagger(t)\tau. \tag{11.11}$$

In the case of the local functions (11.8), invariance to spatial inversion implies

$$(a(\mathbf{x}), \Lambda(t) b(\mathbf{x}')) = (a(-\mathbf{x}), \sigma\Lambda(t)\sigma b(-\mathbf{x}')) \tag{11.12}$$

This condition is more easily expressed in terms of the Fourier transform

$$\begin{aligned} \hat{C}_{AB}(\mathbf{k}, t) &= \int d^3x \exp[i\mathbf{k}\cdot(\mathbf{x} - \mathbf{x}')](a(\mathbf{x}), \Lambda(t) b(\mathbf{x}')) \\ &= \int dX\, \phi(p)\bar{\alpha}(\mathbf{p}) \exp(i\mathbf{k}\cdot\mathbf{q})\, \Lambda(t)\, \beta(\mathbf{p})\delta(\mathbf{q}) \\ &= \int dX\, \phi(p)\bar{\alpha}(\mathbf{p})\,\delta(\mathbf{q})\,\Lambda(t)\,\beta(\mathbf{p}) \exp(-i\mathbf{k}\cdot\mathbf{q}). \end{aligned}$$

It follows from spatial homogeneity that $\exp(-i\mathbf{k}\cdot\mathbf{q})$ is an eigenfunction of Λ in the sense that

$$\Lambda(t) \exp(-i\mathbf{k}\cdot\mathbf{q})\, \beta(\mathbf{p}) = \exp(-i\mathbf{k}\cdot\mathbf{q})\, \hat{\Lambda}(\mathbf{k}, t)\, \beta(\mathbf{p})$$

where $\hat{\Lambda}(\mathbf{k}, t)$ is an operator on functions of momentum. Equation (11.12) gives the condition on $\hat{\Lambda}$

$$\sigma\hat{\Lambda}(\mathbf{k}, t)\sigma = \hat{\Lambda}(-\mathbf{k}, t).$$

The functions usually of interest are such that $\sigma = \tau$. On the corresponding subspace Equation (11.11) shows the adjoint of $\hat{\Lambda}$ to be

$$\hat{\Lambda}^{\dagger}(\mathbf{k},t) = \hat{\Lambda}(-\mathbf{k},t). \tag{11.13}$$

The operator $\Lambda(t)$ plays somewhat the same role as the streaming operator $S(t)$ but is fundamentally different because it acts on single-particle functions. It inherits some, but not all, of the properties of $S(t)$. For example, the group property $S(t_1)S(t_2) = S(t_1 + t_2)$ does not extend to $\Lambda(t)$ because the initial time $t = 0$ plays a special role in the definition of $\mathscr{C}$. The streaming operator is unitary in the sense that

$$\langle AS(t)B\rangle = \langle [S(-t)A]B\rangle = \langle [S^{-1}(t)A]B\rangle$$

but $\Lambda(t)$ need not be unitary.

The density expansion for $\Lambda(t)$ will be developed in the following on a systematic basis but first we discuss its connection to kinetic theory. Consider an ensemble which at time $t = 0$ has the form

$$F_N(0) = F_{N0}\exp(-\Delta Q + \lambda A) \tag{11.14}$$

where F_{N0} denotes the equilibrium ensemble. We suppose A to be a sum of single-particle functions as in (11.1). The constant ΔQ is determined by the normalization condition

$$\exp(\Delta Q) = \langle \exp(\lambda A)\rangle$$

where the brackets denote the equilibrium average. The cumulants of A can be obtained by picking out the corresponding powers of λ.

The ensemble at time t is given by

$$F_N(t) = F_{N0}\,S_N(-t)\exp(-\Delta Q + \lambda A).$$

The one-particle distribution function at time t is then

$$F_1(t) = \sum_{N=1}^{\infty}\frac{1}{(N-1)!}\int dX_2\cdots dX_N F_N(t). \tag{11.15}$$

(The thermodynamic limit should be taken at this point but will be left implicit.) Clearly $F_1(t)$ is a functional of the single-particle function a which determines A according to (11.1). Assuming this functional to be invertible, we solve it for a and substitute the result into $F_1(t)$ to obtain the latter as a functional of its initial value. Thus, within the class of initial conditions having the form (11.14), $F_1(t)$ is determined by its initial value,

$$F_1(t) = R(t)F_1(0).$$

The operator $R(t)$, which need not be linear, satisfies $R(0) = 1$. In addition, since the equilibrium ensemble is time independent, we have $R(t)f_0 = f_0$ where $f_0 = n\phi$. If a kinetic equation, such as the Boltzmann equation, is valid then $R(t)$ is its solution operator. In this case $R(t)$ is a semi-group.

Suppose that B is a second sum of single-particle functions, and let $\langle B\rangle_F$ denote its average over $F_N(t)$:

$$\langle B\rangle_F = \langle BS_N(-t)\exp(-\Delta Q + \lambda A)\rangle.$$

We wish to expand this in powers of λ. Taking into account that

$$\Delta Q = \lambda\langle A\rangle + \dots \tag{11.16}$$

we get

$$\langle B\rangle_F = \langle B\rangle + \lambda\langle BS_N(-t)\Delta A\rangle + \dots = \langle B\rangle + \lambda(a,\Lambda(t)b) + \dots$$

(For simplicity we here assume a and b to be real.) On the other hand $\langle B\rangle_F$ can be expressed in terms of the one-particle distribution function by

$$\langle B\rangle_F = \int dX_1 b(X_1)F_1(X_1,t) = \int dX_1 b(X_1)R(t)F_1(X_1,0).$$

The initial one-particle distribution function is determined by (11.14) and (11.15), and expansion in powers of λ yields

$$F_1(X_1,0) = f_0(p_1)\Big\{1 + \lambda\Big[a(X_1) - \langle a\rangle_M$$
$$+ \int dX_2 f_0(p_2)[g^{(2)}(\mathbf{q}_1 - \mathbf{q}_2) - 1][a(X_2) - \langle a\rangle_M]\Big] + \dots\Big\}.$$

Let $R(t)$ be expanded about equilibrium as

$$R(t)f_0[1 + \lambda\Phi + \dots] = f_0[1 + \lambda T(t)\Phi + \dots]$$

where $T(t)$ is a linear operator. Then we get

$$\langle B\rangle_F - \langle B\rangle = \lambda\int dX_1 f_0(p_1)b(X_1)T(t)\Big\{a(X_1) - \langle a\rangle_M$$
$$+ \int dX_2 f_0(p_2)[g^{(2)}(\mathbf{q}_1 - \mathbf{q}_2) - 1][a(X_2) - \langle a\rangle_M]\Big\} + \dots$$

Using the form (11.7) for the scalar product, we write this as

$$\langle B\rangle_F - \langle B\rangle = \lambda(a,T^+(t)b)$$

where T^+ is the adjoint of T as defined in Equation (11.9). Comparing with (11.16), we see that

$$\Lambda(t) = T^+(t). \tag{11.17}$$

The adjoint occurs because Λ and T act on dual spaces, the former on physical variables and the latter on distribution functions. The time dependence of triple and higher-order cumulants can be obtained from the full operator $R(t)$ by continuing the expansion in powers of λ.

In the Boltzmann approximation $T(t)$ satisfies the linearized Boltzmann equation

$$\left(\frac{\partial}{\partial t} + \mathbf{v}\cdot\nabla + nI\right)T(t) = 0.$$

The initial condition $T(0) = 1$ yields

$$T(t) = \exp\left[-t(\mathbf{v}\cdot\nabla + nI)\right]$$

which is the semi-group denoted by T^t in Chapter 6. In the same approximation $(,)$ can be replaced by $(,)_M$ so the adjoints $\Lambda^\dagger$ and Λ^+ are the same and we get

$$\Lambda(t) = \exp\left[t(\mathbf{v}\cdot\nabla - nI)\right]. \tag{11.18}$$

Actually this holds for positive t while for negative t

$$\Lambda(t) = \exp\left[-t(\mathbf{v}\cdot\nabla + nI)\right], \quad t \le 0$$

as follows from (11.11).

The contraction property of Λ requires that I be non-negative, which is the linearized version of Boltzmann's H-theorem. The contraction property follows from invariance to translations of time, whereas the proof of the H-theorem in the context of the kinetic equation entails detailed balance, or time-reversal invariance, as well. Equation (11.13), which follows from invariance under time reversal and spatial inversion, implies the Hermitian symmetry (4.27) of I.

11.4. CORRELATION FUNCTIONS IN BOLTZMANN APPROXIMATION

In this section we use the Boltzmann approximation (11.18) to analyze the properties of the correlation functions for the conserved densities and fluxes. In this approximation, the potential energy terms in the densities and fluxes can be neglected so we take A, B to be sums of single-particle functions with a, b having the form (11.8). In addition, we use the low-density approximation for the scalar product so the correlation function is

$$\begin{aligned} C_{AB}(\mathbf{x} - \mathbf{x}', t) &= (a(\mathbf{x}), \Lambda(t)b(\mathbf{x}')) \\ &= n\int dX \phi(p)\overline{\alpha}(\mathbf{p})\delta(\mathbf{x} - \mathbf{q})\Lambda(t)\beta(\mathbf{p})\delta(\mathbf{x}' - \mathbf{q}). \end{aligned}$$

With Λ having the particular form (11.18), its transform $\hat{\Lambda}$ is given by

$$\hat{\Lambda}(\mathbf{k}, t) = \exp\left[-tB(\mathbf{k})\right]$$

where $B(\mathbf{k}) = nI - i\mathbf{k}\cdot\mathbf{v}$ is the operator introduced in Equation (6.8). This yields

$$\hat{C}_{AB}(\mathbf{k},t) = n(\alpha, \exp[-tB(\mathbf{k})]\beta)_0$$

where $(,)_0$ is defined by

$$(\alpha,\beta)_0 = \int d^3p\, \phi(p)\bar{\alpha}(\mathbf{p})\beta(\mathbf{p}).$$

This is the scalar product denoted by (,) in Chapter 6.

We assume k to be small and use the perturbation expansion developed in Chapter 6. The hydrodynamic part of the correlation function is obtained by projecting onto the normal solution to the linearized Boltzmann equation. Denoting the hydrodynamic part by C^H_{AB}, we get

$$\hat{C}^H_{AB}(\mathbf{k},t) = n\sum_{\lambda} \exp(-\zeta_\lambda t)(\alpha,g_\lambda)_0(h_\lambda,\beta)_0$$

where $\zeta_\lambda, g_\lambda, h_\lambda$ are the eigenvalues and eigenfunctions of $B(\mathbf{k})$ obtained in Chapter 6. We take A, B to be conserved densities so C is the same as the quantity denoted by G in Chapter 9. To illustrate the calculation, we treat the case $\alpha = \beta = 1$ which corresponds to the density-density correlation function. As in Chapter 6 we assume $\mathbf{k}$ to point in the x-direction; Equations (6.2) and (6.14) give the eigenvalues to second order, and the eigenfunctions to first order, in $\mathbf{k}$. To lowest order in k we get

$$(1,g_1)_0 = -\sqrt{2/5}, \quad (1,g_{2,3})_0 = \sqrt{3/10}, \quad (1,g_{4,5})_0 = 0$$

which yields

$$\hat{G}^H_{nn}(\mathbf{k},t) = n\left\{\frac{2}{5}\exp(-\zeta_1 t) + \frac{3}{10}[\exp(-\zeta_2 t) + \exp(-\zeta_3 t)]\right\}.$$

The remaining elements of G^H can be calculated in a similar way.

The eigenvalues ζ_α are also eigenvalues of the linearized hydrodynamic equations. It follows that G^H satisfies the hydrodynamic equations, in accordance with Onsager's assumption on the regression of fluctuations. The preceding result could have been obtained by solving the hydrodynamic equations subject to the initial condition $G^H(\mathbf{x},0) = g(\mathbf{x})$, but initial slip must be taken into account if the calculation is carried to a higher order in k.

A correlation function for fluxes can be calculated in a similar way. Consider, for example, the correlation function for the energy flux; actually we will use the subtracted energy flux for which $\alpha = \beta = S_x = (\frac{1}{2}mv^2 - \frac{5}{2}\kappa T)v_x$. We find

$$(S_x,g_1)_0 = \sqrt{3}(S_x,g_{2,3})_0 = \sqrt{2/5}(ik/p)(S_x,\mathcal{S}_x)_0 = \sqrt{2/5}(ik\lambda T/n),$$

$$(S_x,g_{4,5})_0 = 0.$$

The hydrodynamic part of the correlation function is

$$\hat{C}^{H}_{S_x,S_x} = -\frac{2}{5}\lambda Tk^2\{\exp(-\zeta_1 t) + \sqrt{3}[\exp(-\zeta_2 t) + \exp(-\zeta_3 t)]\}. \tag{11.19}$$

This could be constructed as a solution to the hydrodynamic equations but with initial conditions which are quite artificial.

The correlation function for the subtracted fluxes at vanishing wave number has no hydrodynamic part. We have $\hat{\Lambda}(0,t) = \exp(-nIt)$ so Equation (9.70) for the thermal conductivity becomes

$$\lambda = \frac{1}{\kappa T^2}\int_0^\infty dt\, n(S_x, \exp(-nIt)S_x) = \frac{1}{\kappa T^2}(S_x, I^{-1}S_x) = \frac{1}{\kappa T^2}(S_x, \mathcal{S}_x)$$

in agreement with Equation (4.21). The viscosity and coefficients of diffusion are treated similarly. Thus the Green-Kubo formulas yield the same results for the transport coefficients as obtained from the Boltzmann equation, if the time dependence of the correlation function is governed by the linearized Boltzmann equation.

11.5. DYNAMICAL CLUSTER EXPANSION

The density expansion for time-dependent correlation functions can be developed by replacing the dynamical N-body problem by the sequence of 1-body, 2-body, 3-body, and so forth, problems. The dynamical cluster expansion will be discussed in this section and then applied to correlation functions in the next section.

Let $S_s(1,\dots,s)$ denote the streaming operator for a set of s particles; for brevity $X_1,\dots,X_s$ will be denoted by the labels $1,\dots,s$ and the time will not be written explicitly. The set of particles will be a subset of the system but the time dependence of S_s is determined as if the remaining particles were absent. Of course the subset could consist of any s particles rather than those which happen to be labeled $1,\dots,s$. $S_s(1,\dots,s)$ acts as the identity on the phases of particles not in the set $1,\dots,s$.

Let the sequence of operators $U_s(1,\dots,s)$ be defined by

$$\begin{aligned} S_1 &= U_1, \\ S_2(1,2) &= U_1(1)U_1(2) + U_2(1,2), \\ S_3(1,2,3) &= U_1(1)U_1(2)U_1(3) + U_1(1)U_2(2,3) \\ &\quad + U_1(2)U_2(1,3) + U_1(3)U_2(1,2) \\ &\quad + U_3(1,2,3), \quad \text{and so forth.} \end{aligned} \tag{11.20}$$

In general the equation for S_s is obtained by partitioning the set $1,\ldots,s$ into disjoint subsets; for each partition there is a product of U's, one for each subset, and the sum is taken over all partitions. The U-operators are analogous to the Ursell-Mayer functions of equilibrium statistical mechanics. They can be expressed in terms of the streaming operators by solving the preceding equations successively,

$$\begin{aligned} U_1 &= S_1, \\ U_2(1,2) &= S_2(1,2) - S_1(1)S_1(2), \\ U_3(1,2,3) &= S_3(1,2,3) - S_1(1)S_2(2,3) - S_1(2)S_2(1,3) - S_1(3)S_2(1,2) \\ &\quad + 2S_1(1)S_1(2)S_1(3), \quad \text{and so forth.} \end{aligned} \tag{11.21}$$

The relation between the U-operators and the streaming operators is algebraically similar to that between cumulants and moments. Let $G(u)$ denote the functional

$$G(u) = 1 + \sum_{s=1}^{\infty} \frac{1}{s!} \int dX_1 \cdots dX_s\, S_s(1,\ldots,s)\, u(1) \cdots u(s)$$

where u is an arbitrary function of the phase X. If we write

$$\ln G(u) = \sum_{s=1}^{\infty} \frac{1}{s!} \int dX_1 \cdots dX_s\, U_s(1,\ldots,s) u(1) \cdots u(s)$$

then it can be shown that the U_s are the same as defined by (11.20). The logarithmic transformation is analogous to the transformation in statistics from the characteristic function to the generating function.

The U_s have a cluster property which can be described as follows. Consider a set of s particles and suppose their initial phases $X_1,\ldots,X_s$ to be such that during the time t they move without interaction. (For this to be possible it is necessary that the force vanish beyond a finite range.) Thus each particle moves as if the others were absent, so in effect

$$S_s(1,\ldots,s) = S_1(1) \cdots S_1(s).$$

Then it is readily checked that except for U_1 the U-operators vanish:

$$U_r(1,\ldots,r) = 0, \quad r = 2,\ldots,s.$$

More generally let $\{s\}$ denote a set of s particles with arbitrary labels. Suppose their initial phases to be such that $\{s\}$ is the union of two disjoint subsets $\{q\}$ and $\{r\}$ with the property that during time t there are no interactions of particles in $\{q\}$ with those in $\{r\}$. Then each of the two subsets, or clusters, moves independently of the other so

$$S_s(\{s\}) = S_q(\{q\})S_r(\{r\}). \tag{11.22}$$

Let $\{\bar{s}\}$, $\{\bar{q}\}$, $\{\bar{r}\}$ be subsets of $\{s\}$, $\{q\}$ and $\{r\}$ respectively with $\{\bar{s}\}$ being the union of $\{\bar{q}\}$ and $\{\bar{r}\}$, and suppose in addition that for any such subsets

$$S_{\bar{s}}(\{\bar{s}\}) = S_{\bar{q}}(\{\bar{q}\})S_{\bar{r}}(\{\bar{r}\}). \tag{11.23}$$

Then again it follows that

$$U_{\bar{s}}(\{\bar{s}\}) = 0, \quad \bar{s} = 2,\ldots,s. \tag{11.24}$$

In words, a U-operator vanishes if the set of particles on which it acts divides into noninteracting clusters.

That the product property (11.22), (11.23) implies the cluster property (11.24) for the U's is called the cluster theorem. It should be kept in mind that the product property is really a property of the initial phases. For $t = 0$ the U's have the cluster property for all values of the phases.

As an illustration, consider the set of particles 1,2,3. Let one cluster be 1,2 and the other 3. If there are no interactions between particle 3 and either of 1 or 2 during time t then

$$\begin{aligned} S_3(1,2,3) &= S_1(3)S_2(1,2), \\ S_2(1,3) &= S_1(1)S_1(3), \quad S_2(2,3) = S_1(2)S_1(3). \end{aligned} \tag{11.25}$$

It is then readily checked that

$$U_3(1,2,3) = U_2(1,3) = U_2(2,3) = 0.$$

However $U_2(1,2)$ need not vanish.

It is not the case that (11.22) implies (11.23) since the motion of a subset will in general be different if the governing equations of motion are those for $\{s\}$ or those for $\{\bar{s}\}$. In the illustration the first of Equations (11.25) does not imply the other two because in the motion described by, say, $S_2(1,3)$ there could be a collision between 1 and 3 which in the motion described by $S_3(1,2,3)$ is prevented by an earlier collision between 1 and 2.

The cluster theorem has the following simple corollary. Let f be a function which is independent of the phases X_α for particles in a set $\{s\}$ but may depend on the phases of other particles. Then

$$U_s(\{s\})f = 0, \quad s \geq 2. \tag{11.26}$$

11.6. CLUSTER EXPANSION FOR CORRELATION FUNCTIONS

We now apply the cluster expansion to a time-dependent correlation function

$$\begin{aligned} C_{AB}(t) &= (a, \Lambda(t)b) \\ &= \lim_{V\to\infty} \Bigg\{ \Big\langle \sum_{\alpha=1}^{N} \bar{a}(\alpha) S_N(1, \ldots, N) \sum_{\beta=1}^{N} b(\beta) \Big\rangle \\ &\qquad - \Big\langle \sum_{\alpha=1}^{N} \bar{a}(\alpha) \Big\rangle \Big\langle S_N(1, \ldots, N) \sum_{\beta=1}^{N} b(\beta) \Big\rangle \Bigg\}. \end{aligned} \tag{11.27}$$

In the last term S_N could be replaced by 1, but the desired expansion is obtained more directly if it is retained as written.

Let S_N be expanded in U-operators. Each factor $U_1(\alpha) = S_1(\alpha)$ can be replaced by 1 when operating on $b(\beta)$ unless $\beta = \alpha$. Furthermore, by (11.26) each factor $U_p(\{p\})$, $p \geq 2$, vanishes when operating on $b(\beta)$ unless β is in the set $\{p\}$. Collecting the remaining terms together, we get

$$S_N(1, \ldots, N) \sum_{\beta=1}^{N} b(\beta) = \sum_{\beta} U_1(\beta)b(\beta) + \frac{1}{2}\sum_{\beta\gamma} U_2(\beta, \gamma)[b(\beta) + b(\gamma)]$$
$$+ \frac{1}{3!}\sum_{\beta\gamma\mu} U_3(\beta, \gamma, \mu)[b(\beta) + b(\gamma) + b(\mu)] + \ldots$$

This is to be substituted into Equation (11.27) and after some reduction we obtain

$$(a, \Lambda(t)b) = \lim_{V\to\infty} \Big\{ \langle N\bar{a}(1)U_1(1)b(1)\rangle + \langle N(N-1)\bar{a}(1)U_1(2)b(2)\rangle$$
$$- \langle N\bar{a}(1)\rangle\langle NU_1(2)b(2)\rangle + \langle N(N-1)\bar{a}(1)U_2(1,2)[b(1) + b(2)]\rangle$$
$$+ \frac{1}{2}\Big\langle N(N-1)(N-2)\bar{a}(1)\Big(U_3(1,2,3)\sum_{\beta=1}^{3} b(\beta) + U_2(2,3)\sum_{\beta=2}^{3} b(\beta)\Big)\Big\rangle$$
$$- \frac{1}{2}\langle N\bar{a}(1)\rangle\Big\langle N(N-1)U_2(2,3)\sum_{\beta=2}^{3} b(\beta)\Big\rangle + \ldots \Big\}.$$

We next introduce the reduced distribution functions to find

$$(a, \Lambda(t)b) = n\int dX_1\phi(1)\bar{a}(1)U_1(1)b(1) + n^2\int dX_1 dX_2\phi(1)\phi(2)\bar{a}(1)$$
$$\times \{g^{(2)}(1,2) - 1]U_1(2)b(2) + g^{(2)}(1,2)U_2(1,2)[b(1) + b(2)]\}$$
$$+ \sum_{s=3}^{\infty} \frac{1}{(s-1)!} n^2 \int dX_1 \cdots dX_s\phi(1)\cdots\phi(s)\bar{a}(1)$$
$$\times \Big\{g^{(s)}(1,\ldots,s)U_s(1,\ldots,s)\sum_{\beta=1}^{s} b(\beta)$$
$$+ [g^{(s)}(1,\ldots,s) - g^{(s-1)}(2,\ldots,s)]U_{s-1}(2,\ldots,s)\sum_{\beta=2}^{s} b(\beta)\Big\}.$$

To convert the preceding equation into a power series in the density, we use the expansion (11.6) for $g^{(s)}$. This yields

$$(a, \Lambda(t)b) = n\int dX_1\phi(1)\bar{a}(1)U_1(1)b(1)$$

$$+ n^2\int dX_1 dX_2\phi(1)\phi(2)\bar{a}(1)\{g_0^{(2)}(1,2) - 1]U_1(2)b(2)$$

$$+ g_0^{(2)}(1,2)U_2(1,2)[b(1) + b(2)]\}$$

$$+ \sum_{p=3}^{\infty} n^p \sum_{s=2}^{p} \frac{1}{(s-1)!}\int dX_1 \cdots dX_s\phi(1)\cdots\phi(s)\bar{a}(1)$$

$$\times \left\{ g_{p-s}^{(s)}(1,\ldots,s)U_s(1,\ldots,s)\sum_{\beta=1}^{s} b(\beta) \right\}$$

$$+ [g_{p-s}^{(s)}(1,\ldots,s) - g_{p-s}^{(s-1)}(2,\ldots,s)]U_{s-1}(2,\ldots,s)\sum_{\beta=2}^{s} b(\beta)\}. \tag{11.28}$$

To identify $\Lambda(t)$ it is necessary to also expand the scalar product in powers of the density:

$$(a,b) = n\int dX_1\phi(1)\bar{a}(1)b(1) + n^2\int dX_1 dX_2\phi(1)\phi(2)\bar{a}(1)[g_0^{(2)}(1,2) - 1]b(2)$$

$$+ \sum_{p=3}^{\infty} n^p \int dX_1 dX_2\phi(1)\phi(2)g_{p-2}^{(2)}(1,2)\bar{a}(1)b(2). \tag{11.29}$$

Let the expansion for $\Lambda(t)$ be written as

$$\Lambda(t) = \sum_{s=0}^{\infty} \Lambda_s(t)\, n^s. \tag{11.30}$$

Then by substituting (11.29) and (11.30) into the left side of (11.28) and comparing with the right side we obtain

$$\Lambda_0 = U_1 = S_1,$$

$$\Lambda_1 b(1) = \int dX_2\phi(2)g_0^{(2)}(1,2)U_2(1,2)[b(1) + b(2)],$$

$$\Lambda_2 b(1) = \frac{1}{2}\int dX_2 dX_3\phi(2)\phi(3)\left\{ g_0^{(3)}(1,2,3)U_3(1,2,3)\sum_{\beta=1}^{3} b(\beta) \right. \tag{11.31}$$

$$\left. + [g_0^{(3)}(1,2,3) - g_0^{(2)}(2,3)]U_2(2,3)[b(2) + b(3)] \right\}$$

$$- \int dX_2\phi(2)\{[g_0^{(2)}(1,2) - 1](\Lambda_1 b)(2) - g_1^{(2)}(1,2)U_2(1,2)[b(1) + b(2)]\}.$$

The general term can be determined from

$$\Lambda_s b(1) = \sum_{r=2}^{s} \frac{1}{r!} \int dX_2 \cdots dX_{r+1} \phi(1) \cdots \phi(r+1)$$

$$\times \left\{ g_{s-r}^{(r+1)}(1, \ldots, r+1) U_{r+1}(1, \ldots, r+1) \sum_{\beta=1}^{r+1} b(\beta) \right.$$

$$\left. + [g_{s-r}^{(r+1)}(1, \ldots, r+1) - g_{s-r}^{(r)}(2, \ldots, r+1)] U_r(2, \ldots, r+1) \sum_{\beta=2}^{r+1} b(\beta) \right\}$$

$$- \int dX_2 \phi(2) \left\{ [g_0^{(2)}(1, 2) - 1](\Lambda_{s-1} b)(2) + \sum_{r=2}^{s-1} g_{r-1}^{(2)}(1, 2)(\Lambda_{s-r} b)(2) \right.$$

$$\left. - g_{s-1}^{(2)}(1, 2) U_2(1, 2)[b(1) + b(2)] \right\} \tag{11.32}$$

where $s = 3, 4 \ldots$.

The expansion can be done in a slightly different way by starting from

$$(a, \Lambda(t) b) = \sum_{N=1}^{\infty} \frac{1}{N!} \exp(-Q + \nu N) \int dX_1 \cdots dX_N$$
$$\times \overline{A} S_N(t) \exp(-\beta H_N) \Delta B. \tag{11.33}$$

This is the same as (11.27) because H_N is a constant of the motion, but the resulting expressions for Λ_s are somewhat different in form. For Λ_1, Λ_2 we find

$$\phi(1)\Lambda_1 b(1) = \int dX_2 \Big\{ U_2(1,2)\phi(1)\phi(2) g_0^{(2)}(1,2)[b(1) + b(2)]$$

$$+ \phi(1)\phi(2) \Big[S_1(1)[g_0^{(2)}(1,2) - 1] b(2) - [g_0^{(2)}(1,2) - 1] S_1(2) b(2) \Big] \Big\}, \tag{11.34}$$

$$\phi(1)\Lambda_2 b(1) = \frac{1}{2} \int dX_2 dX_3 \Big\{ U_3(1, 2, 3)\phi(1)\phi(2)\phi(3) g_0^{(3)}(1, 2, 3) \sum_{\beta=1}^{3} b(\beta)$$

$$+ 2b(3) U_2(1, 2)\phi(1)\phi(2)\phi(3)[g_0^{(3)}(1, 2, 3) - g_0^{(2)}(1, 2)] \Big\}$$

$$+ \int dX_2 \Big\{ U_2(1, 2)\phi(1)\phi(2) g_1^{(2)}(1, 2)[b(1) + b(2)]$$

$$+ \phi(1)\phi(2) \Big[[g_0^{(2)}(1, 2) - 1](\Lambda_1 b)(2) - g_1^{(2)}(1, 2) S_1(2) b(2)$$

$$+ S_1(1) g_1^{(2)}(1, 2) b(2) \Big] \Big\}.$$

Mixtures can be treated by a straightforward modification of the preceding expansion; here we only list the results for the case of self-diffusion. Let n_1 denote the density of the labeled component and n that of the remainder; we assume $n_1 << n$. Let A, B be sums of single-particle functions as in (11.1) but with the sum extending only over the labeled particles. The second term in Equation (11.7) can be neglected so

$$(a, b) = n_1 \int dX\, \phi(p)\bar{a}(X)b(X).$$

We write the time-dependent correlation function as $C_{AB}(t) = (a, \Lambda^D(t)b)$ and find

$$(a, \Lambda^D(t)b) = n_1 \int dX_1 \phi(1)\bar{a}(1)\Big\{S_1(1)$$
$$+ \sum_{s=2}^{\infty} \frac{1}{(s-1)!} n^{s-1} \int dX_2 \ldots dX_s\, \phi(2)\ldots\phi(s)g^{(s)}(1, \ldots, s)U_s(1, \ldots, s)\Big\}b(1).$$

If Λ^D is expanded as

$$\Lambda^D = \sum_{s=0}^{\infty} \Lambda_s^D n^s$$

then the coefficients are found to be

$$\Lambda_0^D = S_1,$$

$$\Lambda_s^D b(1) = \sum_{r=2}^{s+1} \frac{1}{(r-1)!} \int dX_2 \ldots dX_r \phi(2)\ldots\phi(r)g_{s+1-r}^{(r)}(1, \ldots, r)U_r(1, \ldots, r)b(1).$$

Similar results apply to a dilute solution, in which n_1 is the density of the dilute component and n of the solvent.

11.7. REARRANGEMENT OF EXPANSION

The series obtained in the previous section cannot be used directly for large times because it does not converge uniformly in the time. This can be shown by an argument like that given in Chapter 7 but is also evident from the Boltzmann approximation (11.18). There the density occurs multiplied by t so the expansion in powers of n is also an expansion in powers of t. Of course in general the coefficient of a given power of n is a more complicated function of t; (11.18) is only asymptotically valid for long times and it also involves a low-density approximation which corresponds, roughly speaking, to expanding the exponent through linear terms in n.

A valid description of long-time behavior is provided by kinetic theory, as discussed in Chapter 7. The long-time behavior of $\Lambda(t)$, beyond the low-density approximation, could be obtained by following the procedure leading to Equation (11.18) but using the generalized Boltzmann equation. However, this introduces an error for times of the order of a collision time or less, and the Green-Kubo formulas contain an integral over time extending down to 0. Of course this does not mean that the transport coefficients cannot be determined from the generalized Boltzmann equation. To the contrary, transport coefficients apply in the hydrodynamic stage, when kinetic theory is valid. One can follow the Chapman-Enskog procedure to construct the normal solution to the generalized Boltzmann equation and thereby obtain density-dependent corrections to the transport coefficients. Effects which occur at small times in the Green-Kubo formulas are included, but in a different way. However, if it is desired to apply the Green-Kubo formulas, then it is important to treat small times properly.

Thus, what is needed for $\Lambda(t)$ is a representation which is valid over the full range of values for t. Such a representation, introduced by Zwanzig, can be obtained from the Laplace transform of Λ:

$$\mathsf{G}(\mathsf{s}) = \int_0^\infty dt e^{-\mathsf{s}t} \Lambda(t).$$

Since $\Lambda(t)$ is a contraction, its Laplace transform exists as a bounded operator in the right half-plane $\mathrm{Re}\,\mathsf{s} > 0$. In the Boltzmann approximation (11.18) we have

$$\mathsf{G}(\mathsf{s}) = (\mathsf{s} - \mathbf{v}\cdot\nabla + nI)^{-1}.$$

As a generalization to higher values of the density, let an operator $B(\mathsf{s})$ be defined by

$$\mathsf{G}(\mathsf{s}) = [\mathsf{s} + B(\mathsf{s})]^{-1}. \tag{11.35}$$

The notation is somewhat inconsistent since B corresponds to the adjoint of the quantity with the same name introduced in the discussion of the linearized Boltzmann equation.

Let $B(\mathsf{s})$ be expanded as

$$B(\mathsf{s}) = \sum_{p=0}^{\infty} n^p B_{p+1}(\mathsf{s}). \tag{11.36}$$

The subscript has been written as indicated since then, as will be seen, B_2 is a binary collision operator, B_3 a triple collision operator, and so on. This expansion is to be substituted into Equation (11.35) and the result expanded in powers of n. The coefficients $B_{p+1}(\mathsf{s})$ can be obtained by comparing with the Laplace transform of (11.30), where the coefficients will be denoted by $\mathsf{G}_p(\mathsf{s})$:

$$\mathsf{G}_p(\mathsf{s}) = \int_0^\infty dt e^{-\mathsf{s}t} \Lambda_p(t).$$

In lowest order we have

$$[\mathsf{s} + B_1(\mathsf{s})]^{-1} = \mathsf{G}_0(\mathsf{s}).$$

It is readily shown that $\Lambda_0(t) = \exp(t\mathbf{v}\cdot\nabla)$ so $\mathsf{G}_0(\mathsf{s}) = (\mathsf{s} - \mathbf{v}\cdot\nabla)^{-1}$ which yields

$$B_1(\mathsf{s}) = -\mathbf{v}\cdot\nabla.$$

The next two terms in the density expansion yield

$$\mathsf{G}_0 B_2 \mathsf{G}_0 = \mathsf{G}_1, \ \mathsf{G}_0(B_2\mathsf{G}_0 B_2 - B_3)\mathsf{G}_0 = \mathsf{G}_2. \tag{11.37}$$

Insofar as the long-time behavior or the transport coefficients are concerned, we are interested in $B(\mathsf{s})$ for s near 0. The limiting values will be denoted by

$$I_2 = \lim_{\mathsf{s}\to 0} B_2(\mathsf{s}),$$

$$I_3 = \lim_{\mathsf{s}\to 0} B_3(\mathsf{s}), \quad \text{and so forth.}$$

In the spatially homogeneous case we expect I_2 to reduce to the linearized Boltzmann collision operator I. The analysis is similar to that given in Chapter 7, but we nevertheless provide the details. We assume the force between particles to vanish beyond a finite range D, and to be repulsive so no bound states exist. We have

$$B_2(\mathsf{s}) = -(\mathsf{s} - \mathbf{v}\cdot\nabla)\mathsf{G}_1(\mathsf{s})(\mathsf{s} - \mathbf{v}\cdot\nabla)$$

and the terms in $\mathbf{v}\cdot\nabla$ drop out because of spatial homogeneity. Using Equation (11.31) we get

$$\begin{aligned} B_2 b(1) = & -\mathsf{s}^2\int dX_2\,\phi(2)\,g_0^{(2)}(1,2) \\ & \times \int_0^\infty dt\,e^{-\mathsf{s}t}[S_2(1,2,t) - S_0(1,2,t)][b(1) + b(2)]. \end{aligned} \tag{11.38}$$

The integrand vanishes if the two particles do not come within their range of interaction at any time $t \geq 0$. To describe the conditions leading to an interaction, let $\mathbf{q}_1, \mathbf{v}_1$ and $\mathbf{v}_2$ be fixed and consider a cylinder with radius D and with axis through $\mathbf{q}_1$ parallel to $\boldsymbol{g} = \mathbf{v}_2 - \mathbf{v}_1$. For an interaction to occur, particle 2 must lie within this cylinder, either on the approaching side with $(\mathbf{q}_2 - \mathbf{q}_1)\cdot\boldsymbol{g} \leq 0$ or within the sphere of interaction $|\mathbf{q}_2 - \mathbf{q}_1| \leq D$. Let t_1 be the time when interaction first occurs (so $t_1 = 0$ if $|\mathbf{q}_2 - \mathbf{q}_1| \leq D$) and t_2 the time when the particles have passed beyond the range of interaction. The lower limit of the time integral can be replaced by t_1. For $t \geq t_2$ the particles move freely so

$$S_2(t)S_0(-t) = S_2(t_2)S_0(-t_2), \ t \geq t_2.$$

Taking the limit $t \to \infty$ we get

$$S_2(t_2)S_0(-t_2) = \Omega \tag{11.39}$$

where

$$\Omega = \lim_{t\to\infty} S_2(t)S_0(-t).$$

Equation (11.39) is just the statement that the collision is completed at time t_2. For $t \geq t_2$ the time integral can be carried out to give

$$\mathsf{s}\int_0^\infty dt\, e^{-\mathsf{s}t}[S_2(1,2,t) - S_0(1,2,t)]$$

$$= \mathsf{s}\int_{t_1}^{t_2} dt\, e^{-\mathsf{s}t}[S_2(1,2,t) - S_0(1,2,t)] + \exp(-\mathsf{s}t_2)(\Omega - 1)S_0(t_2).$$

Because of spatial homogeneity S_0 reduces to 1 and the equation for B_2 becomes

$$B_2 b(1) = -\int dX_2\, \phi(p_2) g_0^{(2)}(1,2)\Big\{\mathsf{s}^2\int_{t_1}^{t_2} dt\, e^{-\mathsf{s}t}[S_2(1,2,t) - 1]$$

$$+ \mathsf{s}\exp(-\mathsf{s}t_2)(\Omega - 1)\Big\}[b(1) + b(2)].$$

For the integral over $\mathbf{q}_2$ we use cylindrical polar coordinates with origin at $\mathbf{q}_1$ and z-axis parallel to $\boldsymbol{g}$. The volume element is $d^3q_2 = b\,db\,d\varphi\,dz$ and the condition for interaction to occur at a positive time is $b \leq D$, $z \leq (D^2 - b^2)^{1/2}$. The first term in brackets can be shown to vanish in the limit $\mathsf{s} \to 0$ because of the factor s^2. The remaining term contains a factor s so the contribution from any finite region vanishes in the limit $\mathsf{s} \to 0$; the collision operator is determined by configurations where the particles are far apart. For z large and negative we have $t_2 \to -z/\boldsymbol{g}$ and may use Abel's theorem

$$\lim_{\mathsf{s}\to 0} \mathsf{s}\int_{-\infty}^{0} dz \exp(\mathsf{s}z/\boldsymbol{g})\, f(z) = \boldsymbol{g} \lim_{z\to -\infty} f(z).$$

For $z \to -\infty$, $g_0^{(2)}$ takes on its limiting value of 1, showing that initial correlations have no effect. We are left with

$$I_2 b(\mathbf{p}_1) = -\int d^3p_2 \int \boldsymbol{g}\, b\,db\, d\varphi\, (\Omega - 1)[b(\mathbf{p}_1) + b(\mathbf{p}_2)].$$

The wave operator converts the momenta to those after the collision, so the linearized Boltzmann collision operator is obtained. The binary collision operator in the spatially nonhomogeneous case will be worked out later for hard spheres.

The representation (11.35) has a variation which is sometimes useful. Assume that $B(\mathrm{s}) - B_1(\mathrm{s})$ is the Laplace transform of a function $M(t)$:

$$B(\mathrm{s}) - B_1(\mathrm{s}) = \int_0^\infty dt\, e^{-\mathrm{s}t} M(t).$$

Then it is straightforward to show that

$$\frac{\partial \Lambda}{\partial t} - \mathbf{v} \cdot \nabla \Lambda = -\int_0^t d\tau\, M(\tau)\Lambda(t - \tau).$$

The memory function $M(t)$ can be obtained by an expansion in powers of the density and Λ then determined by solving the preceding equation. For hard spheres a slightly different definition of the memory function is necessary since the left-hand side above does not vanish at $t = 0$.

Notes on Chapter 11

Here only the method of cluster expansions is discussed, but there are several other approaches to the many-body problem. Diagrammatic techniques have been extensively developed and applied by Prigogine and coworkers; for example, see Prigogine [1962] and Balescu [1975].

11.2 Reduced distribution functions are discussed in de Boer [1948], Hill [1960], Cohen [1962c], and Mayer and Mayer [1977].

11.3 The relation between time-dependent correlation functions and the kinetic equation has been discussed by a number of authors. In addition to those cited later, see Nelkin and Ghatak [1964], van Leeuwen and Yip [1965], Nelkin and Yip [1966], Ranganathan and Yip [1966], Lebowitz, Percus, and Sykes [1969], and Chappell [1970].

11.4 The proof that for a dilute gas the Green-Kubo formulas reduce to the same results as obtained from the Boltzmann equation is given in Green [1954].

11.5 For the cluster expansion in the context of time-dependent correlation functions, see Ono and Shizume [1963], Weinstock [1963a], Ernst [1966a, b], Dorfman and Cohen [1972, 1975a], Dorfman [1973, 1975b], and Eu [1971]. Convergence of the cluster expansion for finite t is discussed in Gallavotti [1969], and Gallavotti, Lanford, and Lebowitz [1969].

11.6 Time-dependent correlation functions and transport coefficients have been calculated by a variety of techniques in addition to the cluster expansion. Early calculations used the generalized master equation; see Chester and Thellung [1959], Verboven [1960], Swenson [1962, 1963], and McLennan and Swenson [1963b]. In solids other methods than the density expansion are needed; for example, see Hardy, Swenson, and Schieve [1965a], Hardy [1965b], Ranninger [1965], DeVault and McLennan [1965a, b], DeVault and Hardy [1967a], DeVault [1967b], Goyal and Sharma [1980], and Sahu and Sharma [1983].

11.7 The method used here was introduced by Zwanzig [1963, 1964]. One can also proceed by relating $\Lambda(t)\, b$ to a one-particle distribution function as in Section 3, then working out the kinetic equation satisfied by the distribution function. The two methods are reviewed and compared in Ernst, Haines, and Dorfman [1969].

The memory function was introduced in Zwanzig [1961] and Mori [1965a, b]. Much work on dense systems is based on the memory function; for example, see Lebowitz, Percus, and Sykes [1969], Mazenko [1972a, 1974], Mazenko, Wei, and Yip [1972b], Mazenko and Yip [1977], Lindenfeld [1977], and Yip [1979].

12

DENSITY EXPANSION FOR TRANSPORT COEFFICIENTS

12.1. INTRODUCTION

We now apply the discussion of the previous chapter to the evaluation of the transport coefficients of a monatomic gas. To lowest (or zero'th) order in the density, the results are the same as those obtained from the Boltzmann equation. At the next order, analysis of the transport coefficients leads to the triple collision operator. This will be discussed in detail only for hard spheres, for which the results agree, to a good approximation, with those obtained from the Enskog equation.

Persistent correlations, or long-time tails, appear at the second order in density. These have the effect that the transport coefficients are not analytic functions of density, and the coefficient of n^2 obtained from the formal expansion does not exist. To handle this divergence an additional rearrangement of the series is necessary, which leads to the ring operator. When described by the ring operator, the integrand of a Green-Kubo formula decreases for large time as $t^{-3/2}$ which is sufficient for convergence of the integral, but the dependence on density is not analytic. An additional consequence is that Burnett (and higher-order) coefficients do not exist; the Chapman-Enskog expansion is not convergent when persistent correlations are included.

12.2. TRIPLE COLLISION OPERATOR

Equation (11.37) yields for the triple collision operator

$$B_3 = (\mathsf{s} - \mathbf{v} \cdot \nabla)[\mathsf{G}_1(\mathsf{s})(\mathsf{s} - \mathbf{v} \cdot \nabla)\mathsf{G}_1(\mathsf{s}) - \mathsf{G}_2(\mathsf{s})](\mathsf{s} - \mathbf{v} \cdot \nabla). \tag{12.1}$$

We limit attention to the spatially homogeneous case. In addition for simplicity we assume a and b to have vanishing Maxwellian averages. The formulas (11.34) then reduce to

$$\phi(1)\,\Lambda_1 b(1) = \int dX_2\, U_2(1,2)\phi(1)\phi(2)W(1,2)[b(1) + b(2)],$$

$$\phi(1)\,\Lambda_2 b(1) = \int dX_2 dX_3\,\{U_3(1,2,3)\,W_3(1,2,3) \qquad (12.2)$$
$$+ U_2(1,2)\,W(1,2)\,\mathsf{f}(1,3)\,\mathsf{f}(2,3)\}\,\phi(1)\phi(2)\phi(3)\sum_{\beta=1}^{3} b(\beta).$$

Here we have introduced the abbreviations

$$W(1,2) = g_0^{(2)}(1,2),$$
$$W_3(1,2,3) = g_0^{(3)}(1,2,3) = W(1,2)W(1,3)W(2,3) \qquad (12.3)$$

and f denotes the Mayer function

$$\mathsf{f} = W - 1.$$

In addition we have used the formula

$$g_1^{(2)}(1,2) = W(1,2)\int d^3q_3\,\mathsf{f}(1,3)\,\mathsf{f}(2,3). \qquad (12.4)$$

The operators G_1 and G_2 are obtained from the Laplace transform of (12.2) and these are then to be substituted into (12.1). The result for B_3 is found to be

$$\phi(1)\,B_3(\mathsf{s})\,b(1)$$
$$= -\mathsf{s}^2\int dX_2 dX_3\Big\{\frac{1}{2}\mathcal{U}_3(1,2,3)W_3(1,2,3) + \mathcal{U}_2(1,2)W(1,2)\mathsf{f}(1,3)\mathsf{f}(2,3)$$
$$- \mathcal{U}_2(1,2)W(1,2)\Big[\mathsf{s} - \sum_{\alpha=1}^{3}\mathbf{v}_\alpha\cdot\nabla_\alpha\Big][\mathcal{U}_2(1,3)\,W(1,3) + \mathcal{U}_2(2,3)\,W(2,3)]\Big\}$$
$$\times\;\phi(1)\phi(2)\phi(3)\sum_{\beta=1}^{3} b(\beta).$$

Here $\mathcal{U}_2, \mathcal{U}_3$ denote the Laplace transforms of U_2, U_3.

The Laplace transform of the streaming operator, usually referred to as the propagator, will be denoted by G_s:

$$G_s(\mathsf{s}) = \int_0^\infty dt\,e^{-\mathsf{s}t}\,S_s(t).$$

For $s = 0$ in particular we have

$$G_0(1,2,3,\mathsf{s}) = \int_0^\infty dt\, e^{-\mathsf{s}t}\, S_0(1,2,3,t) = (\mathsf{s} - \mathcal{L}_0)^{-1}$$

where $S_0(1,2,3) = S_1(1)S_1(2)S_1(3)$ and $\mathcal{L}_0 = \Sigma_\alpha\, \mathbf{v}_\alpha \cdot \nabla_\alpha$. We also use the abbreviation

$$G_2(1,2) = \int_0^\infty dt\, e^{-\mathsf{s}t}\, S_2(1,2,t)S_1(3,t)$$

since the presence of the third particle as a freely moving spectator can be inferred from context. The Laplace transforms of the Ursell operators are

$$\mathcal{U}_2 = G_2 - G_0, \quad \mathcal{U}_3 = G_3 - \sum_m G_2(m) + 2G_0$$

where m denotes a pair and the sum is over the three pairs which can be formed from the three particles. The transition operator $T_+(m)$ for the pair m is defined by

$$G_2(m) = G_0 + G_0T_+(m)G_0 \tag{12.5}$$

and satisfies

$$-\theta(m)G_2(m) = T_+(m)\, G_0. \tag{12.6}$$

Here $\theta(m)$ denotes the operator (7.20) referred to the pair m. Thus, if m consists of particles α, β then

$$\theta(m) = \frac{\partial U(\mathbf{q}_\alpha - \mathbf{q}_\beta)}{\partial \mathbf{q}_\alpha} \cdot \left(\frac{\partial}{\partial \mathbf{p}_\alpha} - \frac{\partial}{\partial \mathbf{p}_\beta}\right).$$

The operator related to T_+ by time reversal will be denoted by T_-.

After some reduction we obtain the form for B_3

$$\begin{aligned}
\phi(1)\, &B_3(\mathsf{s})b(1) \\
&= -\mathsf{s}^2 \int dX_2 dX_3 \left\{\frac{1}{2}(G_3 - G_0)\, W_3 - \frac{1}{2} G_0 \sum_m T_+(m)\, G_0 W_3 \right. \\
&\qquad + G_0T_+(1,2)\, G_0W(1,2)\, \mathsf{f}(1,3)\, \mathsf{f}(2,3) \\
&\left. - G_0T_+(1,2)\, G_0W(1,2)[T_+(1,3)\, G_0W(1,3) + T_+(2,3)\, G_0W(2,3)]\right\} \\
&\qquad \times \phi(1)\phi(2)\phi(3) \sum_{\beta=1}^{3} b(\beta).
\end{aligned} \tag{12.7}$$

To proceed further it is necessary to come to grips with the three-body problem. Furthermore, we will specialize to the case of hard spheres, whose peculiar dynamics needs a special discussion. We take a detour to deal with these two subjects before returning to the triple collision operator.

12.3. HARD SPHERE DYNAMICS

The dynamics of hard spheres was discussed briefly in Chapter 7 in the context of the Enskog equation. Here we give a more complete treatment, following a discussion by Ernst et al.

Consider first a system of two hard spheres, with diameter D. Let $X = \mathbf{q}_1, \mathbf{p}_1, \mathbf{q}_2, \mathbf{p}_2$ denote a point in the two-particle phase space; we write $\mathbf{r} = \mathbf{q}_1 - \mathbf{q}_2$, $\boldsymbol{g} = \mathbf{v}_1 - \mathbf{v}_2$. As in Chapter 7 we treat the discontinuity in velocity on collision by the provision that if the spheres are just brought into contact at time t, then their velocities are unchanged. The motion is defined following an initial configuration X provided it is nonoverlapping ($r \geq D$); the streaming operator $S_2(t)$ is defined on functions $f(X)$ which vanish in the overlapping region ($r < D$).

Overlapping configurations are nonphysical and do not actually occur. Nevertheless, it is convenient to extend the definition of S_2 to functions which may be nonvanishing in the overlapping region. Let an extension S_2^e be defined by

$$S_2^e - S_0 = W(S_2 - S_0).$$

Thus, points for which the particles overlap lead to free motion. The propagator

$$G_2^e(\mathrm{s}) = \int_0^\infty dt\, e^{-st}\, S_2^e(t)$$

can be calculated as follows. Let f be a function of the relative variables $\mathbf{r}, \boldsymbol{g}$. If $r < D$, or if the conditions are otherwise such that a collision does not occur for positive t, then $(S_2^e(t)f)(\mathbf{r}, \boldsymbol{g}) = f(\mathbf{r} + \boldsymbol{g}t, \boldsymbol{g})$. As described in Chapter 7, a collision occurs for positive t if $r \geq D$, $\mathbf{r} \cdot \boldsymbol{g} \leq 0$ and the impact parameter satisfies $b \leq D$. The value of $\mathbf{r}$ on collision can be written $\mathbf{r} = D\mathbf{e}$ where $\mathbf{e}$ is a unit vector. The time t_c of the collision satisfies

$$g t_c = -\mathbf{r} \cdot \boldsymbol{g} - (D^2 - b^2)^{1/2}$$

and when the conditions for a collision are met we have

$$\begin{aligned} S_2^e(t) f(\mathbf{r}, \boldsymbol{g}) &= f(\mathbf{r} + \boldsymbol{g}t, \boldsymbol{g}), && t \leq t_c, \\ &= f(\mathbf{r} + \boldsymbol{g}t_c + \boldsymbol{g}'(t - t_c), \boldsymbol{g}'), && t > t_c \end{aligned}$$

where $\boldsymbol{g}' = \boldsymbol{g} - 2\mathbf{e}\mathbf{e}\cdot\boldsymbol{g}$ is the relative velocity after the collision.

Writing out the conditions for a collision to occur or not, we get

$$G_2^e(\mathsf{s})f(\mathbf{r},\boldsymbol{g}) = \int_0^{t_c} dt\, e^{-\mathsf{s}t} f(\mathbf{r}+\boldsymbol{g}t,\boldsymbol{g}) + \int_{t_c}^{\infty} dt\, e^{-\mathsf{s}t} f(\mathbf{r}+\boldsymbol{g}t_c+\boldsymbol{g}'(t-t_c),\boldsymbol{g}'),$$

$$r \geq D,\ b \leq D,\ \mathbf{r}\cdot\boldsymbol{g} \leq 0,$$

$$= \int_0^{\infty} dt\, e^{-\mathsf{s}t} f(\mathbf{r}+\boldsymbol{g}t,\boldsymbol{g}), \tag{12.8}$$

$$r < D,\ \text{or}\ b > D,\ \text{or}\ r \geq D,\ b \leq D,\ \mathbf{r}\cdot\boldsymbol{g} \geq 0.$$

The operator G_0 has the form of the last expression independently of the conditions, so

$$[G_2^e(\mathsf{s}) - G_0(\mathsf{s})]f = \int_{t_c}^{\infty} dt\, e^{-\mathsf{s}t}[f(\mathbf{r}+\boldsymbol{g}t_c+\boldsymbol{g}'(t-t_c),\boldsymbol{g}') - f(\mathbf{r}+\boldsymbol{g}t,\boldsymbol{g}],$$

$$r \geq D,\ b \leq D,\ \mathbf{r}\cdot\boldsymbol{g} \leq 0, \tag{12.9}$$

$$= 0 \quad \text{otherwise.}$$

In the present context $G_0 = (\mathsf{s} - \boldsymbol{g}\cdot\nabla_r)^{-1}$ and the arguments of f are such that $\boldsymbol{g}\cdot\nabla_r f = \partial f/\partial t$. Hence, we get

$$(G_2^e - G_0)(\mathsf{s} - \boldsymbol{g}\cdot\nabla_r)f = \exp(-\mathsf{s}t_c)[f(\mathbf{r}+\boldsymbol{g}t_c,\boldsymbol{g}') - f(\mathbf{r}+\boldsymbol{g}t_c,\boldsymbol{g}],$$

$$r \geq D,\ b \leq D,\ \mathbf{r}\cdot\boldsymbol{g} \leq 0,$$

$$= 0, \quad \text{otherwise.}$$

By definition of $t_c, \mathbf{r} + \boldsymbol{g}t_c = D\mathbf{e}$. Hence, we can write the preceding equation as

$$(G_2^e - G_0)G_0^{-1}f = 0, \qquad z < 0,$$

$$= \exp(-\mathsf{s}z/\boldsymbol{g})[f(D\mathbf{e},\boldsymbol{g}') - f(D\mathbf{e},\boldsymbol{g})], \qquad z \geq 0 \tag{12.10}$$

where $z = \boldsymbol{g}t_c$.

The transition operator is determined by Equation (12.5) or

$$T_+ = G_0^{-1}(G_2^e - G_0)G_0^{-1}. \tag{12.11}$$

Using $\boldsymbol{g}\cdot\nabla_r\mathbf{e} = 0$, $\boldsymbol{g}\cdot\nabla_r z = -\boldsymbol{g}$ it can be seen that the only contribution to T_+ comes from the discontinuity at $z = 0$ in (12.10) so we get

$$T_+ f = \boldsymbol{g}\,\delta(z)[f(D\mathbf{e},\boldsymbol{g}') - f(D\mathbf{e},\boldsymbol{g})].$$

For small z, $r = D + z|\mathbf{e}\cdot\hat{\boldsymbol{g}}|$ and

$$\delta(z) = -\mathbf{e}\cdot\hat{\boldsymbol{g}}\,\delta(r - D), \qquad \mathbf{e}\cdot\boldsymbol{g} \leq 0,$$

$$= 0, \qquad \mathbf{e}\cdot\boldsymbol{g} \geq 0.$$

Therefore, T_+ can be written

$$T_+ f = -\mathbf{e}\cdot\boldsymbol{g}\,\delta(r - D)[f(D\mathbf{e},\boldsymbol{g}') - f(D\mathbf{e},\boldsymbol{g})], \quad \mathbf{e}\cdot\boldsymbol{g} \le 0, \qquad (12.12)$$
$$= 0, \quad \mathbf{e}\cdot\boldsymbol{g} \ge 0.$$

It is noteworthy that T_+ is independent of the Laplace transform variable s, which has come about because of the impulsive nature of the hard-sphere interaction.

It follows by direct calculation from Equation (12.12) that

$$T_+\, G_2^e = T_+\, G_0. \qquad (12.13)$$

Equation (12.5) then shows that

$$T_+\, G_0 T_+ = 0 \qquad (12.14)$$

This is intuitively clear, since it means that two hard spheres cannot interact more than once without the intervention of other particles. Since G_2^e satisfies $W(G_2^e - G_0) = G_2^e - G_0$ we also obtain

$$W\, G_0\, T_+ = G_0\, T_+ \qquad (12.15)$$

Inversion of the Laplace transform shows that

$$S_2^e(t) = S_0(t) + \int_0^t ds\, S_0(t - s) T_+\, S_0(s).$$

It follows from Equation (12.13) that $T_+\, S_2^e = T_+\, S_0$ so an alternative form is

$$S_2^e(t) = S_0(t) + \int_0^t ds\, S_0(t - s)\; T_+\, S_2^e(s).$$

This yields the equation of motion

$$\partial S_2^e/\partial t = (\mathcal{L}_0 + T_+) S_2^e$$

so T_+ plays the role of $-\theta$.

The preceding equations apply to positive time, but the equations for negative time can be obtained by a time reversal. For example, the time-reversed transition operator is given by

$$T_- f = \mathbf{e}\cdot\boldsymbol{g}\,\delta(r - D)[f(\mathbf{q}_1,\mathbf{v}_1',\mathbf{q}_1 + D\mathbf{e},\mathbf{v}_2')$$
$$- f(\mathbf{q}_1,\mathbf{v}_1,\mathbf{q}_1 + D\mathbf{e},\mathbf{v}_2)], \quad \mathbf{e}\cdot\boldsymbol{g} \ge 0,$$
$$= 0, \quad \mathbf{e}\cdot\boldsymbol{g} \le 0.$$

A different extension of the streaming operator can be obtained by writing

$$\overline{S}_2^e - S_0 = (S_2 - S_0)W.$$

Since $S_2W = WS_2$ we have

$$\overline{S}_2^e - S_2^e = WS_0 - S_0W.$$

The corresponding equation for the Laplace transform is

$$\overline{G}_2^e - G_2^e = WG_0 - G_0W.$$

Calculation gives $(\mathsf{s} - \mathcal{L}_0)W - W(\mathsf{s} - \mathcal{L}_0) = -\mathbf{e}\cdot\boldsymbol{g}\,\delta(r - D)$ and this yields

$$WG_0 - G_0W = -G_0\mathbf{e}\cdot\boldsymbol{g}\,\delta(r - D)G_0. \tag{12.16}$$

The operator $\overline{T}_+$ obtained from $\overline{S}_2^e$ in the manner of (12.11) is given by

$$\overline{T}_+ = T_+ - \mathbf{e}\cdot\boldsymbol{g}\,\delta(r - D) \tag{12.17}$$

which works out to

$$\begin{aligned} \overline{T}_+ f &= -\mathbf{e}\cdot\boldsymbol{g}\,\delta(r - D)\, f(D\mathbf{e},\boldsymbol{g}'), \quad \mathbf{e}\cdot\boldsymbol{g} \le 0, \\ &= -\mathbf{e}\cdot\boldsymbol{g}\,\delta(r - D)\, f(D\mathbf{e},\boldsymbol{g}), \quad \mathbf{e}\cdot\boldsymbol{g} \ge 0. \end{aligned} \tag{12.18}$$

Corresponding to Equations (12.14) and (12.15), $\overline{T}_+$ satisfies

$$\overline{T}_+ G_0 \overline{T}_+ = 0, \quad \overline{T}_+ G_0 W = \overline{T}_+ G_0. \tag{12.19}$$

The equation of motion for $\overline{S}_2^e$ is

$$\partial\overline{S}_2^e/\partial t = (\mathcal{L}_0 + \overline{T}_+)\overline{S}_2^e.$$

Let $\mathcal{H}$ denote the Hilbert space of functions which are square integrable on $\mathbf{r},\boldsymbol{g}$. Using Equation (12.8) it can be shown that for functions f in $\mathcal{H}$, $(G_0)^{-1}\,G_2^e f = f$ which implies $\mathcal{L}_2 G_2^e = \mathcal{L}_0 G_2^e$. G_2^e transforms $\mathcal{H}$ onto the domain of $\mathcal{L}_2$ so it follows that $\mathcal{L}_2$ is a restriction of $\mathcal{L}_0$ to functions having the form of the right-hand side of Equation (12.8). Such functions can be characterized by a boundary condition on the surface $r = D$. In other words, for f in $\mathcal{H}$, $S_2^e(t)f$ satisfies the same differential equation as $S_0(t)f$ but has a different boundary condition. The transition operators incorporate the boundary condition into the equations of motion. Although $T_\pm$ are not proper operators on $\mathcal{H}$, we may define formal conjugates $T_\pm^\dagger$ by $(f, T_\pm h) = (T_\pm^\dagger f, h)$ where for the moment (,) denotes the integral which gives the scalar product on $\mathcal{H}$ and f, h are any functions for which both sides of the equation are well defined. Then it can be shown that $T_\pm^\dagger = \overline{T}_\mp$.

To generalize to a system of N particles, let extensions of the N-body streaming operator be defined by

$$S_N^e - S_0 = W_N(S_N - S_0), \quad \overline{S}_N^e - S_0 = (S_N - S_0)W_N.$$

Here W_N denotes the characteristic function for the nonoverlapping region in the N-particle phase space; W_N vanishes when any pair of particles overlap and otherwise is equal to 1. The equations of motion for $S_N^e, \overline{S}_N^e$ are natural generalizations of the two-particle equations, so we will state them without derivation:

$$\partial S_N^e/\partial t = \left[\mathcal{L}_0 + \sum_m T_+(m)\right] S_N^e, \quad \partial \overline{S}_N^e/\partial t = \left[\mathcal{L}_0 + \sum_m \overline{T}_+(m)\right]\overline{S}_N^e.$$

The operators $T_+(m), \overline{T}_+(m)$ are as defined previously but referred to the m'th pair. These equations hold for positive t, while if t is negative $T_+, \overline{T}_+$ are replaced by $T_-, \overline{T}_-$.

The Liouville equation can be obtained as follows. Let an ensemble F_N be set up at time $t = 0$. The average value of a quantity A can be calculated two ways:

$$\langle A \rangle = \int d\Gamma_N\, F_N(0)A(t) = \int d\Gamma_N\, F_N(t)A(0).$$

The ensemble will contain a factor W_N so the first of these can be written

$$\langle A \rangle = \int d\Gamma_N F_N W_N S_N(t) A = \int d\Gamma_N F_N S_N^e(t) A.$$

This shows that

$$F_N(t) = (S_N^e(t))^\dagger F_N(0)$$

where the dagger has the same meaning as before but generalized to the phase space for N particles. The equation of motion for S_N^e then yields the Liouville equation

$$\frac{\partial F_N}{\partial t} + \left[\mathcal{L}_0 - \sum_m \overline{T}_-(m)\right]F_N = 0.$$

The BBGKY hierarchy can be obtained directly from the Liouville equation. Using the definition (7.4) for the reduced distribution functions, we get

$$\left(\frac{\partial}{\partial t} + \mathbf{v}_1 \cdot \nabla_1\right) F_1(X_1,t) = n\int dX_2 \overline{T}_-(1,2)\, F_2(X_1,X_2,t),$$

$$\left(\frac{\partial}{\partial t} + \mathbf{v}_1 \cdot \nabla_1 + \mathbf{v}_2 \cdot \nabla_2 - \overline{T}_-(1,2)\right) F_2(X_1,X_2,t)$$
$$= n\int dX_3[\overline{T}_-(1,3) + \overline{T}_-(2,3)]\, F_3(X_1,X_2,X_3,t).$$

Succeeding equations follow a similar pattern.

Using the preceding results, we work out the operator B_2 for the case when spatial variations are present. Equation (11.38) yields

$$B_2 b = -\int dX_2\, \phi(2)(G_0)^{-1}\, W[G_2 - G_0](G_0)^{-1}[b(1) + b(2)]$$
$$= -\int dX_2\, \phi(2)\, T_+[b(1) + b(2)].$$

Using Equation (12.12) for T_+ we get

$$B_2 b = D^2 \int d^3p_2 \int_- d^2e\, |\mathbf{e}\cdot\mathbf{g}|[b(\mathbf{q}_1,\mathbf{v}_1) + b(\mathbf{q}_1 + D\mathbf{e},\mathbf{v}_2) - b(\mathbf{q}_1,\mathbf{v}_1') - b(\mathbf{q}_1 + D\mathbf{e},\mathbf{v}_2')].$$

The Fourier-transformed operator is

$$\hat{B}_2\beta = D^2\int d^3p_2\phi(p_2)\int d^2e|\mathbf{e}\cdot\mathbf{g}|\,\{\beta(\mathbf{v}_1) - \beta(\mathbf{v}_1') + \exp(-iD\mathbf{e}\cdot\mathbf{k})[\beta(\mathbf{v}_2) - \beta(\mathbf{v}_2')]\}. \tag{12.20}$$

The operator B_2 is related to Enskog's collision operator as follows. Consider the operator on the right-hand side of Equation (7.30), that is, Enskog's operator without the collision-rate factor χ, and let I_E denote the corresponding linearized collision operator. Then $B_2 = I_E^+$ in agreement with Equation (11.17).

12.4. BINARY COLLISION EXPANSION

One way of making progress with the three- (and more) body problem is by the binary collision expansion. This expresses the many-body dynamics as a sum of sequences of two-body interactions, which is possible if all forces are sums of two-body forces.

Consider the time evolution operator $S_N(t)$ for a system of N particles. With continuous two-body interactions, its generator is given by

$$\mathcal{L}_N = \mathcal{L}_0 - \sum_m \theta(m).$$

Let G_N denote the Laplace transform of S_N:

$$G_N = \left[\mathrm{s} - \mathcal{L}_0 + \sum_m \theta(m)\right]^{-1}. \tag{12.21}$$

Then G_N satisfies

$$G_N - G_0 = -G_0 \sum_m \theta(m) G_N. \tag{12.22}$$

In Equation (12.21) separate out a particular pair, say the m'th:

$$G_N = \left[\mathsf{s} - \mathcal{L}_0 + \theta(m) + \sum_{n \neq m} \theta(n) \right]^{-1}.$$

Then it is straightforward to show that

$$G_N - G_2(m) = -G_2(m) \sum_{n \neq m} \theta(n)\, G_N \tag{12.23}$$

where

$$G_2(m) = [\mathsf{s} - \mathcal{L}_0 + \theta(m)]^{-1}.$$

In $G_2(m)$ only the pair m interacts while the remaining particles move freely, and so $G_2(m)$ can be worked out from the solution to the two-body problem. Combining the preceding equations we get

$$G_N = G_0 - G_0 \sum_m \theta(m)\, G_2(m) + G_0 \sum_m \sum_{n \neq m} \theta(m)\, G_2(m)\, \theta(n) G_N.$$

Continued application of (12.23) leads to the series

$$G_N = G_0 - G_0 \sum_m \theta(m)\, G_2(m) + G_0 \sum{}' \theta(m)\, G_2(m)\, \theta(n)\, G_2(n)$$

$$- G_0 \sum{}' \theta(m)\, G_2(m)\, \theta(n)\, G_2(n)\, \theta(p)\, G_2(p) + \ldots \tag{12.24}$$

Here the sums extend over all pairs with the restriction, indicated by the prime, that no pair occurs twice consecutively. Thus, the third term has the restriction $n \neq m$ and $n \neq p$ but terms with $m = p$ are to be included. If there are only two particles, then this restriction causes the series to terminate after the second term.

Equation (12.24) is the desired binary collision expansion. It expresses the dynamics of the N-body system as a series in which no particles interact, one pair interacts, two pairs interact, and so on. An equivalent series for the streaming operator can be obtained by inverting the Laplace transform:

$$S_N(t) = S_0(t) - \int_0^t dt_1 \sum_m S_0(t_1)\, \theta(m)\, S_2(m, t - t_1)$$

$$+ \int_0^t dt_1 \int_0^{t_1} dt_2 \sum{}' S_0(t - t_1)\, \theta(m)\, S_2(m, t_1 - t_2)\, \theta(n)\, S_2(n, t_2) + \ldots$$

Here $S_2(m)$ denotes the streaming operator in which only the pair m interacts.

For hard spheres the operator θ is not defined, so it is necessary to write the binary collision expansion in a different form. If we use Equation (12.6) the result

$$G_N = G_0 + G_0 \sum_m T_+(m)\, G_0 + G_0 \sum{}' T_+(m)\, G_0 T_+(n)\, G_0 + \ldots \tag{12.25}$$

contains no reference to θ. However, this still does not apply to hard spheres since overlapping configurations have not been excluded, nor has G_N been extended to overlapping configurations. As shown by Ernst and colleagues, all that is necessary is to multiply by W_N and the binary collision expansion becomes

$$W_N\, G_N = W_N\Big\{ G_0 + G_0 \sum_m T_+(m)\, G_0 + G_0 \sum{}' \, T_+(m)\, G_0 T_+(n)\, G_0 + \ldots \Big\}. \tag{12.26}$$

An alternative form is

$$\begin{aligned} G_N W_N &= W_N G_N \\ &= \Big\{ G_0 + G_0 \sum_m \overline{T}_+(m)\, G_0 + G_0 \sum{}' \overline{T}_+(m)\, G_0 \overline{T}_+(n)\, G_0 + \ldots \Big\} W_N. \end{aligned} \tag{12.27}$$

Actually the sums can be extended over all pairs without restriction since in view of Equation (12.14) terms where the same pair occurs twice consecutively are zero anyway. Hence, we may write

$$\begin{aligned} W_N(G_N - G_0) &= W_N\Big\{ G_0 + G_0 \sum_m T_+(m)\, G_0 + \ldots \Big\} \sum_n T_+(n)\, G_0 \\ &= W_N G_N \sum_n T_+(n)\, G_0. \end{aligned} \tag{12.28}$$

Similarly, we obtain

$$(G_N - G_0)\, W_N = G_0 \sum_m \overline{T}_+ (m)\, G_N\, W_N. \tag{12.29}$$

12.5. HARD-SPHERE TRIPLE COLLISION OPERATOR

We now return to Equation (12.7) for the triple collision operator and reduce it further for hard spheres. It is convenient to first write $G_N W_N = W_N G_N$ and substitute Equation (12.28) into (12.29) to obtain

$$(G_N - G_0)\, W_N = G_0 \sum_m \overline{T}_+(m)\, W_N\Big[1 + G_N \sum_n T_+(n) \Big]\, G_0. \tag{12.30}$$

The advantage of this form is that factors G_0 appear on the extreme right and left, where in (12.7) they may be replaced by s^{-1}. This equation (with $N = 3$) is to be substituted into (12.7), and the result can then be reduced with the aid of the relations obtained previously for the hard-sphere T-matrices. We write the result as

$$B_3 = B_3^E + B_3^S$$

where

$$\phi(1)B_3^E b(1) = \int dX_2 dX_3 \overline{T}_+(1,2)\mathsf{f}(1,3)\mathsf{f}(2,3)\phi(1)\phi(2)\phi(3)\sum_{\beta=1}^{3} b(\beta),$$

$$\phi(1)B_3^S b(1) = -\int dX_2 dX_3 \overline{T}_+(1,2)\Big\{W_3 G_3 \sum_m T_+(m) \tag{12.31}$$

$$- G_0[T_+(1,3) + T_+(2,3)]\Big\}\phi(1)\phi(2)\phi(3)\sum_{\beta=1}^{3} b(\beta).$$

In B_3^E there is no remaining dependence on the Laplace transform variable s so we write $I_3^E = B_3^E$. Furthermore, the integral of the term containing $b(3)$ vanishes. The integral over X_3 can be carried out with the aid of (12.4) and the result is

$$\phi(1)I_3^E b(1) = -\int dX_2 \overline{T}_+(1,2) g_1^{(2)}(1,2)\phi(1)\phi(2)[b(1) + b(2)].$$

From Equation (12.18) for $\overline{T}_+$ it is seen that $g_1^{(2)}$ is evaluated at contact, and so according to Equation (7.31) contributes a factor $\frac{5}{12}\pi D^3$. The remaining integration yields the linearized Boltzmann collision operator. Thus, we get

$$I_3^E = \tfrac{5}{12}\pi D^3 I, \quad I_2 + nI_3^E = \chi I. \tag{12.32}$$

This incorporates Enskog's correction to the collision frequency.

The operator B_3^S has been analyzed by Sengers and coworkers and is found to give corrections to the Enskog theory which amount to about 5 to 10 percent. In their analysis the binary collision expansion is used for G_3. This expansion actually terminates at the fifth term, as in a system of three hard spheres more than four collisions cannot occur. There are additional restrictions which correspond to other dynamically forbidden collision sequences. We denote the limit of B_3^S by I_3^s and (12.31) is found to reduce to

$$\phi(1)I_3^S b(1) = -\lim_{\mathsf{s}\to 0}\int dX_2 dX_3 \overline{T}_+(1,2)\Big\{\mathsf{f}(1,3)G_0 T_+(2,3)$$

$$+ \mathsf{f}(2,3)G_0 T_+(1,3) + G_0\sum_{m,n} T_+(m)G_0 T_+(n)$$

$$\times\Big[1 + G_0\sum_p T_+(p)\Big]\Big\}\phi(1)\phi(2)\phi(3)\sum_{\beta=1}^{3} b(\beta).$$

The first two terms correspond to a sequence of two collisions with the colliding particles being forced by the Mayer function to overlap; these are referred to as single overlap terms with the Enskog contribution being called double overlap. The remaining terms correspond to sequences of three and four collisions. It is possible to reduce I_3^S somewhat further by analytical methods, but the geometry of the collision sequences is so involved that ultimately it is necessary to resort to numerical techniques. It is then found that the four-collision sequences give a negligible contribution, being only about one millionth of the Enskog correction. The contributions from the single-overlap and three-collision sequences have about the same magnitude. It is noteworthy that I_3^S is not, like the Enskog correction, proportional to I and so different transport coefficients are affected in different ratios. The numerical results for the transport coefficients will be quoted later.

12.6. HARD-SPHERE TRANSPORT COEFFICIENTS

The Green-Kubo formulas have the general form

$$L = \int_0^\infty dt \, \langle JJ(t)\rangle .$$

In general the flux J is a sum of two contributions, $J = K + U$. The kinetic contribution K is a sum of single-particle terms, and the potential contribution U is a sum over pairs. Consequently, L is the sum

$$L = L_{KK} + L_{KU} + L_{UK} + L_{UU}$$

where

$$L_{KK} = \int_0^\infty dt \, \langle KK(t)\rangle ,$$

$$L_{KU} = \int_0^\infty dt \, \langle KU(t)\rangle ,$$

$$L_{UK} = \int_0^\infty dt \, \langle UK(t)\rangle ,$$

$$L_{UU} = \int_0^\infty dt \, \langle UU(t)\rangle .$$

The last three are contributions due to collisional transfer. There are no such contributions to the coefficients of diffusion since the associated fluxes do not contain two-body terms.

It follows from time-reversal invariance that $L_{KU} = L_{UK}$. These terms are proportional to the density (at low density) while L_{UU} is quadratic in the density. We will consider only those corrections to L which are linear in the density, so L_{KU} can be evaluated to its lowest nonvanishing order and L_{UU} can be neglected.

The evaluation of L_{KK} to first order in the density involves the triple collision operator. For the kinetic contributions to the thermal conductivity and viscosity we have

$$\lambda_{KK} = \frac{1}{\kappa T^2}\int d^3p\,\phi(p)S_x(I + nI_3)^{-1}S_x,$$

$$\eta_{KK} = \frac{1}{\kappa T}\int d^3p\,\phi(p)T_{xy}(I + nI_3)^{-1}T_{xy}.$$

For hard spheres in Enskog's approximation, the results are particularly simple since Equation (12.32) yields

$$\lambda_{KK} = \chi^{-1}\lambda_0, \quad \eta_{KK} = \chi^{-1}\eta_0 \tag{12.33}$$

where the subscript 0 denotes the Boltzmann approximation. Through terms linear in the density, this is the same as

$$\lambda_{KK} = \left(1 - \frac{5}{12}\pi n D^3\right)\lambda_0, \quad \eta_{KK} = \left(1 - \frac{5}{12}\pi n D^3\right)\eta_0. \tag{12.34}$$

Use of the full triple collision operator yields corrections to the Enskog results. The analysis of Sengers and colleagues obtained

$$\begin{aligned}\lambda_{KK} &= \left(1 - 0.963\frac{5}{12}\pi n D^3\right)\lambda_0,\\ \eta_{KK} &= \left(1 - 0.972\frac{5}{12}\pi n D^3\right)\eta_0.\end{aligned} \tag{12.35}$$

We also quote the result for the coefficient of self-diffusion,

$$D = \left(1 - 0.918\frac{5}{12}\pi n D^3\right)D_0. \tag{12.36}$$

We next consider the effect on the transport coefficients of collisional transfer. A cluster expansion can be developed for L_{KU} but it is simpler to follow a less direct approach. Consider, for example, the shear viscosity, for which the Green-Kubo formula can be written

$$\eta = \frac{1}{\kappa T}\int_0^\infty dt\,\langle t_{xy}T_{xy}(t)\rangle \tag{12.37}$$

where t_{xy} is the momentum flux. Its volume integral T_{xy} is given by

$$T_{xy} = \frac{1}{m}\sum_\alpha p_{\alpha x}p_{\alpha y} + \frac{1}{2}\sum_{\alpha,\beta}(q_{\beta x} - q_{\alpha x})\partial U_{\alpha\beta}/\partial q_{\alpha y}.$$

As noted by Ernst, this is the time derivative $T_{xy} = dV_{xy}/dt$ where

$$V_{xy} = \sum_\alpha q_{\alpha x}p_{\alpha y}.$$

The kinetic contribution to t_{xy} is

$$t_{xy}^{K} = \frac{1}{m} \sum_{\alpha} p_{\alpha x} p_{\alpha y} \delta(\mathbf{q}).$$

If these two expressions are used in (12.37), the result is $\eta_{KK} + \eta_{KU}$. The time integral can be carried out to give

$$\eta_{KK} + \eta_{KU} = \frac{1}{\kappa T} \lim_{t\to\infty} \langle t_{xy}^{K} V_{xy}(t) \rangle. \tag{12.38}$$

(The integrand is easily seen to vanish at $t = 0$.) The advantage of this formulation is that both t_{xy} and V_{xy} are sums of single-particle functions, so the time dependence is governed by the operator $\Lambda(t)$ as worked out previously.

In terms of G, Equation (12.38) is

$$\eta_{KK} + \eta_{KU} = \frac{1}{m\kappa T} \lim_{\mathsf{s}\to 0} \mathsf{s}\, (p_x p_y \delta(\mathbf{q}), \mathsf{G}(\mathsf{s}) q_x p_y).$$

Expressing η_{KK} in a similar form, we get

$$\eta_{KU} = \frac{1}{m\kappa T} \lim_{\mathsf{s}\to 0} (p_x p_y\, \delta(\mathbf{q}),\ \mathsf{G}(\mathsf{s})[\mathsf{s} q_x - \frac{1}{m} p_x] p_y).$$

The scalar product here is as defined in Equation (11.7) but for the purpose of the present calculation it is sufficient to use the leading term, $(a, b) = n\int dX\, \phi(p)\bar{a}(X) b(X)$. For G we use Equation (11.35) and to the order of interest B need be expanded only to first order in density. Thus, we get

$$\eta_{KU} = \frac{1}{m\kappa T} \lim_{\mathsf{s}\to 0} (p_x p_y\, \delta(\mathbf{q}), (\mathsf{s} - \mathbf{v}\cdot\nabla + nB_2)^{-1} \times (\mathsf{s} q_x - \frac{1}{m} p_x) p_y). \tag{12.39}$$

This can be evaluated by using the Fourier representation. We have

$$\begin{aligned}(\mathsf{s} - \mathbf{v}\cdot\nabla + n\, B_2)^{-1} q_x p_y &= i\frac{\partial}{\partial k_x}(\mathsf{s} - \mathbf{v}\cdot\nabla + n\, B_2)^{-1} \exp(-i\mathbf{k}\cdot\mathbf{q}) p_y\big|_{\mathbf{k}=0} \\ &= i\frac{\partial}{\partial k_x}\exp(-i\mathbf{k}\cdot\mathbf{q})[\mathsf{s} + i\mathbf{v}\cdot\mathbf{k} + \hat{B}_2(\mathbf{k})]^{-1} p_y\big|_{\mathbf{k}=0} \\ &= \mathsf{s}^{-1}[q_x p_y - i(\mathsf{s} + nI)^{-1}(iv_x + n\beta_x) p_y]\end{aligned}$$

where

$$\beta_x = \partial \hat{B}_2(\mathbf{k})/\partial k_x\big|_{\mathbf{k}=0}$$

and we have used $\hat{B}(0) = I$ and $I\mathbf{p} = 0$. Equation (12.39) reduces to

$$\eta_{KU} = -i\frac{n}{m\kappa T}\lim_{s\to 0}\,(p_x p_y\,\delta(\mathbf{q}), (s + nI)^{-1}\,\beta_x p_y).$$

We now specialize to hard spheres and use Equation (12.20) for $\hat{B}_2(\mathbf{k})$. This yields

$$\begin{aligned}\beta_x p_y &= iD^3\int d^3p_2\phi(p_2)\int_- d^2e\,\mathbf{e}\cdot\mathbf{g}\,e_x(p_{2y} - p'_{2y})\\ &= i\frac{4}{15}\pi D^3(p_x p_y/m).\end{aligned}$$

The result for η_{KU} is

$$\eta_{KU} = \frac{4}{15}\pi n D^3\frac{1}{\kappa T}\int d^3p\,\phi(p)\,T_{xy}\,I^{-1}\,T_{xy} = \frac{4}{15}\pi n D^3\,\eta_0. \tag{12.40}$$

The thermal conductivity can be treated in a similar way. The volume integral of the energy flux, as given by Equation (8.9), can be written as the time derivative

$$S_i = \frac{d}{dt}\sum_\alpha\left(\frac{1}{2m}p^2 + \sum_\beta U_{\alpha\beta}\right).$$

This is not a sum of single-particle functions, but the two-particle term gives a contribution to the thermal conductivity which is second order in the density and can be neglected. Hence, the analysis is similar to that for the viscosity, and the result is found to be

$$\lambda_{KU} = \frac{2}{5}\pi n D^3\,\lambda_0.$$

By combining Equations (12.34) and (12.40), we obtain for the viscosity in Enskog's approximation

$$\begin{aligned}\eta_E &= \eta_{KK} + 2\,\eta_{KU}\\ &= \left(1 + \frac{7}{60}\pi n D^3\right)\eta_0.\end{aligned}$$

Similarly, the thermal conductivity is

$$\lambda_E = \left(1 + \frac{23}{60}\pi n D^3\right)\lambda_0.$$

Using the numerical results (12.35) instead of Enskog's approximation, we get

$$\eta = \left(1 + 1.101\frac{7}{60}\pi n D^3\right)\eta_0, \quad \lambda = \left(1 + 1.040\frac{23}{60}\pi n D^3\right)\lambda_0.$$

The increased collision rate reduces the transport coefficients, but this is more than compensated for by the collisional transfer. Hence, the effect of increasing the density is to increase the thermal conductivity and viscosity but to decrease the coefficients of diffusion.

12.7. DIAGRAMMATIC EXPANSION

It is clear that on going to higher orders of the density expansion, the complexity of the analysis increases rapidly. However, progress can be made by an intuitive selection of interesting contributions. The representation of higher-order terms by diagrams is useful for this purpose and will be discussed in this section.

A particular problem to be addressed is the divergence of the density expansion, which first appears when four-body collisions are considered. (In two dimensions the divergence occurs one order sooner, that is, at the level of the triple collision operator.) Treatment of the divergence requires a resummation of selected terms from the density expansion, to all orders in the density. Such a resummation was first carried out by Kawasaki and Oppenheim; the resulting collision operator is called the ring operator since the terms contributing to it, when represented by diagrams, have the appearance of rings.

Consider Equation (11.32) for Λ_s, and take its Laplace transform to obtain G_s. Then G_s involves the Laplace transforms $\mathcal{U}_r$ of the cluster operators U_r, $r = 2,\ldots,s+1$. Each U_r is given in terms of streaming operators S_p, $p = 1,\ldots,r$ by Equation (11.21). The streaming operators can be written in the form

$$S_p(t) = \exp\left\{t\left[\mathcal{L}_0 - \sum_m \theta(m)\right]\right\}. \tag{12.41}$$

(For hard spheres, θ is to be replaced by $-T_+$.) The Laplace transform of S_p is the propagator G_p given by Equation (12.21) (for $N = p$). The products of streaming operators in U_r refer to disjoint sets of particles and so can be combined into an exponential form like that previously shown. Thus, for example,

$$S_2(1,2)S_2(3,4) = \exp\{t[\mathcal{L}_0 - \theta(1,2) - \theta(3,4)]\}.$$

Let the Laplace transform of this quantity be denoted by $G(1,2;3,4)$; it is given by

$$G(1,2;3,4) = [\mathsf{s} - \mathcal{L}_0 + \theta(1,2) + \theta(3,4)]^{-1}. \tag{12.42}$$

Other products of streaming operators can be treated in a similar way. Thus $\mathcal{U}_r$ is equal to G_r plus a linear combination of modified G_r's in which interactions between clusters are omitted.

Let each G-operator in $\mathcal{U}_r$ be expanded in a binary collision expansion. The expansion for a modified G-operator such as (12.42) is easily obtained from that for G_r by eliminating the T_+-operators which connect different clusters. Each term in the resulting expansion can be depicted by a diagram in the following way. Draw a horizontal line for each of the particles. For each T_+-operator, draw a vertical line connecting the two horizontal lines for the pair of particles which interact. The vertical lines are to be drawn in the same order as the T_+-operators occur in the term. The horizontal line segments between vertical lines, and at the extreme right and left, stand for the free-particle propagators G_0. Thus, a diagram represents an ordered sequence of interactions among labeled particles. (Note that adjacent vertical lines never connect the same pair of horizontal lines because of the corresponding restriction on the binary collision expansion.) The contribution to $\mathsf{G}_s b$ is obtained by applying the sequence of G_0- and T_+-operators associated with the diagram to $\Sigma\, b$, multiplying by the statistical factors (and by $1/s!$) in Equation (11.21), and then integrating over the coordinates and momenta of all particles except number 1.

These diagrams are similar to the Mayer diagrams which are familiar in equilibrium statistical mechanics. However, it is now more convenient to represent the particles by lines rather than points.

A diagram is connected if there is a path consisting of vertical and horizontal line segments between any two horizontal lines. Thus, in the set of diagrams for G_3, the diagram for $G_0 T_+ (1,2) G_0 T_+ (2,3) G_0$ is connected while the one for $G_0 T_+ (1,2) G_0$ is not. If a diagram is not connected, then the particles involved can be divided into two clusters such that there are no interactions of particles in one cluster with those in the other. It follows from the cluster property of U_r that only connected diagrams contribute to $\mathcal{U}_r$; the unconnected contributions to the G-operators cancel when all terms in $\mathcal{U}_r$ are collected together. (This point is easily illustrated by writing out the contributions to $\mathcal{U}_3$.) Thus, we have $\mathcal{U}_r = [G_r]_c$ where the subscript c means that only connected diagrams are included. Note that a connected diagram for r particles must contain at least $r - 1$ vertical lines.

The operator G_s involves correlations $g_p^{(r)}$ in addition to the cluster operators. Their inclusion in diagrams is cumbersome and will not be attempted.

It is readily verified that $T_+(\alpha,\beta)\, F = 0$ if F is independent of X_α, X_β. In G_s, a cluster operator $\mathcal{U}_r$ operates on the sum of single-particle functions

Σb. Hence, the first interaction on the right, say $T_+(\alpha,\beta)$, results in a function of the coordinates and momenta of particles α, β only. The next interaction to the left must therefore involve either particle α or β to give a nonzero contribution. Continuing, we see that any vertical line must connect at most one new particle beyond those already connected to the right of it.

12.8. GRAD LIMIT

As an exercise in the use of diagrams, we rederive the low-density form for Λ by taking the Grad limit. This is the limit of increasing n and decreasing D, with the combination nD^2 held fixed. It may seem paradoxical that n increases, but the point is that the dimensionless parameter nD^3 becomes vanishingly small; the mean free path is large in comparison to the average distance between particles.

Each T_+ operator contains a factor $\delta(r - D)$ which, after integration over the coordinates, gives a factor D^2. A correlation $g_p^{(r)}$, $p \geq 1$, or $g_0^{(r)} - 1$, restricts a particle's coordinates to a region of the order of D^3. Hence, in the Grad limit interactions are favored over correlations. It is then not difficult to see that the Laplace transform of Equation (11.32) reduces to

$$\mathsf{G}_s b(1) = \frac{1}{s!}\int dX_2 \cdots dX_{s+1}\, \phi(2) \cdots \phi(s+1)\mathcal{U}_{s+1}(1,\ldots,s+1) \times \sum_{\beta=1}^{s+1} b(\beta). \tag{12.43}$$

To obtain the lowest order in D^2 we want the smallest number of interactions, and for connected diagrams this smallest number is s. Thus in the Grad limit we have

$$\mathsf{G}_s b(1) = \frac{1}{s!}\int dX_2 \cdots dX_{s+1}\, \phi(2) \cdots \phi(s+1) \times \mathsf{G}_0 \sum T_+(m_1)\, G_0 \cdots \sum T_+(m_s)\, G_0 \sum_{\beta=1}^{s+1} b(\beta) \tag{12.44}$$

where $m_1,\ldots,m_s$ denote the s pairs which interact to provide connected diagrams.

For $D \to 0$ the difference in position of two colliding particles can be ignored so I_2 reduces to the linearized Boltzmann collision operator I; furthermore, we have

$$\int dX_\alpha dX_\beta \phi(\alpha)\, \phi(\beta)\, T_+(\alpha,\beta) F = 0. \tag{12.45}$$

Consequently, the first interaction on the left in (12.44) must involve particle 1; suppose the other particle to be 2. Then the next interaction must be between 1 or 2 and a third particle, say 3. Proceeding to the right, any vertical line must connect at most one new particle in addition to those already connected to the left of it. Because of the requirement that the diagram be connected with the minimum number, s, of interactions, there can be no repetition of interactions; a contribution such as $G_0 T_+(1,2)\, G_0 T_+(2,3) \times G_0 T_+(1,2)\ldots$ will not occur. In summary, the diagrams contributing in the Grad limit have a typical form with the first interaction on the left being between particles 1 and 2, the next between 1 or 2 and 3, the next between 1, 2 or 3 and 4, and so forth. To account for permutations of the particles $2,\ldots,s+1$ we multiply by $s!$ which cancels the leading factor in Equation (12.44).

With the diagrams classified in this way, the last interaction on the right is the only one involving particle $s+1$ and we may immediately integrate over its coordinates and momenta. The result is

$$\begin{aligned}\int dX_{s+1}\, \phi(s+1) \sum_{r=1}^{s} G_0 T_+(r,s+1)\, G_0[b(r) + b(s+1)] \\ = -\sum_{r=1}^{s} (G_0 I G_0 b)(r).\end{aligned} \tag{12.46}$$

The next interaction can be treated in a similar way, and continuing on down the line we eventually obtain

$$\mathsf{G}_s b = G_0(-IG_0)^s b. \tag{12.47}$$

Summing up, we get

$$\begin{aligned}\mathsf{G} &= G_0 + \sum_{s=1}^{\infty} n^s \mathsf{G}_s = G_0(1 + nIG_0)^{-1} \\ &= (\mathsf{s} - \mathbf{v}\cdot\nabla + nI)^{-1}\end{aligned} \tag{12.48}$$

which is the desired low-density result.

The Grad limit can also be used to derive the full nonlinear Boltzmann equation from the hierarchy. Mathematically this is perhaps the most straightforward derivation.

12.9. REDUCIBLE DIAGRAMS

We next consider an approximation in which correlations and the difference in position of colliding particles are neglected as in the Grad limit, but in which diagrams beyond those of the previous section are retained. We will not consistently retain all contributions of a given order, but otherwise the analysis is much more complicated.

A connected diagram is said to be reducible if there is a subset of particles, not counting particle 1, such that all interactions involving particles in the subset occur to the right of any other interactions, and the subset is connected to the remainder of the diagram by just one particle which is called the node. Although particle 1 may not be in the subset, it may serve as the node. As an example, the diagram corresponding to

$$\cdots G_0 T_+(3,4) G_0 T_+(4,5) G_0 T_+(3,4) G_0,$$

where the unwritten factors on the left do not involve particles 4 or 5, is reducible and particle 3 is the node. A reducible diagram will fall into two pieces if cut at the node. Of course an irreducible diagram is one which is not reducible.

The concept of reducibility gives a useful distinction only for diagrams with three or more particles. However, the single connected two-particle diagram is conventionally classified as irreducible.

If a reducible diagram is cut at the node, each of the resulting subdiagrams can be looked at as a diagram with a smaller number of particles and classified as reducible or irreducible. For this purpose the node is included in the subdiagram and plays the role of particle 1 in the definition of reducibility. We can continue until the diagram is completely broken down into irreducible subdiagrams. Insofar as G_s is concerned, the product of operators acts on Σb and it follows that a nonvanishing contribution is obtained only if the node particle is involved in the first interaction to the left of the subdiagram for which it is the node. Furthermore, because of Equation (12.45), the node must be involved in the left-most interaction of its subdiagram. Thus, adjacent subdiagrams are connected together in the shortest possible manner.

Consider a diagram for $s + 1$ particles which provides a contribution to G_s. Break it down into irreducible diagrams, and suppose there are i of these. Let r_j, $j = 1, \ldots, i$, be the number of particles in the j'th subdiagram not counting its node; we number the subdiagrams from right to left. The r_j satisfy

$$1 \leq r_j \leq s, \quad \sum_{j=1}^{i} r_j = s. \tag{12.49}$$

(For an irreducible diagram, $i = 1$ and $r_1 = s$.) Form the product of operators associated with the first subdiagram, up to but not including the free-particle propagator on its left. The coordinates and momenta of the r_1 particles (excepting the node) can be integrated out, and the result is a function of X_β where β denotes the node. Since this function is linear in b, we write it as $(Y_{r_1} G_0 b)(\beta)$. Any of the $s + 1 - r_1$ particles not in the subset might be the node, so including the contributions of all diagrams which differ only in this node we obtain

$$\sum_{\beta} (Y_{r_1} G_0 b)(\beta). \tag{12.50}$$

Thus, on peeling off the subdiagram we replace the sum over $s + 1$ single-particle functions b by the sum over $s + 1 - r_1$ functions of the form just given, and have a simpler diagram left to deal with.

Proceeding on through the diagram, we eventually obtain a contribution

$$(G_0 Y_{r_i} G_0 \cdots Y_{r_1} G_0 b)(1). \tag{12.51}$$

Each Y_r corresponds to a particular irreducible subdiagram. In general there will be other irreducible subdiagrams for the same particles, which may differ by permutations of the r particles (not counting the node) or in their structure. Let the sum of the Y_r for all such irreducible diagrams be denoted by Z_r; thus $(1/r!) G_0 Z_r G_0$ is the same as G_r except that contributions from reducible diagrams are left out. By summing (12.51) over all such possibilities we obtain

$$(G_0 Z_{r_i} G_0 \cdots Z_{r_1} G_0 b)(1).$$

There will still be other diagrams which differ from the ones included so far by different partitions of the s particles into i subsets. The number of these partitions is $s!\,(\Pi_j r_j!)^{-1}$. The contribution to G_s of all diagrams which decompose into irreducible subdiagrams with r_j, $j = 1, \ldots, i$ particles in each is then

$$G_0 \prod_{j=1}^{i} \frac{1}{r_j!} Z_{r_j} G_0.$$

Finally to obtain the total contribution to G_s we must sum over all possible ways of forming diagrams from irreducible parts. This yields

$$\mathsf{G}_s = G_0 \sum \prod_{j=1}^{i} \frac{1}{r_j!} Z_{r_j} G_0$$

where the sum and product extend over all values of i and r_j which are consistent with (12.49).

The operator G is given by

$$\mathsf{G} = G_0 + \sum_{s=1}^{\infty} n^s G_0 \sum \prod_{j=1}^{i} \frac{1}{r_j!} Z_{r_j} G_0.$$

This is the expansion of

$$\begin{aligned} \mathsf{G} &= G_0 \left(1 - \sum_{s=1}^{\infty} \frac{n^s}{s!} Z_s G_0\right)^{-1} \\ &= \left(G_0^{-1} - \sum_{s=1}^{\infty} \frac{n^s}{s!} Z_s\right)^{-1}. \end{aligned}$$

Comparing with Equations (11.35), (11.36) we see that

$$B_{s+1} = -\frac{1}{s!} Z_s.$$

Thus, in the approximation considered (that is, neglect of correlations and of the difference in position of colliding particles) there is a simple diagrammatic representation for B_{s+1} obtained by summing contributions from all irreducible diagrams with $s + 1$ particles. (Note that diagrams for B_s or Z_s do not contain horizontal line segments on the extreme right or left since the corresponding G_0-operators have been factored out.)

12.10. RING OPERATOR

Our purpose next is to introduce ring diagrams and obtain their contribution to B. However, a certain amount of background is desirable before commencing the calculations.

The processes involved in the triple collision operator B_3 give indications of a breakdown of the separation of time scales for collisions and for hydrodynamics. It is necessary to consider irreducible sequences of three binary collisions, and such sequences may occupy long intervals of time. Consider, for example, the sequence (1,2), (2,3) and (1,2) for hard spheres. (Here, for instance, (1,2) denotes a collision between particles 1 and 2.) It is not difficult to see that particle 3 may strike 2 in such a way that 2 catches back up with 1 only very slowly, so that the time between the two (1,2) collisions is very long. Thus, triple collisions may occupy long time intervals and may spread over large distances. Similar remarks apply to collisions among four and more particles.

It could be that the initial conditions leading to extended collision sequences occupy a negligibly small region of phase space, so detailed calculations are necessary to determine the consequences. We have already seen that the triple collision operator (in three dimensions) presents no divergences. However when four-body collisions are included it is found that, because of extended collision sequences, the long time limit which should determine a collision operator, or the limit $\mathsf{s} \to 0$ of $B(\mathsf{s})$, does not exist in any useful sense. The transport coefficients cannot be evaluated by a series expansion in powers of the density, but instead the formal manipulations based on the density expansion lead to divergent coefficients. The correlation functions which appear in the Green-Kubo formulas, when calculated by including collisions among four and more particles, do not decay with time fast enough for convergence of the time integral.

It is to be emphasized that the divergence is associated only with the long time limit. For finite t there is no indication that the correlation functions are not analytic in the density (in a neighborhood of $n = 0$), and indeed convergence of the density expansion has been proven for some cases.

The way to a proper treatment of the divergent terms can be described as follows. A triple (or higher-order) collision which takes a long time to be completed will, in a real gas, be interrupted by collisions with other particles. Suppose, for example, that particles 1, 2 and 3 have initial conditions which would cause them, in isolation, to undergo a (1,2), (2,3), (1,2) collision sequence. If the distance between any two successive binary collisions is significantly greater than the mean free path (as determined by the binary collision operator), the chances are that the collision sequence will actually not proceed as indicated. The probability that a particle will move a distance L without undergoing a collision is roughly proportional to $\exp(-L/\ell)$ where ℓ is the mean free path, and such a factor should be associated with the motion of a particle between successive binary collisions. Hence, there is a mean free path cutoff, which fixes an effective upper bound on the extent of higher-order collisions. Since the mean free path is inversely proportional to the density, the incorporation of the cutoff into the collision operator requires a sum of appropriate contributions from the cluster expansion, to all orders in the density. The cutoff is a many-body effect which cannot be obtained from any finite order of the density expansion.

The diagrams to be summed up are the ring diagrams. It has been noted that a connected diagram of $s + 1$ particles must have at least s interactions; it is easily seen that (except for the two-particle case) an irreducible diagram must have at least $s + 1$ interactions. Ring diagrams are the simplest irreducible diagrams, which are the ones with just $s + 1$ interactions. The contribution to G of the simplest connected diagrams was seen to correspond to the Grad limit. Since the simplest irreducible diagrams have one more T_+ their inclusion amounts, roughly speaking, to the Grad limit plus certain corrections of order D^2.

The interactions in a ring diagram can be described as follows. The first one on the left involves particle 1 plus another particle which we take to be 2. The next interaction involves either particle 1 or 2 and a third particle, say 3. There will not be any repetition of interactions until the very end since otherwise the limited number, $s + 1$, of interactions would be insufficient to make the diagram irreducible. Hence, each interaction introduces one new particle in addition to those which have already interacted. The final particle, taken to be number $s + 1$, will become involved at the s'th interaction, and the last interaction is then drawn to make the diagram irreducible. The typical form of a ring diagram is like that of the connected diagrams which occur in the Grad limit, plus one additional interaction on the end. An example is $(1,2), (2,3), \ldots, (s, s + 1), (s + 1, 1)$.

The factors associated with the last two interactions on the right are

$$\sum_{\alpha=1}^{s} T_+(\alpha,s+1)\,G_0 \sum_m T_+(m) \sum_{\beta=1}^{s+1} b(\beta) = \sum_{\alpha\neq\lambda} T_+(\alpha,s+1)\,G_0 \\ \times \{T_+(s+1,\lambda)[b(s+1)+b(\lambda)] + T_+(\alpha,\lambda)[b(\alpha)+b(\lambda)]\}. \tag{12.52}$$

(We are considering diagrams for B and so factors G_0 on the extreme ends are not included.) Particle $s+1$ does not appear elsewhere and so can be integrated out. This gives

$$\int dX_{s+1}\,\phi(s+1) \sum_{\alpha=1}^{s} T_+(\alpha,s+1)\,G_0 \sum_m T_+(m) \sum_{\beta=1}^{s+1} b(\beta) = \\ -\sum_{\alpha=2}^{s}\sum_{\lambda=1}^{\alpha-1} [I(\alpha)+I(\lambda)]G_0 T_+(\alpha,\lambda)[b(\alpha)+b(\lambda)].$$

Here $I(\alpha)$ denotes the linearized Boltzmann collision operator on the momentum of the α'th particle. Continuing to the left, we can treat each interaction similarly and eventually we obtain

$$(-1)^{s-1}\int dX_2\,\phi(2)\,T_+(1,2)\,G_0\{[I(1)+I(2)]G_0\}^{s-1} \\ \times T_+(1,2)[b(1)+b(2)]. \tag{12.53}$$

This is to be multiplied by $s!$ to account for permutations of the particles $2,\ldots,s+1$. The ring contribution to B, which we denote by R, is then given by

$$Rb(1) = n\int dX_2\phi(2)\,T_+(1,2)\{G_0^{-1} + n[I(1)+I(2)]\}^{-1} \\ \times T_+(1,2)[b(1)+b(2)]. \tag{12.54}$$

Including the two lowest-order terms, we obtain for B in the ring approximation

$$B_R = -\mathbf{v}\cdot\nabla + nI + R.$$

It is straightforward (in fact slightly simpler) to modify the analysis for the case of self-diffusion. Only the term $b(1)$ is retained, and $I(1)$ is replaced by the operator $I_S(1)$ given in Equation (4.47). Thus, the ring operator for self-diffusion is given by

$$R_S b(1) = n\int dX_2\phi(2)\,T_+(1,2)\{G_0^{-1} + n[I(1)+I_S(2)]\}^{-1}\,T_+(1,2)b(1).$$

The factor between the two T_+ operators in (12.54) describes a diffusive motion of the two particles according to the linearized Boltzmann equation. The two particles collide, diffuse through the surrounding gas, then collide again. The replacement of free-particle motion by propagation through the medium is called renormalization. Of course R would vanish without renormalization since the two particles would not be able to collide twice.

As mentioned, the reason for summing ring diagrams is to eliminate a divergence which occurs in the expansion of transport coefficients in powers of the density. Here we will not give the arguments which demonstrate the existence of the divergence, nor derive the dependence on density which follows from the ring operator. Detailed arguments provide evidence that ring diagrams give the most divergent contributions to the power series so the summation of ring diagrams is the proper approach to handle the divergence. Analysis of ring contributions to the transport coefficients reveals a logarithmic dependence on density such that the viscosity, for example, has the expansion

$$\eta = \eta_0 + \eta_1 n + (\eta_2' + \eta_2'' \ln n)n^2 + \ldots \qquad (12.55)$$

The ring operator itself only includes ring contributions to triple and higher-order collisions and so does not reproduce the value of η_1 exactly, but this coefficient can be determined from the triple collision operator. Evaluation of η_2' requires solution of the four-body problem (beyond the ring approximation) and so is quite difficult.

12.11. LONG-TIME TAILS

One consequence of extended collision sequences is a slow decay of the time dependent correlation functions which appear in the Green-Kubo formulas. When four-body collisions are included, the correlation functions are found to decrease only as $1/t$ for large t so the time integrals do not converge. The ring summation corrects this to a $t^{-3/2}$ decay which is rapid enough for convergence of the integral but is still slower than the exponentially bounded decay obtained from the linearized Boltzmann equation (in the case of hard-core potentials). In this section we will obtain the $t^{-3/2}$ long-time tail from the ring operator.

Consider the autocorrelation for a function A which is a sum of single-particle functions a of momentum. Supposing the Maxwellian average of a to vanish, we write the correlation function as

$$C(t) = n\frac{1}{V}\int dX\,\phi\, a\Lambda(t)a.$$

Its Laplace transform is

$$\mathscr{C} = n\frac{1}{V}\int dX\,\phi\, a\mathrm{G}a$$

where in the ring approximation

$$\mathsf{G} = (\mathsf{s} - \mathbf{v}\cdot\nabla + nI + R)^{-1}.$$

We treat R as a perturbation and expand,

$$\mathsf{G} = \mathsf{G}_B - \mathsf{G}_B R \mathsf{G}_B + \ldots \tag{12.56}$$

Here G_B denotes the Boltzmann approximation (12.48). Since a is independent of position, the term $\mathbf{v}\cdot\nabla$ drops out. We are concerned with the long-time behavior and so consider small s; nothing of interest is lost if at this point we take the limit for $\mathsf{s}\to 0$ of G_B in the second term of Equation (12.56). This yields

$$\begin{aligned}\delta\mathscr{C} &= \mathscr{C} - \mathscr{C}_B \\ &= -\frac{1}{n}\int d^3p\,\phi\mathscr{A}R\mathscr{A}\end{aligned}$$

where $\mathscr{C}_B$ denotes the Boltzmann approximation and $\mathscr{A}$ is determined by

$$I\mathscr{A} = a. \tag{12.57}$$

We next introduce Equation (12.54) for R and write the result as

$$\begin{aligned}\delta\mathscr{C} = \frac{1}{2}\int d^3p_1\,dX_2\,\phi(1)\,\phi(2)\,V^{\dagger}(1,2)& \\ \times\,(G_0^{-1} + n[I(1) + I(2)])^{-1}V(1,2)&\end{aligned} \tag{12.58}$$

where

$$V(1,2) = T_+(1,2)[\mathscr{A}(1) + \mathscr{A}(2)],\; V^{\dagger}(1,2) = T_+^{\dagger}(1,2)[\mathscr{A}(1) + \mathscr{A}(2)].$$

The long-time behavior is dominated by hydrodynamic contributions to the Boltzmann propagation in the ring operator. To separate out the hydrodynamic part, we first Fourier-transform the spatial dependence. Since $\mathscr{A}$ depends only on momentum, V depends on position only through the difference $\mathbf{q}_1 - \mathbf{q}_2$ so we may write

$$V(1,2) = \int d^3k \exp\left[i\mathbf{k}\cdot(\mathbf{q}_1 - \mathbf{q}_2)\right]\alpha(1,2). \tag{12.59}$$

Here α is a function of the momenta; it also depends on $\mathbf{k}$ but this will not be explicitly indicated. When (12.59) is introduced into (12.58), the result can be written

$$\begin{aligned}\delta\mathscr{C} = -\frac{1}{2}\int d^3k\int d^3p_1 d^3p_2\,\phi(1)\,\phi(2)\,\alpha^{\dagger}(1,2)& \\ \times\,[\mathsf{s} + B(1) + B^{\dagger}(2)]^{-1}\alpha(1,2).&\end{aligned} \tag{12.60}$$

Here $\alpha^\dagger$ is defined by

$$\alpha^\dagger(1,2) = (2\pi)^{-3} \int d^3 q_2 \exp\left[i\mathbf{k}\cdot(\mathbf{q}_1 - \mathbf{q}_2)\right] V^\dagger(1,2)$$

and $B, B^\dagger$ are the operators defined in Equations (6.8), (6.10).

The quantity α can be obtained by inverting the transform in (12.59) and using the explicit form (12.12) for T_+. Our concern is with values of k small in comparison to $1/D$ so the k dependence associated with T_+ can be neglected. We then find

$$\alpha(1,2) = (2\pi)^{-3} D^2 \int_- d^2 e\, \mathbf{e}\cdot\mathbf{g}[\mathcal{A}(\mathbf{p}_1) + \mathcal{A}(\mathbf{p}_2) - \mathcal{A}(\mathbf{p}_1') - \mathcal{A}(\mathbf{p}_2')]. \quad (12.61)$$

In the small k limit we have $\alpha^\dagger = \alpha$.

To obtain the hydrodynamic contribution, we project onto the hydrodynamic eigenfunctions obtained in Chapter 6. It is necessary to use a product of eigenfunctions, one each for B and $B^\dagger$, whose hydrodynamic eigenfunctions are g_α and h_α respectively. We recall that $h_\alpha(\mathbf{k}) = g_\alpha(-\mathbf{k}) = \bar{g}_\alpha(\mathbf{k})$, and furthermore $(h_\alpha, g_\beta) = \delta_{\alpha\beta}$. Let α_h denote the hydrodynamic projection of α defined by

$$\alpha_h(1,2) = \sum_{\alpha,\beta=1}^{5} C_{\alpha\beta}\, g_\alpha(1)\, h_\beta(2)$$

where the coefficients $C_{\alpha\beta}$ are given by

$$C_{\alpha\beta} = \int d^3p_1\, d^3p_2\, \phi(1)\,\phi(2)\, g_\alpha(1)\, h_\beta(2)\, \alpha(1,2).$$

When Equation (12.61) is used, the result after some manipulation is found to be

$$C_{\alpha\beta} = -(2\pi)^{-3} \int d^3p\, \phi\, \mathcal{A}\, L(g_\alpha, h_\beta) \quad (12.62)$$

where L is the bilinear operator (4.50). We note that $C_{\alpha\beta} = \bar{C}_{\beta\alpha}$ and $C_{\alpha\beta} = \tau\, C_{\beta\alpha}$ where τ is the parity of $\mathcal{A}$ under time reversal.

Replacing α by α_h in Equation (12.60) we obtain the hydrodynamic contribution to $\delta\mathcal{C}$:

$$\delta\mathcal{C}_h = -4\pi^3 \int d^3k \sum_{\alpha,\beta=1}^{5} C_{\alpha\beta}^2 \left[\mathsf{s} + \zeta_\alpha + \bar{\zeta}_\beta\right]^{-1}$$

where $\zeta_\alpha, \bar{\zeta}_\beta$ are the eigenvalues of $B, B^\dagger$. When the Laplace transform is inverted, the result is

$$\delta C_h(t) = -4\pi^3 \int d^3k \sum_{\alpha,\beta=1}^{5} C_{\alpha\beta}^2 \exp\left[-(\zeta_\alpha + \bar{\zeta}_\beta)t\right]. \quad (12.63)$$

For large t the exponential limits the integration to small k, which confirms that the long-time behavior is dominated by the hydrodynamic contributions. We may replace $C_{\alpha\beta}$ by its value for $\mathbf{k} = 0$. The summational invariants ψ_α can be shown to satisfy $L(\psi_\alpha, \psi_\beta) = -I(\psi_\alpha \psi_\beta)$. Using Equation (12.57) we reduce Equation (12.62) to

$$C^0_{\alpha\beta} = (2\pi)^{-3} \int d^3p\, \phi\, a\, g^0_\alpha g^0_\beta \tag{12.64}$$

where the superscript 0 denotes the value for $\mathbf{k} = 0$.

The eigenfunctions g^0_α depend on the direction of $\mathbf{k}$. To make this explicit, we rewrite the formulas of Chapter 6 as

$$g^0_1 = \sqrt{\frac{2}{5}}\left(\frac{\epsilon}{\kappa T} - \frac{5}{2}\right),$$

$$g^0_2 = \sqrt{\frac{2}{15}}\frac{\epsilon}{\kappa T} - \sqrt{\frac{m}{2\kappa T}}\ \hat{\mathbf{k}}\cdot\mathbf{v},$$

$$g^0_3 = \sqrt{\frac{2}{15}}\frac{\epsilon}{\kappa T} + \sqrt{\frac{m}{2\kappa T}}\ \hat{\mathbf{k}}\cdot\mathbf{v},$$

$$g^0_4 = \sqrt{\frac{m}{\kappa T}}\,\mathbf{i}\cdot\mathbf{v},\quad g^0_5 = \sqrt{\frac{m}{\kappa T}}\ \mathbf{j}\cdot\mathbf{v}.$$

Here $\hat{\mathbf{k}}$ is a unit vector in the direction of $\mathbf{k}$, while $\mathbf{i}$ and $\mathbf{j}$ are unit vectors orthogonal to $\mathbf{k}$ and to each other. The direction of $\mathbf{k}$ enters into (12.63) only through $C_{\alpha\beta}$ since the eigenvalues ζ_α depend only on its magnitude. Hence, we may replace $C^2_{\alpha\beta}$ in Equation (12.63) by its average over the direction of $\mathbf{k}$. This average can be calculated with the aid of the equations

$$\begin{aligned}
\frac{1}{4\pi}\int d^2\hat{k}\, k_i k_j &= \frac{1}{3}\delta_{ij}k^2,\\
\frac{1}{4\pi}\int d^2\hat{k}\, k_i k_j k_k k_l &= \frac{1}{15}k^4(\delta_{ij}\delta_{kl} + \delta_{ik}\delta_{jl} + \delta_{il}\delta_{jk}),\\
\frac{1}{4\pi}\int d^2\hat{k}\, k_i k_j i_k i_l &= \frac{1}{30}k^2(4\delta_{ij}\delta_{kl} - \delta_{ik}\delta_{jl} - \delta_{il}\delta_{jk})
\end{aligned} \tag{12.65}$$

plus the similar equations obtained by permuting $\mathbf{i}, \mathbf{j}, \hat{\mathbf{k}}$.

To proceed further, it is necessary to specify the quantity a whose correlation is being calculated. To illustrate the calculations we take $a = p_x p_y/m$ so the correlation function is the one in the Green-Kubo formula for the shear viscosity. It is straightforward to carry out the integration in Equation

(12.64) and the results are

$$C_{1\alpha}^{0} = 0, \ \alpha = 1,\ldots,5,$$

$$C_{22}^{0} = C_{33}^{0} = -\ C_{23}^{0} = \mu \hat{k}_x \hat{k}_y,$$

$$C_{24}^{0} = -\ C_{34}^{0} = -\ \frac{1}{\sqrt{2}}\mu(\hat{k}_x i_y + \hat{k}_y i_x),$$

$$C_{44}^{0} = 2\mu i_x i_y, \ C_{45}^{0} = \mu(i_x j_y + i_y j_x)$$

where μ is the abbreviation $\mu = \kappa T/(2\pi)^3$. The equations for C_{25}^0, C_{35}^0 and C_{55}^0 are obtained from those for C_{24}^0, C_{34}^0 and C_{44}^0 by replacing **i** by **j**. We now substitute the previous equations into (12.63) and use Equations (12.65) to average over the direction of **k**. Using Equation (6.2) for the eigenvalues ζ_α, we find

$$\delta C_h(t) = -\ \frac{2\kappa^2 T^2}{15\pi^3} \int d^3k \,\{[1 + \cos 2ckt] \exp(-2\Gamma k^2 t) + 6 \cos ckt \exp(-\nu k^2 t) + 7 \exp(-2\nu k^2 t)\}.$$

It is straightforward to carry out the integration over **k**. The dominant contribution for large t is then found to be

$$C(t) \rightarrow -\ \frac{2\kappa^2 T^2}{15(2\pi t)^{3/2}} [\Gamma^{-3/2} + 7\nu^{-3/2}].$$

This exhibits the $t^{-3/2}$ long-time tail.

To obtain the ring contributions to the transport coefficients, it is necessary to evaluate the correlation functions over the full range of values for t. For small t it is not sufficient to consider only the hydrodynamic contribution, but large values of **k** must be included. This is the case in particular for the calculation of the coefficient of $\ln n$ in Equation (12.55).

If the ring operator is used to determine the Burnett-order transport coefficients, it is found that these diverge. Indications of this can be found in the formulas for these coefficients obtained in Chapter 9. Certain terms contain the same correlation functions as in the Green-Kubo formulas, but multiplied by the time. The $t^{-3/2}$ decay then gives a net dependence on t, for large t, as $t^{-1/2}$ which is not integrable. This divergence (and the corresponding divergence of the Navier-Stokes transport coefficients in two dimensions) is due to a fault with the Chapman-Enskog expansion. In the linear case the difficulty can be stated as a lack of analyticity in the wavenumber, while for the nonlinear case there is a lack of analyticity in the amplitude of the disturbance.

It is to be emphasized that the long-time tail is associated with spatially homogeneous correlation functions and is to be distinguished from the similar dependence on time of correlation functions which depend on position. The latter have a hydrodynamic part which displays directly the slow diffusion of heat, vorticity, and so on. The slow decay of the former occurs because of an interference of hydrodynamics with the (otherwise rapid) relaxation in momentum space.

The $t^{-3/2}$ decay was discovered in computer experiments by Alder and Wainwright on the velocity autocorrelation. It could be seen that the slow decay of velocity was associated with a vortex which forms around the particle and then carries the particle along. An analytical description (essentially the one given in Chapter 2 for the case of Brownian motion) was provided on the basis of the Navier-Stokes equations for the motion of the vortex. Subsequently, Dorfman and Cohen extracted the slow decay from the ring operator, showing that it has a common dynamical origin with the singular dependence of transport coefficients on density.

Notes on Chapter 12

12.3 The dynamics of hard spheres is discussed in Ernst, Dorfman, Hoegy, and van Leeuwen [1969b], Konijnendijk and van Leeuwen [1973], and Dormaradzki [1977].

12.4 The binary collision expansion was derived in Siegert and Teramoto [1958]. See also Weinstock [1962], Zwanzig [1963], Haines, Dorfman, and Ernst [1966], Dorfman [1967], and Blum and Lebowitz [1969].

12.6 Calculations of L_{KU} are given in McLennan [1963c], Kawasaki and Oppenheim [1964], Imam-Rahajoe and Curtiss [1967], and Herman and McLennan [1969].

12.7 Diagrams including the effect of correlations are discussed in van Beijeren and Ernst [1979].

12.8 See Grad [1958].

12.10 The divergence in the density expansion of a transport coefficient was discovered by several people at about the same time; reviews are given in Ernst, Haines, and Dorfman [1969a], Brush [1972], and Dorfman [1981]. For the logarithmic term in the density expansion, see the review by Kan, Dorfman, and Sengers [1977], and Kamgar-Parsi and Sengers [1983].

The ring operator was constructed in Kawasaki and Oppenheim [1965]. See also Pomeau [1968, 1971a, 1971b], Dorfman and Cohen [1972, 1975a], and the review by Keyes and Masters [1985].

The analysis is somewhat more complete for a Lorentz gas; see van Leeuwen and Weyland [1967a, b], Hauge and Cohen [1969], Hoegy [1969], and van Beijeren [1982].

12.11 See Alder and Wainwright [1969, 1970], and Alder, Gass, and Wainwright [1971]. The subsequent literature is extensive; in addition to the papers cited in Chapter 2, see for example Wood [1975], Wood and Erpenbeck [1976], Evans [1980], Erpenbeck and Wood [1981], Fox [1983], Machta, Ernst, van Beijeren, and Dorfman [1984], Kirkpatrick [1984], Harrison and Zwanzig [1986], Kirkpatrick and Nieuwoudt [1986b], Machta [1986a], Machta and Reinhold [1986b], and the reviews by Pomeau and Resibois [1975] and Keyes [1977].

The divergence of the Burnett coefficients, because of long-time tails, is discussed in Alle and Pomeau [1972], de Schepper, van Biejeren, and Ernst [1974], and Brey [1983]. This requires a different kind of expansion than that of Chapman-Enskog. An expansion in fractional powers of the wavenumber has been obtained by Ernst and Dorfman [1972, 1975]; see also the review by Pomeau and Resibois [1975]. There are related complications in the behavior of the Knudsen layer near a boundary; see Wolynes [1976].

For uniform shear, a nonanalytic dependence of stress on strain rate was obtained by Kawasaki and Gunton [1973], Yamada and Kawasaki [1975], and Ernst, Chichoki, Dorfman, Sharma, and van Beijeren [1978]. For other discussions dealing with large strain rate, see Zwanzig [1979, 1981], Eu [1983], and Dufty [1983, 1984a, 1984b]. Computer experiments by Erpenbeck [1984] on a hard-sphere gas at high density and high strain rate have shown a transition to a phase with a two-dimensional lattice; the particles then channel through the lattice with a drop in viscosity. See Kirkpatrick and Nieuwoudt [1986a], and Lutsko and Dufty [1986].

There is also interest in effects of large wavenumber; see de Schepper and Cohen [1982], Yip, Alley, and Alder [1982], Leutheusser, Yip, Alder, and Alley [1983], Alley and Alder [1983a], Alley, Alder, and Yip [1983b], and Kirkpatrick [1985].

13

INTERNAL STATES

13.1. HIERARCHY IN QUANTUM MECHANICS

In this chapter the kinetic equations will be discussed for a gas whose particles have internal degrees of freedom. As an example, the particles may be polyatomic molecules but the methods are not limited to this case.

The variety of particles with internal structure is huge, and the description of their collisions is complex. Our discussion will not include detailed calculations, which are generally tailored to specific systems, but instead will be limited to a survey of the kinetic equations and their general features.

The approach will be based on the BBGKY hierarchy, as it is formulated in quantum theory. However, a parallel treatment can be given in terms of correlation functions.

In quantum kinetic theory there are density-dependent corrections similar to those in the classical theory, which are characterized by the dimensionless parameter nD^3 where D is the force range. Quantum mechanics introduces degeneracy as an additional density-dependent effect, which is measured by $n\lambda^3$ where λ is the average de Broglie wavelength. Here we will be concerned only with corrections of the first kind, and so deal with a nondegenerate gas.

In the case of identical particles without internal structure and interacting by pair-wise additive forces, the Hamiltonian has the form (7.1)

$$H = \sum_{\alpha} (p_\alpha^2/2m) + \frac{1}{2}\sum_{\alpha\beta} U_{\alpha\beta}$$

where the coordinates and momenta are now operators satisfying the basic commutation relation $[q,p] = i\hbar$. The simplest internal degree of freedom is spin angular momentum. Spin can be possessed by a particle which is elementary, that is, one which is not, or for the purposes at hand need not be considered to be, a cluster of other particles bound together. A state of the gas is described on a Hilbert space of functions of the external variables (position or momentum) and the spin variables for each particle; the effect of spin-dependent forces can be included in the potential energy. For a gas of composite particles, such as molecules, two levels of description are possible. At the more basic level molecules are viewed as bound states of two or more elementary particles, and they then occur as a consequence of the dynamics (assuming, of course, that the force between elementary particles supports bound states). The Hamiltonian includes the energy of all of the elementary particles, and takes a form like that just given. In the other description, molecules are put in "by hand," with the numbers of molecules of various kinds being assigned definite values. The internal energy is represented by a diagonal matrix H_{int} acting on a vector space called the internal space. The Hamiltonian for a molecule is $H_1 = K + H_{\text{int}}$ where K is the kinetic energy of its center-of-mass motion, and the Hamiltonian for the gas is

$$H = \sum_{\alpha} H_{1\alpha} + \frac{1}{2}\sum_{\alpha\beta} U_{\alpha\beta}.$$

The potential energy now describes an effective force between molecules, and may depend on the quantum numbers for the internal states. A discussion of the reduction of the description in terms of elementary particles to the one on the internal space will be given in the following.

Statistical mechanics is introduced through a statistical operator, or density matrix, F on the Hilbert space for the system. The density matrix is normalized according to

$$\operatorname{Tr} F = 1 \tag{13.1}$$

where Tr denotes the trace, and the average of a physical variable A is

$$\langle A \rangle = \operatorname{Tr} AF. \tag{13.2}$$

If the number of particles N is variable as is the case with the grand canonical ensemble, then the trace may be interpreted to include a sum over its values. The time dependence of F is governed by the von Neumann equation

$$i\hbar\, \partial F/\partial t + [F,H] = 0.$$

The identity of particles requires that the state of the gas be described on a Hilbert space $\mathcal{H}_{\mathscr{E}}$ of functions with the proper exchange symmetry. Thus, if Ψ is a function in $\mathcal{H}_{\mathscr{E}}$ and $\Pi(\alpha,\beta)$ is the operator which exchanges particles α and β, then $\Pi(\alpha,\beta)\Psi = \epsilon\Psi$ where $\epsilon = 1(-1)$ for bosons (fermions). As a practical matter, it is sometimes more convenient to work with the Hilbert space $\mathcal{H}$ of functions for which the requirement of symmetry or antisymmetry is waived. The space $\mathcal{H}_{\mathcal{E}}$ is a subspace of $\mathcal{H}$ which can be obtained by the projection $\mathcal{H}_{\mathcal{E}} = \mathcal{E}\mathcal{H}$ where

$$\mathcal{E} = \frac{1}{N!}\sum_{\Pi} \epsilon^{\Pi}\,\Pi. \tag{13.3}$$

Here Π denotes a permutation of the N particles, the sum extends over all permutations, and $\epsilon^{\Pi} = 1$ for bosons while for fermions $\epsilon^{\Pi} = \pm 1$ depending on whether the permutation is even or odd. A density matrix on $\mathcal{H}_{\mathcal{E}}$ can be extended to $\mathcal{H}$ by $F = \mathcal{E}F\mathcal{E}$ and the extension satisfies the symmetry condition $F = F\mathcal{E} = \mathcal{E}F$. The trace $\mathrm{Tr}_{\mathcal{E}}$ on $\mathcal{H}_{\mathcal{E}}$ is obtained from the trace Tr on $\mathcal{H}$ by projection, $\mathrm{Tr}_{\mathcal{E}}A = \mathrm{Tr}\,A\mathcal{E}$. Because of the symmetry condition on F, Equations (13.1) and (13.2) can be written with either Tr or $\mathrm{Tr}_{\mathcal{E}}$. Let F^B be any density matrix on $\mathcal{H}$, which we assume to commute with $\mathcal{E}$ (as must any physically meaningful quantity) but not necessarily to satisfy the symmetry condition. Then $F = F^B\mathcal{E}$ is a density matrix on $\mathcal{H}_{\mathcal{E}}$. If the effects of exchange are negligible, the system can be described by Boltzmann statistics. In this case only the identity permutation gives a contribution to the trace and the normalization condition for F yields $\mathrm{Tr}\,F^B = N!$ where Tr now refers to $\mathcal{H}$ rather than $\mathcal{H}_{\mathcal{E}}$. However, it is customary to replace F^B by $F^B/N!$ so

$$\mathrm{Tr}\,F^B = 1, \quad \langle A\rangle = \mathrm{Tr}\,A\,F^B.$$

The derivation of the hierarchy follows the same lines as in the classical case. To make the number of particles N explicit we write F_N instead of F, or $F_N(1, \ldots, N)$ when it is desired to include the particle labels. Reduced density matrices F_s are defined by a relation similar to (7.4):

$$F_s(1, \ldots, s) = V^s\,\mathrm{Tr}(s + 1, \ldots, N)F_N(1, \ldots, N)$$

where $\mathrm{Tr}(s + 1, \ldots, N)$ denotes the trace over the variables for particles $s + 1, \ldots, N$. The symmetry condition on F_N requires that $F_s = \mathcal{E}_s F_s = F_s\mathcal{E}_s$ where $\mathcal{E}_s$ is the projection (13.3) for the system of s particles; this condition on F_s is dropped when the system is described by Boltzmann statistics. In the thermodynamic limit, the hierarchy is

$$i\hbar\,\frac{\partial F_s(1, \ldots, s)}{\partial t} + [F_s(1, \ldots, s), H_s] = n\,\mathrm{Tr}(s + 1)[U^{(s)}, F_{s+1}]$$

where $U^{(s)}$ has the same meaning as classically,

$$U^{(s+1)}(1, \ldots, s + 1) = \sum_{\alpha=1}^{s} U_{\alpha,s+1}.$$

The time development operator on wavefunctions for an isolated system of s particles with and without interactions among them will be denoted by V_s, V_0:

$$V_s(t) = \exp(-iH_s t/\hbar), \quad V_0(t) = \exp(-iKt/\hbar)$$

where K is the kinetic energy. The solution to the von Neumann equation for F_N is given by

$$F_N(t) = V_N(t - t_0)F_N(t_0)V_N(t_0 - t).$$

The solution to the hierarchy can be expressed in the form given in Chapter 7 but with the integrals replaced by traces and $S_s(-t)F_s$ replaced by

$$S_s(-t)F_s = V_s(t)F_sV_s(-t).$$

The kinetic equation is obtained by construction of the time-independent functional $F_2(F_1)$ which determines F_2 in terms of F_1. When this is substituted into the first equation of the hierarchy, the result is

$$\frac{\partial F_1}{\partial t} + (1/i\hbar)[F_1, H_1] = nC$$

where C denotes the collision operator

$$C = (1/i\hbar)\,\mathrm{Tr}(2)[U, F_2(F_1)].$$

Generally it is convenient to use momentum representation for the density matrices and other operators. The one-particle density matrix, for example, will be represented by $F_1(\mathbf{p}|\mathbf{p}')$ defined so that if ψ is a function in momentum space, then

$$(F_1\psi)(\mathbf{p}) = \int d^3p'\, F_1(\mathbf{p}|\mathbf{p}')\psi(\mathbf{p}').$$

The trace of an operator A is given by $\mathrm{Tr}A = \int d^3p A(\mathbf{p}|\mathbf{p})$.

A closer correspondence with the classical case can be obtained by using the Wigner function instead of the one-particle density matrix. This is a function of position and momentum, defined by

$$W(\mathbf{x}, \mathbf{p}) = (n/h^3)\int d^3p'\, F_1(\mathbf{p} + \frac{1}{2}\mathbf{p}'|\mathbf{p} - \frac{1}{2}\mathbf{p}')\exp(i\mathbf{p}' \cdot \mathbf{x}/\hbar). \qquad (13.4)$$

If A is a sum of single-particle functions of momentum, $A = \Sigma a(\mathbf{p}_\alpha)$, its average can be calculated from W by

$$\langle A\rangle = \mathrm{Tr}AF_N = n\mathrm{Tr}a(\mathbf{p}_1)F_N = (N/V)\int d^3p\, a(\mathbf{p})F_1(\mathbf{p}|\mathbf{p})$$

$$= \int d^3p\, d^3x\, a(\mathbf{p})W(\mathbf{x}, \mathbf{p}).$$

Thus, W has the same normalization as the function f used in the theory of the Boltzmann equation. The local density n, for example, is given by

$$n(\mathbf{x}) = \int d^3p\, W(\mathbf{x}, \mathbf{p}). \tag{13.5}$$

Internal degrees of freedom can be included by allowing W to be a matrix on the internal space.

If the particles have no internal energy and if no external field is present the one-body Hamiltonian is $H_1 = p^2/2m$. This yields

$$(n/h^3)\int d^3p'\,(1/i\hbar)[F_1, H_1](\mathbf{p} + \frac{1}{2}\mathbf{p}'|\mathbf{p} - \frac{1}{2}\mathbf{p}')\exp(i\mathbf{p}' \cdot \mathbf{x}/\hbar) = (1/m)\mathbf{p} \cdot \nabla W.$$

Hence the first equation of the hierarchy can be written

$$\partial W/\partial t + \mathbf{v} \cdot \nabla W = J$$

where $\mathbf{v} = \mathbf{p}/m$ and J is obtained by transforming C in the manner of (13.4).

13.2. BOLTZMANN EQUATION

To illustrate the derivation of a kinetic equation in quantum theory, we first consider the simplest case of particles without internal structure and interacting by a repulsive force. The Boltzmann equation holds in this case, but the details of the derivation are different in quantum mechanics than in classical mechanics.

Initially we assume the particles to satisfy Boltzmann statistics, so the molecular chaos condition has the same form as classically:

$$F_2(1,2,t_0) = F_1(1,t_0)F_1(2,t_0). \tag{13.6}$$

In constructing the functional relation between F_2 and F_1 to lowest order, we may take the time dependence to be that given by the hierarchy in lowest order

$$F_1(1,t) = S_1(1,t_0 - t)F_1(1,t_0), \quad F_2(1,2,t) = S_2(t_0 - t)F_2(1,2,t_0) \tag{13.7}$$

which yields

$$F_2(1,2,t) = S_2(t_0 - t)\, S_0(t - t_0)\, F_1(1,t_0)\, F_1(2,t_0). \tag{13.8}$$

As in the classical argument, the next step is to move the initial time t_0 back into the remote past. The quantum mechanical wave operators are defined by

$$\Omega_{\pm} = \lim_{t\to\pm\infty} V_2(t)\, V_0(-t). \tag{13.9}$$

At the moment we assume the two particles not to form bound states, and then the adjoint of $\Omega_\pm$ is given by the limit

$$\Omega_\pm^\dagger = \lim_{t \to \pm\infty} V_0(t)\, V_2(-t). \tag{13.10}$$

The wave operator now of interest is Ω_+ which we denote by Ω. Taking the limit $t_0 \to -\infty$ in Equation (13.8) we obtain the time-independent functional

$$F_2(1,2) = \Omega\, F_1(1)\, F_1(2)\, \Omega^\dagger \tag{13.11}$$

which yields the binary collision operator

$$C_2(1) = (1/i\hbar)\, \mathrm{Tr}(2)[U(1,2), \Omega\, F_1(1)\, F_1(2)\, \Omega^\dagger].$$

This result, due to Snider, is the quantum mechanical analogue of Bogoliubov's binary collision operator.

The transition operator, or T-matrix, is defined by $T = U\Omega$. (Specifically, we have to do with $T_+ = U\Omega_+$.) In terms of T the collision operator is

$$C_2(1) = -(in/\hbar)\, \mathrm{Tr}(2)[T\, F_1(1)\, F_1(2)\Omega^\dagger - \Omega F_1(1) F_1(2)\, T^\dagger]. \tag{13.12}$$

As in the classical case the simplest form is obtained when the spatial extent of a collision is neglected. This means that in the collision operator, the variation of W with $\mathbf{x}$ is neglected, which in view of (13.4) is the same as taking F_1 to have the form

$$F_1(\mathbf{p}|\mathbf{p}') = (h^3/n)\, W(\mathbf{p})\, \delta(\mathbf{p} - \mathbf{p}'). \tag{13.13}$$

Let $\mathbf{P}, \mathbf{p}$ denote the total and relative momenta

$$\mathbf{P} = \mathbf{p}_1 + \mathbf{p}_2, \quad \mathbf{p} = \frac{1}{2}(\mathbf{p}_1 - \mathbf{p}_2).$$

The wave operator and T-matrix commute with the total momentum so their matrix elements contain a δ-function for conservation of momentum. To simplify the notation, we use the same symbol for the operator on the two-particle space and on the space for relative motion, and write the matrix elements as

$$\Omega(\mathbf{p}_1, \mathbf{p}_2 | \mathbf{p}_1', \mathbf{p}_2') = \delta(\mathbf{P} - \mathbf{P}')\, \Omega(\mathbf{p}|\mathbf{p}'),$$
$$T(\mathbf{p}_1, \mathbf{p}_2 | \mathbf{p}_1', \mathbf{p}_2') = \delta(\mathbf{P} - \mathbf{P}')\, T(\mathbf{p}|\mathbf{p}').$$

Conservation of momentum and (13.13) yield a factor $\delta(\mathbf{p}_1 - \mathbf{p}_1')$ in the collision operator C. The corresponding collision operator in the kinetic equation for W is

$$J_2(\mathbf{p}_1) = 2\pi i h^2 \int d^3p_2\, d^3\mathsf{p}'\, W(\mathbf{p}_1')\, W(\mathbf{p}_2') \times [\Omega(\mathsf{p}|\mathsf{p}')\, T^\dagger(\mathsf{p}'|\mathsf{p}) - T(\mathsf{p}|\mathsf{p}')\,\Omega^\dagger(\mathsf{p}'|\mathsf{p})]. \tag{13.14}$$

Here the momenta $\mathbf{p}_1', \mathbf{p}_2', \mathsf{p}'$ of the intermediate state satisfy

$$\mathsf{p}' = \frac{1}{2}(\mathbf{p}_1' - \mathbf{p}_2'), \quad \mathbf{p}_1' + \mathbf{p}_2' = \mathbf{P} = \mathbf{p}_1 + \mathbf{p}_2.$$

A more compact way of writing J_2 is

$$J_2(1) = 2\pi i h^2 \mathrm{tr}(2)\, [\Omega\, W(1)\, W(2)\, T^\dagger - T W(1)\, W(2)\, \Omega^\dagger] \tag{13.15}$$

where tr means to integrate over the diagonal matrix elements but leave out the overall momentum-conserving δ-function.

The wave operator can be written in terms of the T-matrix as

$$\Omega(\mathsf{p}|\mathsf{p}') = \delta(\mathsf{p} - \mathsf{p}') - 2\pi i \delta_+(E - E')T(\mathsf{p}|\mathsf{p}') \tag{13.16}$$

where E, E' denote the energies of relative motion in the two states

$$E = \mathsf{p}^2/m, \quad E' = \mathsf{p}'^2/m. \tag{13.17}$$

In addition we use the notation

$$\delta_\pm(x) = \pm(1/2\pi i)[x \mp i0]^{-1}$$

for which we have $\delta_+(x) + \delta_-(x) = \delta(x)$. When (13.16) is used in (13.14), the result is

$$J_2(\mathbf{p}_1) = -2\pi i h^2 \int d^3p_2 d^3\mathsf{p}'\, W(\mathbf{p}_1')\, W(\mathbf{p}_2') \times \{\delta(\mathsf{p} - \mathsf{p}')[T(\mathsf{p}|\mathsf{p}) - \overline{T}(\mathsf{p}|\mathsf{p})] + 2\pi i \delta(E - E')|T(\mathsf{p}|\mathsf{p}')|^2\} \tag{13.18}$$

where the bar denotes the complex conjugate. In the first term we use the optical theorem,

$$T(\mathsf{p}|\mathsf{p}) - \overline{T}(\mathsf{p}|\mathsf{p}) = -2\pi i \int d^3\mathsf{p}'\, \delta(E - E')|T(\mathsf{p}|\mathsf{p}')|^2.$$

The T-matrix is evaluated on the energy shell where it may be expressed in terms of the scattering amplitude, $T(\mathsf{p}|\mathsf{p}') = -(1/\pi m h) f(\mathsf{p}' \to \mathsf{p})$. The scattering cross section is given by $\sigma = |f|^2$ and the result for J_2 is

$$J_2(\mathbf{p}_1) = (4/m^2) \int d^3p_2\, d^3\mathsf{p}\, \delta(E - E')\sigma(\mathsf{p}' \to \mathsf{p}) \times [W(\mathbf{p}_1')W(\mathbf{p}_2') - W(\mathbf{p}_1)W(\mathbf{p}_2)]. \tag{13.19}$$

Changing momentum to velocity, we find this to be the same as the Boltzmann collision operator in the form (3.15) but with the cross section being calculated by quantum mechanics.

Exchange symmetry requires a modification of the molecular chaos assumption to satisfy $F_2 = F_2\mathscr{E}_2$. In place of (13.6) we write

$$F_2(1,2,t_0) = F_1(1,t_0)F_1(2,t_0)[1 \pm \Pi(1,2)].$$

where the + or − sign corresponds to particles which are bosons or fermions, and the operator $\Pi(1,2)$ exchanges particles 1 and 2. Written out explicitly in terms of matrix elements, this is

$$F_2(\mathbf{p}_1, \mathbf{p}_2|\mathbf{p}_1', \mathbf{p}_2') = F_1(\mathbf{p}_1|\mathbf{p}_1')F_1(\mathbf{p}_2|\mathbf{p}_2') \pm F_1(\mathbf{p}_1|\mathbf{p}_2')F_1(\mathbf{p}_2|\mathbf{p}_1') \qquad (13.20)$$

where the time argument has been dropped for brevity. It is straightforward to carry the preceding derivation through with the exchange term included. Interchanging the momenta $\mathbf{p}_1$ and $\mathbf{p}_2$ is equivalent to changing the sign of $\mathbf{p}$, so Equation (13.18) becomes

$$J_2(\mathbf{p}_1) = -2\pi i h^2 \int d^3p_2\, d^3\mathrm{p}'\, W(\mathbf{p}_1')W(\mathbf{p}_2') \\ \times \{\delta(\mathbf{p} - \mathbf{p}')[T_{\mathscr{E}}(\mathbf{p}|\mathbf{p}) - \overline{T}_{\mathscr{E}}(\mathbf{p}|\mathbf{p})] + 2\pi i \delta(E - E')T(\mathbf{p}|\mathbf{p}')\overline{T}_{\mathscr{E}}(\mathbf{p}|\mathbf{p}')\}$$

where $T_{\mathscr{E}}$ is the exchange modified T-matrix,

$$T_{\mathscr{E}}(\mathbf{p}|\mathbf{p}') = T(\mathbf{p}|\mathbf{p}') \pm T(-\mathbf{p}|\mathbf{p}').$$

Interchanging the two particles and averaging the result with the preceding expression, we get a form with the exchange terms included in a symmetrical manner.

$$J_2(\mathbf{p}_1) = -2\pi i h^2 \int d^3p_2\, d^3\mathrm{p}'\, W(\mathbf{p}_1')\mathrm{W}(\mathbf{p}_2') \\ \times \{\delta(\mathbf{p} - \mathbf{p}')[T_{\mathscr{E}}(\mathbf{p}|\mathbf{p}) - \overline{T}_{\mathscr{E}}(\mathbf{p}|\mathbf{p})] + \pi i \delta(E - E')|T_{\mathscr{E}}(\mathbf{p}|\mathbf{p}')|^2\}$$

The optical theorem for $T_{\mathscr{E}}$ is

$$T_{\mathscr{E}}(\mathbf{p}|\mathbf{p}) - \overline{T}_{\mathscr{E}}(\mathbf{p}|\mathbf{p}) = -\pi i \int d^3\mathrm{p}'\, \delta(E - E')|T_{\mathscr{E}}(\mathbf{p}|\mathbf{p}')|^2. \qquad (13.21)$$

In terms of the scattering angle θ, the relation between $T_{\mathscr{E}}$ and the scattering amplitude is $T_{\mathscr{E}}(\mathbf{p}|\mathbf{p}') = (-1/2\pi m h)[f(\theta) \pm f(\pi - \theta)]$. The collision operator takes the form (13.19) with σ replaced by $\frac{1}{2}|f(\theta) \pm f(\pi - \theta)|^2$. The cross section, as it is usually defined, is twice as much since the incident and target particles contribute equally to the counting rate.

Of course this description is incomplete for fermions which always have a nonvanishing spin angular momentum.

13.3. WALDMANN-SNIDER EQUATION

Suppose a particle of the gas to have internal degrees of freedom, which at this point we describe on the internal space. The density matrix F_1 and Wigner function are matrices on the internal space; F_2 is a matrix on the product of two internal spaces.

If exchange is neglected, the molecular chaos condition is formally the same as (13.6). The calculation of the collision operator follows the same lines as before and C has the form (13.12) but with the trace including a sum over the diagonal elements in internal space. Again neglecting the spatial extent of a collision, we obtain

$$J = -2\pi i h^2 \int d^3p_2\, d^3\mathbf{p}'\, \mathrm{Sp}(2)\, \{T(\mathbf{p}|\mathbf{p}')\, W_1(\mathbf{p}_1')\, W_2(\mathbf{p}_2')\Omega^\dagger(\mathbf{p}'|\mathbf{p}) - \Omega(\mathbf{p}|\mathbf{p}')\, W_1(\mathbf{p}_1')\, W_2(\mathbf{p}_2')\, T^\dagger(\mathbf{p}'|\mathbf{p})\} \tag{13.22}$$

where for brevity the subscript 2 on J has been dropped. Here Sp denotes the trace on the internal space and the subscripts 1, 2 on the W's label the internal spaces for particles 1 and 2 respectively. When the commutator $[F_1, H_1]$ is transformed as in Equation (13.4), there is now an additional contribution from the internal Hamiltonian so the kinetic equation for W is

$$\partial W/\partial t + \mathbf{v}\cdot\nabla W + (1/i\hbar)[W, H_{\mathrm{int}}] = J.$$

The wave and transition operators are matrices on the product of the two internal spaces. Their matrix elements will be written $\Omega_{im,jn}$, $T_{im,jn}$ where i,j refer to matrix elements on one internal space and m,n the other. The wave operator can be written in terms of the transition operator by an equation similar to (13.16):

$$\Omega_{im,jn}(\mathbf{p}|\mathbf{p}') = \delta(\mathbf{p}-\mathbf{p}')\,\delta_{ij}\delta_{mn} - 2\pi i\,\delta_+(E_{im} - E_{jn}')\, T_{im,jn}(\mathbf{p}|\mathbf{p}'). \tag{13.23}$$

Here E_{im} is defined by

$$E_{im} = E + \epsilon_i + \epsilon_m$$

where ϵ_i is the energy of the i'th internal state and E is again the energy of the center-of-mass motion (13.17). Equation (13.23) is to be substituted into (13.22). We write the result as the sum of two terms which correspond to the gain and loss terms in the Boltzmann collision operator,

$$J = G + L. \tag{13.24}$$

The loss term, obtained from the first term in (13.23), is given by

$$L = -2\pi i h^2 \int d^3p_2\, \mathrm{Sp}\,(2) \times [T(\mathbf{p}|\mathbf{p})\, W_1(\mathbf{p}_1)\, W_2(\mathbf{p}_2) - W_1(\mathbf{p}_1)\, W_2(\mathbf{p}_2)\, T^\dagger(\mathbf{p}|\mathbf{p})]. \tag{13.25}$$

The gain term with all subscripts written out is

$$G_{ij} = (2\pi h)^2 \int d^3p_2\, d^3\mathbf{p}' \sum_{mnpqr} [\delta_+(E_{im} - E_{pq}') + \delta_-(E_{jm} - E_{nr}')] \times T_{im,pq}(\mathbf{p}|\mathbf{p}')\, W_{pn}(\mathbf{p}_1')\, W_{qr}(\mathbf{p}_2')\, \overline{T}_{jm,nr}(\mathbf{p}|\mathbf{p}'). \tag{13.26}$$

In general the δ_+ and δ_- functions in G_{ij} do not combine to give an overall energy-conserving delta function. Furthermore, T need not commute with the W's so the loss term does not contain the difference $T - T^\dagger$ needed to apply the optical theorem. However, the collision operator can be simplified by imposing special conditions on the Wigner function. We assume first that W is diagonal in the internal quantum numbers, $W_{mn} = W_m\delta_{mn}$, and calculate the diagonal elements of the collision operator. For the loss term we get

$$L_{ii} = -2\pi i h^2 \int d^3p_2 \sum_j [T_{ij,ij}(\mathbf{p}|\mathbf{p}) - \overline{T}_{ij,ij}(\mathbf{p}|\mathbf{p})]\, W_i(\mathbf{p}_1)\, W_j(\mathbf{p}_2)$$

which can be handled as before. In the gain term the functions δ_+ and δ_- now have the same arguments and so combine to a single δ-function. The result is

$$G_{ii} = (2\pi h)^2 \int d^3p_2\, d^3\mathrm{p}' \sum_{jpq} \delta(E_{ij} - E'_{pq})|T_{ij,pq}(\mathbf{p}|\mathbf{p}')|^2\, W_p(\mathbf{p}_1')\, W_q(\mathbf{p}_2').$$

The T-matrix can be put in terms of the scattering amplitude, and the square of its magnitude then in terms of the cross section. Collecting terms together, we obtain for the collision operator

$$J_{ii} = (4/m^2) \int d^3p_2\, d^3\mathrm{p}' \sum_{jpq} \sigma(\mathbf{p}', pq \to \mathbf{p}, ij)\, \delta(E_{ij} - E'_{pq}) \times [W_p(\mathbf{p}_1')\, W_q(\mathbf{p}_2') - W_i(\mathbf{p}_1) W_j(\mathbf{p}_2)].$$

The initial and final states need not have the same internal energies, so the possibility of inelastic collisions is included. This description of the effects of inelastic collisions is due to Wang Chang, Uhlenbeck and de Boer.

The assumption that W is diagonal in the internal quantum numbers does not imply the same property for the collision operator; collisions induce off-diagonal elements. The assumption provides a good approximation if the internal energy levels are widely spaced and nondegenerate. However, rotational levels are degenerate in the magnetic quantum number, and for these Waldmann and Snider made the weaker assumption that the Wigner function is diagonal in the internal energy: $W_{ij} = 0$ if $\epsilon_i \neq \epsilon_j$. Restricting ourselves to matrix elements of the collision operator which are also diagonal in the internal energy, we obtain for the gain term

$$G_{ij} = (2\pi h)^2 \int d^3p_2\, d^3\mathrm{p}' \sum_{mnpqr} \delta(E_{im} - E'_{pq}) \times T_{im,pq}(\mathbf{p}|\mathbf{p}')\, W_{pn}(\mathbf{p}_1')\, W_{qr}(\mathbf{p}_2') \overline{T}_{jm,nr}(\mathbf{p}'|\mathbf{p}), \quad \epsilon_i = \epsilon_j.$$

Suppressing the internal quantum numbers, we write the collision operator as

$$J = (2\pi h)^2 \int d^3p_2 \,\mathrm{Sp}(2) \Big\{ (1/2\pi i)[T(\mathbf{p}|\mathbf{p})\, W_1(\mathbf{p}_1)\, W_2(\mathbf{p}_2) - W_1(\mathbf{p}_1)\, W_2(\mathbf{p}_2)\, T^{\dagger}(\mathbf{p}|\mathbf{p})] + \int d^3\mathbf{p}' \; T(\mathbf{p}|\mathbf{p}')\; \delta(E - E')\; W_1(\mathbf{p}_1')\; W_2(\mathbf{p}_2')\; T^{\dagger}(\mathbf{p}'|\mathbf{p}) \Big\}. \tag{13.27}$$

The kinetic equation with collision operator given by Equations (13.24) through (13.26) can be viewed as the Waldmann-Snider equation in its most general form, but as the term is normally used the collision operator has the simpler form just given.

The assumption that W is diagonal either in the internal energy or in the internal quantum numbers implies that the commutator $[W, H_{\text{int}}]$ vanishes. However, the interaction of a molecular electric or magnetic moment with an external field can be described by a Hamiltonian H_{ext} which is a matrix on the internal space and the Waldmann-Snider equation in this case is

$$\partial W/\partial t + \mathbf{v} \cdot \nabla W + (1/i\hbar)[W, H_{\text{ext}}] = J.$$

The T-matrix can be expressed in terms of scattering amplitudes for the various elastic and inelastic collisions which occur. In general the collision operator (13.27) does not reduce to a form which is determined by cross sections but instead depends on the relative phases of the scattering amplitudes. In scattering experiments, the phase relations between different scattering amplitudes are responsible for a variety of polarization and correlation effects. To obtain a description in terms of cross sections, it is necessary to eliminate the coherence between amplitudes by the introduction of some degree of phase randomization. For example, the incident beam may be assumed to be "unpolarized" and a sum taken over all final states. At the other extreme, one may consider a "complete" scattering experiment in which the incident beam is prepared in a definite internal state and measurements are made on the scattered particles to select a definite final state. Either approach introduces elements which are external to the dynamical equations of motion. In a gas, as described by the Waldmann-Snider equation, there is no automatic selection of states which would reduce the description to one in terms of cross sections so the collision-induced coherence remains in the collision operator. (The Boltzmann collision operator is determined by the cross section because at low density the large distance between particles selects a definite momentum between collisions. This does not occur at higher density; the triple collision operator is not determined by the three-body cross section.)

In the description of radiative transitions, it is necessary to retain those elements of the density matrix which connect states of different energies. The collision operator must then be used in the general form given by Equations (13.24) through (13.26). The resulting kinetic equation provides a basis for the description of collisional effects on spectra and the interaction with radiation.

Exchange symmetry can be included in the Waldmann-Snider equation in a way similar to that used in Section 2. The molecular chaos assumption is formally the same as (13.20) but the operator $\Pi(1,2)$ must exchange the internal states as well as the momenta (or positions) of the two particles. The case of fermions with spin will be treated in detail later.

13.4. TRANSPORT COEFFICIENTS

Internal energy levels lead to a nonvanishing bulk viscosity and also have an effect on the thermal conductivity and viscosity. These effects appear in the normal solution to the equation of Wang Chang, Uhlenbeck and de Boer, as was shown by these authors. We will briefly consider the normal solution to the Waldmann-Snider equation; it turns out that the results are the same as obtained from the equation of Wang Chang, Uhlenbeck, and de Boer.

The derivation of the conservation laws follows the same general lines as for the Boltzmann equation, but in calculating the average of a density or flux the trace over the internal space must be included. Thus the momentum density, for example, is given by $\mathbf{g} = \int d^3p\,\mathbf{p}\,\mathrm{Sp}\,W$. An additional difference is that the summational invariant for energy is now $\frac{1}{2}mv^2 + H_{\mathrm{int}}$ so the energy density and flux are given by

$$\epsilon = \int d^3p\ \mathrm{Sp}\left(\frac{p^2}{2m} + H_{\mathrm{int}}\right) W, \quad \mathbf{s} = \int d^3p\ \mathrm{Sp}\left(\frac{p^2}{2m} + H_{\mathrm{int}}\right)\frac{\mathbf{p}}{m}\, W.$$

The fluxes obtained this way are purely convective. Fluxes across the molecule itself are also possible, but to include them in the theory it is necessary to describe the molecule as a bound state of elementary particles.

The Wigner function in equilibrium is

$$W_0(p) = n\phi(p)\, Z^{-1} \exp\left(-H_{\mathrm{int}}/\kappa T\right)$$

where ϕ is the Maxwellian momentum distribution and Z is the partition function

$$Z = \mathrm{Sp}\,\exp\left(-H_{\mathrm{int}}/\kappa T\right).$$

The equilibrium Wigner function for a gas in motion is obtained by including the flow velocity $\mathbf{u}$ in ϕ in the same way as before, and local equilibrium is described by allowing the thermodynamic variables to depend on position and time. The expansion about local equilibrium is

$$W = W_L[1 + \Phi^{(1)} + \ldots].$$

By following the Chapman-Enskog procedure, we obtain an integral equation for $\Phi^{(1)}$ which formally has the same appearance as (4.12),

$$I\Phi^{(1)} = -\frac{1}{n\kappa T^2}\mathbf{S}\cdot\nabla T - \frac{1}{n\kappa T}T_{ij}\frac{\partial u_i}{\partial x_j} \tag{13.28}$$

where I is obtained by linearizing the operator (13.27) about equilibrium. In this case $\mathbf{S}$ and $\mathbf{T}$ work out to

$$\mathbf{S} = \left\{\frac{1}{2}mv^2 - \frac{5}{2}\kappa T + [H_{\text{int}} - \langle H_{\text{int}}\rangle]\right\}\mathbf{v},$$

$$T_{ij} = m\left(v_i v_j - \frac{1}{3}\delta_{ij}v^2\right)$$

$$- \delta_{ij}\left\{\left(\gamma - \frac{5}{3}\right)\left(\frac{1}{2}mv^2 - \frac{3}{2}\kappa T\right) + (\gamma - 1)(H_{\text{int}} - \langle H_{\text{int}}\rangle)\right\}.$$

The brackets $\langle\rangle$ denote the equilibrium average and γ is the ratio of specific heats. The inhomogeneous term in (13.28) is diagonal in the internal quantum numbers, and as a consequence the solution is determined by the cross sections for inelastic collisions. Thus, the transport coefficients are the same whether obtained from the Waldmann-Snider equation or the equation of Wang Chang, Uhlenbeck and de Boer. However, this is not the case if the effects of an external field are considered, when the results depend on the phase relations between scattering amplitudes.

The solution to Equation (13.28) can be written

$$\Phi^{(1)} = -\frac{1}{n\kappa T^2}\mathcal{S}_i\frac{\partial T}{\partial x_i} - \frac{1}{n\kappa T}\mathcal{T}_{ij}\frac{\partial u_i}{\partial x_j}$$

where $\mathcal{S}$ and $\mathcal{T}$ are solutions to

$$I\mathcal{S}_i = S_i, \quad I\mathcal{T}_{ij} = T_{ij}.$$

The thermal conductivity and coefficients of viscosity are then given by

$$\lambda = \frac{1}{\kappa T^2}(S_x, \mathcal{S}_x), \quad \eta = \frac{1}{\kappa T}(T_{xy}, \mathcal{T}_{xy}), \quad \eta' = \frac{1}{9\kappa T}(T_{ii}, \mathcal{T}_{jj}).$$

Here the scalar product is defined by

$$(A, B) = n^{-1}\int d^3p\, \mathrm{Sp}\, W_0 \overline{A} B.$$

The formal similarity with the results obtained from the Boltzmann equation is evident. Detailed calculations of the transport coefficients, which in this case are much more complex, are available in the literature.

13.5. PARTICLES WITH SPIN

Spin angular momentum is responsible for a variety of interesting effects on the transport properties of a gas. Polarization of the spins, or magnetization, can affect the transport of energy and momentum, and there is a dependence of the thermal conductivity and viscosity on an external magnetic field. Spin diffusion itself is of interest, particularly for magnetic resonance experiments. It was noted by Lhuillier and Laloe that the role of spin can be significant even in the absence of spin-dependent interactions because of the effect of exchange symmetry during a collision.

We consider in particular fermions with spin quantum number $\frac{1}{2}$, which we assume to be the only internal degree of freedom. The spin may be electronic in an atom with one unpaired electron and no nuclear spin, or it may be nuclear spin in an atom with no unpaired electrons. The density matrix and Wigner function are 2×2 matrices in spin space while the wave and transition operators are 4×4 matrices on the space for two spins. Since there is no internal energy, the diagonality assumption used in the derivation of the Waldmann-Snider equation is rigorously satisfied.

For fermions the molecular chaos condition is

$$F_2(1,2,t_0) = F_1(1,t_0)\, F_1(2,t_0)[1 - \Pi(1,2)].$$

The space and spin dependence can be separated in $\Pi(1,2)$ by writing

$$\Pi(1,2) = \rho\tau.$$

Here ρ exchanges the two momenta and so reverses the sign of the relative momentum $\mathbf{p}$. In addition τ is the spin-exchange operator, which is given in terms of the Pauli matrices $\vec{\sigma}_1, \vec{\sigma}_2$ for the two particles by

$$\tau = \frac{1}{2}(1 + \vec{\sigma}_1 \cdot \vec{\sigma}_2).$$

We also have $\tau^2 = 1$, $\tau\vec{\sigma}_1\tau = \vec{\sigma}_2$.

The collision operator is given by

$$J = -2\pi i h^2 \int d^3p_2 \mathrm{Sp}(2)\Big\{[T_{\mathscr{E}}(\mathbf{p}|\mathbf{p})\, W_1(\mathbf{p}_1)\, W_2(\mathbf{p}_2) - W_1(\mathbf{p}_1)\, W_2(\mathbf{p}_2)\, T^{\dagger}_{\mathscr{E}}(\mathbf{p}|\mathbf{p})] + \pi i \int d^3\mathrm{p}'\, T_{\mathscr{E}}(\mathbf{p}|\mathbf{p}')\, \delta(E - E')\, W_1(\mathbf{p}_1')\, W_2(\mathbf{p}_2')\, T^{\dagger}_{\mathscr{E}}(\mathbf{p}'|\mathbf{p})\Big\}$$

where

$$T_{\mathscr{E}}(\mathbf{p}|\mathbf{p}') = T(\mathbf{p}|\mathbf{p}') - \tau T(-\mathbf{p}|\mathbf{p}').$$

The natural expression of $T_{\mathscr{E}}$ is in terms of T-matrices for singlet and triplet states:

$$T_{\mathscr{E}} = P_s T_s + P_t T_t.$$

Here P_s, P_t are the projections on singlet and triplet states:

$$P_s = \frac{1}{4}(1 - \vec{\sigma}_1 \cdot \vec{\sigma}_2), \quad P_t = \frac{1}{4}(3 + \vec{\sigma}_1 \cdot \vec{\sigma}_2)$$

and the singlet and triplet T-matrices are

$$T_s(\mathbf{p}|\mathbf{p}') = T(\mathbf{p}|\mathbf{p}') + T(-\mathbf{p}|\mathbf{p}'), \quad T_t(\mathbf{p}|\mathbf{p}') = T(\mathbf{p}|\mathbf{p}') - T(-\mathbf{p}|\mathbf{p}').$$

For simplicity we assume the interaction to be independent of spin so T_s and T_t are proportional to the unit matrix in spin space.

To express the collision operator in a convenient form, let W and J be expanded in Pauli matrices as

$$W = w_0 + \mathbf{w} \cdot \vec{\sigma}, \quad J = J_0 + \mathbf{J} \cdot \vec{\sigma}.$$

The kinetic equation decomposes into two parts

$$\frac{\partial w_0}{\partial t} + \mathbf{v} \cdot \nabla w_0 = J_0, \quad \frac{\partial \mathbf{w}}{\partial t} + \mathbf{v} \cdot \nabla \mathbf{w} = \mathbf{J}$$

which are coupled together through the collision operators. For J_0 we find

$$J_0 = \frac{1}{2}\mathrm{Sp}\, J = -\pi i h^2 \int d^3p_2\, \mathrm{Sp}(1,2) \Big\{ [T_{\mathscr{E}}(\mathbf{p}|\mathbf{p}) - T^{\dagger}_{\mathscr{E}}(\mathbf{p}|\mathbf{p})]\, W_1(\mathbf{p}_1)\, W_2(\mathbf{p}_2)$$

$$+ \pi i \int d^3\mathbf{p}'\, \delta(E - E')\, T^{\dagger}_{\mathscr{E}}(\mathbf{p}'|\mathbf{p})\, T_{\mathscr{E}}(\mathbf{p}|\mathbf{p}')\, W_1(\mathbf{p}_1')\, W_2(\mathbf{p}_2') \Big\}.$$

The optical theorem for $T_{\mathscr{E}}$ is given by (13.21) but with a trace over spins, which yields

$$J_0 = \pi^2 h^2 \int d^3p_2\, \mathrm{Sp}(1,2) \int d^3\mathbf{p}'\, \delta(E - E')\, T^{\dagger}_{\mathscr{E}}(\mathbf{p}'|\mathbf{p})\, T_{\mathscr{E}}(\mathbf{p}|\mathbf{p}')$$

$$\times [W_1(\mathbf{p}_1')\, W_2(\mathbf{p}_2') - W_1(\mathbf{p}_1)\, W_2(\mathbf{p}_2)].$$

This has the Boltzmann gain-loss form, but the form is deceptive since W is still a matrix in spin space. The trace over spins can be evaluated with the aid of the formulas

$$\mathrm{Sp}(1,2)\, P_s W_1(\mathbf{p}_2)\, W_2(\mathbf{p}_2) = w_0(\mathbf{p}_1) w_0(\mathbf{p}_2) - \mathbf{w}(\mathbf{p}_1) \cdot \mathbf{w}(\mathbf{p}_2),$$

$$\mathrm{Sp}(1,2)\, P_t W_1(\mathbf{p}_1)\, W_2(\mathbf{p}_2) = 3 w_0(\mathbf{p}_1) w_0(\mathbf{p}_2) + \mathbf{w}(\mathbf{p}_1) \cdot \mathbf{w}(\mathbf{p}_2).$$

The result is

$$J_0 = (4/m^2) \int d^3p_2\, d^3\mathbf{p}'\, \delta(E - E')$$

$$\times \Big\{ \frac{1}{4}[\sigma_s(\mathbf{p}' \to \mathbf{p}) + 3\sigma_t(\mathbf{p}' \to \mathbf{p})]\, [w_0(\mathbf{p}_1')\, w_0(\mathbf{p}_2') - w_0(\mathbf{p}_1)\, w_0(\mathbf{p}_2)] \qquad (13.29)$$

$$+ \frac{1}{4}[\sigma_t(\mathbf{p}' \to \mathbf{p}) - \sigma_s(\mathbf{p}' \to \mathbf{p})]\, [\mathbf{w}(\mathbf{p}_1') \cdot \mathbf{w}(\mathbf{p}_2') - \mathbf{w}(\mathbf{p}_1) \cdot \mathbf{w}(\mathbf{p}_2)] \Big\}$$

where σ_s and σ_t are the singlet and triplet cross sections;

$$\sigma_s(\mathbf{p}' \to \mathbf{p}) = (\pi m h)^2 |T_s(\mathbf{p}|\mathbf{p}')|^2, \quad \sigma_t(\mathbf{p}' \to \mathbf{p}) = (\pi m h)^2 |T_t(\mathbf{p}|\mathbf{p}')|^2.$$

The corresponding formula for $\mathbf{J}$ is

$$\begin{aligned}\mathbf{J} = & -\pi i h^2 \int d^3 p_2 \mathrm{Sp}(1,2)\vec{\sigma}_1 \Big\{ T_{\mathscr{E}}(\mathbf{p}|\mathbf{p})\, W_1(\mathbf{p}_1)\, W_2(\mathbf{p}_2) \\ & - W_1(\mathbf{p}_1)\, W_2(\mathbf{p}_2)\, T^{\dagger}_{\mathscr{E}}(\mathbf{p}|\mathbf{p}) \\ & + \pi i \int d^3 \mathrm{p}'\, \delta(E - E')\, T_{\mathscr{E}}(\mathbf{p}|\mathbf{p}')\, W_1(\mathbf{p}_1')\, W_2(\mathbf{p}_2')\, T^{\dagger}_{\mathscr{E}} \mathbf{p}'|\mathbf{p}) \Big\}.\end{aligned}$$

To work out the trace over spins we need the additional formulas

$$\begin{aligned}&\mathrm{Sp}(1,2)\vec{\sigma}_1 P_s W_1(\mathbf{p}_1) W_2(\mathbf{p}_2) P_s = 0, \\ &\mathrm{Sp}(1,2)\vec{\sigma}_1 P_t W_1(\mathbf{p}_1) W_2(\mathbf{p}_2) P_t = 2[w_0(\mathbf{p}_1)\mathbf{w}(\mathbf{p}_2) + w_0(\mathbf{p}_2)\mathbf{w}(\mathbf{p}_1)], \\ &\mathrm{Sp}(1,2)\vec{\sigma}_1 P_s W_1(\mathbf{p}_1) W_2(\mathbf{p}_2) P_t = w_0(\mathbf{p}_2)\mathbf{w}(\mathbf{p}_1) - w_0(\mathbf{p}_1)\mathbf{w}(\mathbf{p}_2) + i\mathbf{w}(\mathbf{p}_1) \times \mathbf{w}(\mathbf{p}_2), \\ &\mathrm{Sp}(1,2)\vec{\sigma}_1 P_t W_1(\mathbf{p}_1) W_2(\mathbf{p}_2) P_s = w_0(\mathbf{p}_2)\mathbf{w}(\mathbf{p}_1) - w_0(\mathbf{p}_1)\mathbf{w}(\mathbf{p}_2) - i\mathbf{w}(\mathbf{p}_1) \times \mathbf{w}(\mathbf{p}_2).\end{aligned}$$

These yield

$$\begin{aligned}\mathbf{J} = & (2/m^2) \int d^3 p_2\, d^3 \mathrm{p}'\, \delta(E - E') \\ & \times \{\sigma_t(\mathbf{p}' \to \mathbf{p})[w_0(\mathbf{p}_1')\mathbf{w}(\mathbf{p}_2') + w_0(\mathbf{p}_2')\mathbf{w}(\mathbf{p}_1') - w_0(\mathbf{p}_1)\mathbf{w}(\mathbf{p}_2) - w_0(\mathbf{p}_2)\mathbf{w}(\mathbf{p}_1)] \\ & + \tau_1(\mathbf{p},\mathbf{p}')[w_0(\mathbf{p}_2')\mathbf{w}(\mathbf{p}_1') - w_0(\mathbf{p}_1')\mathbf{w}(\mathbf{p}_2')] - \tau_2(\mathbf{p},\mathbf{p}')\mathbf{w}(\mathbf{p}_1') \times \mathbf{w}(\mathbf{p}_2')\} \\ & + 2\pi h^2 \int d^3 p_2 \{\mathrm{Im}\,[T_s(\mathbf{p}|\mathbf{p}) + T_t(\mathbf{p}|\mathbf{p})]\,[w_0(\mathbf{p}_2)\mathbf{w}(\mathbf{p}_1) - w_0(\mathbf{p}_1)\mathbf{w}(\mathbf{p}_2)] \\ & \qquad + \mathrm{Re}\,[T_s(\mathbf{p}|\mathbf{p}) - T_t(\mathbf{p}|\mathbf{p})]\,\mathbf{w}(\mathbf{p}_1) \times \mathbf{w}(\mathbf{p}_2)\}. \qquad (13.30)\end{aligned}$$

Here τ_1 and τ_2 are defined by

$$T_s(\mathbf{p}|\mathbf{p}')\,\overline{T}_t(\mathbf{p}|\mathbf{p}') = (\pi m h)^2 [\tau_1(\mathbf{p}|\mathbf{p}') + i\tau_2(\mathbf{p}|\mathbf{p}')]$$

The collision operators J_0 and $\mathbf{J}$ display a dependence on the polarization which in general is quite complex. In the simplest case the gas is unpolarized, or $\mathbf{w} = 0$. Then $\mathbf{J} = 0$ while w_0 is governed by the Boltzmann equation with the spin-averaged cross section $\sigma = \frac{1}{4}(\sigma_s + 3\sigma_t)$. Another simple special case is that of complete spin polarization, say in the z-direction. This is described by taking $w_x = w_y = 0$, $w_z = w_0$. It is easily seen that $J_x = J_y = 0$ while Equations (13.29) and (13.30) yield

$$\begin{aligned}J_0 = J_z = (4/m^2) \int d^3 p_2\, d^3 \mathrm{p}'\, \delta(E - E') \\ \times \sigma_t(\mathbf{p}' \to \mathbf{p})\, [w_0(\mathbf{p}_1')w_0(\mathbf{p}_2') - w_0(\mathbf{p}_1)w_0(\mathbf{p}_2)].\end{aligned}$$

Only the triplet cross section occurs, since with complete polarization the singlet state is not possible. Generally the triplet cross section is small at low energies since S-waves are not available. As a result the mean free path is large at low temperatures, which increases the transport coefficients.

The transport coefficients for a spin-polarized gas were evaluated in detail by Lhuillier and Laloe, who also obtained a solution describing spin waves in the gas.

13.6. BOUND STATES

We now treat the gas as a collection of elementary particles, but allow for the existence of bound states. For simplicity we assume the elementary particles to be identical but use Boltzmann statistics. The modifications which are necessary to deal with a mixture of elementary particles of different kinds, or to take exchange into account, are straightforward.

Bound states, if possible at all, can be formed from the particles available in a variety of ways. A particular arrangement of the set of particles into bound complexes, with some perhaps left over as free particles, is called a channel. (Here "free" is used in the sense of unbound rather than non-interacting.) For example, if the system consists of three hydrogen atoms (viewed as elementary particles), then there are three channels corresponding to the different pairs which may be bound into a molecule, and a fourth channel in which there are no bound pairs. (If exchange symmetry is taken into account, the first three channels mentioned are identical.) One may interpret the term so that bound complexes of the same particles but in different internal states are assigned to different channels. A reaction, or with the latter definition, an inelastic collision, is a transition from an initial channel to a different channel after the collision. As will be seen, the collision operator of kinetic theory divides in a natural way into parts which correspond to transitions among the different scattering channels.

Reactions can occur only if at least three (elementary) particles are present, since in the two-body problem a transition between bound and free states is prevented by conservation of energy and momentum. Nevertheless, we reconsider the derivation of the binary collision operator as there are some important differences when the potential supports bound states.

The motion of particles in a bound state is strongly correlated, so molecular chaos can apply only to scattering states. To project out the bound states, let ξ_i denote an eigenfunction of the Hamiltonian for relative motion of a pair, and let P_i denote the projection on ξ_i (that is, P_i is the product of the projection on ξ_i in the space for relative motion, times the identity in the space for the mass-center motion). The projection on all bound states (which we assume to be finite in number) is $P = \Sigma_i P_i$ and the projection on scattering states is $1 - P$. The corresponding projection on density matrices is $\mathcal{P}F = (1 - P)F(1 - P)$. Then F_2 can be written as a part which acts entirely on the subspace of scattering states plus a remainder:

$$F_2 = \mathcal{P}F_2 + D \tag{13.31}$$

This equation serves as the definition of D, which then satisfies $\mathcal{P}D = 0$. However, F_2 need not be a sum of operators on the orthogonal subspaces so D may have matrix elements which mix bound and scattering states.

We impose molecular chaos at an initial time t_0 but now only on the subspace of scattering states:

$$F_2(1,2,t_0) = \mathcal{P}\, F_1(1,t_0)\, F_1(2,t_0) + D(1,2,t_0). \tag{13.32}$$

No restriction is imposed on D which instead is left as a free variable. Consequently, the concentrations of atoms (by which we mean particles not in bound states) and molecules can be adjusted to any desired initial values.

Proceeding in the same way as before, we first use the hierarchy to lowest order in the density when its solution is given by Equation (13.7). Since P commutes with H_2, D satisfies the same equation as F_2. With the initial condition (13.32), F_2 at time t is

$$F_2(1,2,t) = \mathcal{P}\, S_2(t_0 - t)\, S_0(t - t_0)\, F_1(1,t)\, F_1(2,t) + D(1,2,t).$$

Taking the limit $t_0 \to -\infty$, we may use Equation (13.9). In the presence of bound states the limit (13.10) does not exist on the full two-particle Hilbert space, but the preceding equation contains a projection on the scattering states where the limit does exist:

$$\lim_{t \to \infty} V_0(t)\, V_2(-t)(1 - P) = \Omega^\dagger (1 - P).$$

[Of course $\Omega^\dagger$ is defined as the adjoint of Ω and not by a limiting operation such as (13.10).] The time-independent functional for F_2, now as a functional of D as well as F_1, is

$$F_2(1,2) = \mathcal{P}\,\Omega\, F_1(1)\, F_1(2)\Omega + D(1,2).$$

As the limit of unitary operators, Ω satisfies $\Omega^\dagger\,\Omega = 1$. It follows that $(\Omega\,\Omega^\dagger)^2 = \Omega\,\Omega^\dagger$ so $\Omega\,\Omega^\dagger$ is a projection; indeed it is the projection on scattering states: $\Omega\,\Omega^\dagger = 1 - P$. This implies $\Omega^\dagger P = P\Omega = 0$ and consequently the functional relation can be written

$$F_2(1,2) = \Omega F_1(1)\, F_1(2)\, \Omega^\dagger + D(1,2).$$

This is a natural extension of the time-independent functional (13.11) obtained in the absence of bound states.

When the functional relation is inserted into the first equation of the hierarchy, the result is

$$\frac{\partial F_1(1)}{\partial t} + (1/i\hbar)[F_1(1), H_1]$$
$$= (n/i\hbar)\,\mathrm{Tr}(2)[U(1,2), \Omega F_1(1) F_1(2) \Omega^\dagger + D(1,2)]. \qquad (13.33)$$

This is not a closed equation for F_1 because of the appearance of D, but a closed equation for a modified one-particle density matrix can be obtained as follows. The equation of motion for D at the present order in density is

$$\frac{\partial D}{\partial t} + (1/i\hbar)[D, H_2] = 0. \qquad (13.34)$$

If the trace is taken over the variables for particle 2, the result is

$$\frac{\partial}{\partial t}\mathrm{Tr}(2) D(1,2) + (1/i\hbar)[\mathrm{Tr}(2) D(1,2), H_1(1)] = (1/i\hbar)\,\mathrm{Tr}(2)[U(1,2), D(1,2)].$$

Multiplying by n and subtracting from (13.33) we get

$$\frac{\partial F_A}{\partial t} + (1/i\hbar)[F_A, H_1] = (n/i\hbar)\,\mathrm{Tr}(2)[U(1,2), \Omega F_1(1) F_1(2) \Omega^\dagger] \qquad (13.35)$$

where

$$F_A(1) = F_1(1) - n\,\mathrm{Tr}(2) D(1,2). \qquad (13.36)$$

On the right-hand side of (13.35) the difference between F_1 and F_A can be neglected since it gives a correction of higher order in the density. Thus, F_A satisfies the kinetic equation

$$\frac{\partial F_A}{\partial t} + (1/i\hbar)[F_A, H_1] = (n/i\hbar)\,\mathrm{Tr}(2)[U(1,2), \Omega F_A(1) F_A(2) \Omega^\dagger]. \qquad (13.37)$$

The density matrix F_1 refers to all particles whether they are in bound states or not. The contribution from particles which are in bound states is subtracted out in Equation (13.36) so F_A is the density matrix for atoms. In the present approximation the kinetic equation for atoms is (13.37), while the molecules are governed by the collisionless equation (13.34). The two kinetic equations are independent but become coupled at the next order in density.

13.7. THREE-BODY COLLISIONS

We proceed to work out the next order in the density expansion, which involves interactions in clusters of three particles. The procedure is generally the same as used in Chapter 7. The molecular chaos assumption must be formulated for F_3 as well as F_2; the time dependence of F_3 can be determined to lowest order in the density, but in working out the functional relation for F_2 it is necessary to extend the solution to the hierarchy to the next order in density. The resulting triple collision operators for F_A and D then describe collisions between atoms and (diatomic) molecules, and the former also includes three-body collisions between atoms.

Molecular chaos for F_2 takes basically the same form as (13.32). However, the uncorrelated particles are atoms described by F_A rather than elementary particles described by F_1 so we write

$$F_2(1,2,t_0) = \mathcal{P} F_A(1,t_0) F_A(2,t_0) + D(1,2,t_0).$$

Actually (13.32) and this equation both lead to the same functional relation, but the algebra is somewhat simpler if we use the latter as the starting point.

In a three-body cluster there are, as mentioned, three channels containing a bound pair plus a fourth channel in which all particles are free. The molecular chaos assumption for F_3 is formulated by projecting it onto subspaces for the different channels, then writing a product of density matrices for the particles (whether atoms or molecules) in the channel. The projection on channels is done with the aid of channel wave operators, whose properties we now summarize.

The symbol Υ will be used for the three-particle wave operators to distinguish them from the two-body wave operator Ω. The wave operator for the free-particle channel, denoted by a subscript 0, is defined by

$$\Upsilon_0 = \lim_{t \to \infty} V_3(t) V_0(-t).$$

As a limit of unitary operators, Υ_0 satisfies $\Upsilon_0^\dagger \Upsilon_0 = 1$.

A channel containing a bound pair will be denoted by α which may be viewed as labeling the bound particles or the third particle; $\alpha = 1,2$ and $\alpha = 3$ stand for the same thing. When it is desired to distinguish a particular internal state, the channel will be labeled as αi where i denotes the internal quantum numbers. Let $H_{(\alpha)}$ denote the Hamiltonian for the three-particle system in which only the pair α interacts:

$$H_{(\alpha)} = K + U(\alpha)$$

where as before K denotes the kinetic energy. Thus, $H_{(\alpha)}$ is the "free" Hamiltonian for channel α. The corresponding time development operator will be written $V_{(\alpha)}$:

$$V_{(\alpha)}(t) = \exp\left(-iH_{(\alpha)} t/\hbar\right)$$

and in addition we define

$$S_{(\alpha)}(-t)F = V_{(\alpha)}(t) F V_{(\alpha)}(-t). \tag{13.38}$$

Let $P_{\alpha i}$ denote the projection introduced previously but now referring to the pair α and extended to the three-particle Hilbert space. (In other words $P_{\alpha i}$ is the product of the projection on ξ_i in the space for relative motion of pair α, multiplied by the identities on the spaces for the mass-center motion of pair α and for the third particle.) The wave operator for the αi channel is then defined by

$$\Upsilon_{\alpha i} = \lim_{t\to\infty} V_3(t)\, V_{(\alpha)}(-t)\, P_{\alpha i}.$$

The wave operator for the collection of all bound states for pair α is

$$\Upsilon_\alpha = \sum_i \Upsilon_{\alpha i} = \lim_{t\to\infty} V_3(t)\, V_{(\alpha)}(-t)\, P_\alpha$$

where $P_\alpha = \Sigma_i P_{\alpha i}$. These wave operators are isometric on their corresponding subspaces, and the wave functions for different channels are orthogonal. This can be stated in the form

$$\Upsilon_{\alpha i}^\dagger \Upsilon_{\beta j} = P_{\alpha i}\, \delta_{\alpha\beta}\, \delta_{ij}.$$

Orthogonality of scattering and bound channels implies

$$\Upsilon_0^\dagger \Upsilon_{\alpha i} = 0.$$

The intertwining relations for the channel wave operators are

$$\Upsilon_0 H_0 = H_3 \Upsilon_0, \quad \Upsilon_{\alpha i} H_{(\alpha)} = H_3 \Upsilon_{\alpha i}.$$

It is readily shown that $\Upsilon_0 \Upsilon_0^\dagger$ and $\Upsilon_{\alpha i}\Upsilon_{\alpha i}^\dagger$ are projections, and as a consequence of the intertwining relations they commute with H_3. Thus, they project onto eigenfunctions of H_3 corresponding to the scattering channels. (Actually we should say generalized eigenfunctions, since the presence of a continuous spectrum prevents square integrability.) The completeness of scattering states plus three-body bound states can be stated as

$$\Upsilon_0 \Upsilon_0^\dagger + \sum_{\alpha i} \Upsilon_{\alpha i} \Upsilon_{\alpha i}^\dagger = 1 - Q$$

where Q is the projection on three-body bound states. For present purposes, three-body bound states can be ignored as they affect the kinetic equations only in a higher order than we consider.

Returning to the molecular chaos assumption for F_3, we write its projection on a channel as a product of the density matrices for the atoms and molecules in the channel. As noted, the channel projection operators are $\Upsilon_0 \Upsilon_0^\dagger$ and $\Upsilon_\alpha \Upsilon_\alpha^\dagger$ so

$$F_3(1,2,3,t_0)$$
$$= \Upsilon_0 \Upsilon_0^\dagger G(1,2,3,t_0) \Upsilon_0 \Upsilon_0^\dagger + \sum_\alpha \Upsilon_\alpha \Upsilon_\alpha^\dagger F_A(\alpha,t_0)\, D(\alpha,t_0) \Upsilon_\alpha \Upsilon_\alpha^\dagger. \tag{13.39}$$

Here G is the abbreviation

$$G(1,2,3) = F_A(1)\, F_A(2)\, F_A(3)$$

and α labels both a pair and the remaining third particle. If three-body bound states are taken into account, an additional term similar to the term D in Equation (13.31) should be included.

The generalization of the molecular chaos assumption to other reduced density matrices is straightforward. In general, F_s is to be written as a sum of terms obtained by projecting onto the channels which are available in the s-body problem. Each such projection is then written as a product of the density matrices for the atoms and/or molecules in the channel.

The time-independent functional for F_3 in terms of F_A and D can be obtained directly from (13.39). To lowest order in the density we have

$$\begin{aligned} F_3(1,2,3,t) &= S_3(1,2,3,t_0 - t)\,F_3(1,2,3,t_0) \\ &= S_3(1,2,3,t_0 - t)\,[\Upsilon_0\Upsilon_0^\dagger S_0(1,2,3,t - t_0)\,G(1,2,3,t)\Upsilon_0\Upsilon_0^\dagger \\ &\qquad + \sum_\alpha \Upsilon_\alpha\Upsilon_\alpha^\dagger S_{(\alpha)}(t - t_0)\,F_A(\alpha,t)\,D(\alpha,t)\Upsilon_\alpha\Upsilon_\alpha^\dagger]. \end{aligned}$$

Since S_3 commutes with the channel projection operators, it can be moved in and combined with the operators $S_0, S_{(\alpha)}$; in the limit $t_0 \to -\infty$ the combinations can be written in terms of the wave operators. This yields the time-independent functional for F_3

$$F_3(1,2,3) = \Upsilon_0 G(1,2,3)\Upsilon_0^\dagger + \sum_\alpha \Upsilon_\alpha F_A(\alpha)D(\alpha)\Upsilon_\alpha^\dagger.$$

The kinetic equation for F_2 is obtained by substituting this expression for F_3 into the second equation of the hierarchy. The kinetic equation for $D = (1 - \mathcal{P})F_2$ is then found to be

$$\frac{\partial D}{\partial t} + (1/i\hbar)[D,H_2] = n\,C_M$$

where the collision operator is given by

$$\begin{aligned} C_M = (1/i\hbar)(1 - \mathcal{P})\,\mathrm{Tr}(3)[U^{(3)}, \Upsilon_0 G(1,2,3)\Upsilon_0^\dagger \\ + \sum_\alpha \Upsilon_\alpha F_A(\alpha)\,D(\alpha)\Upsilon_\alpha^\dagger]. \end{aligned} \tag{13.40}$$

In working out the time-independent functional for F_2 we need the solutions to the hierarchy for F_1 and F_2 through terms linear in the density. These are formally the same as their classical counterparts (7.27) and so will be written down without derivation:

$$\begin{aligned} F_1(1,t) &= S_1(1,t_0 - t)\,F_1(1,t_0) \\ &\qquad + n\,\mathrm{Tr}(2)[S_2(1,2,t_0 - t) - S_0(1,2,t_0 - t)]\,F_2(1,2,t_0), \end{aligned}$$

$$\begin{aligned} F_2(1,2,t) &= S_2(1,2,t_0 - t)\,F_2(1,2,t_0) \\ &\quad + n\,\mathrm{Tr}(3)[S_3(1,2,3,t_0 - t) - S_{(3)}(1,2,3,t_0 - t)]\,F_3(1,2,3,t_0). \end{aligned} \tag{13.41}$$

Here $S_{(3)}$ is the operator defined by (13.38). The molecular chaos assumptions for $F_2(t_0)$ and $F_3(t_0)$ are to be substituted into these equations. To eliminate $F_1(t_0)$, we write the first of (13.41) with t, t_0 interchanged and use the lowest-order equation of motion for F_2 to obtain

$$F_1(1,t_0) = S_1(1,t - t_0)F_1(1,t) + n\mathrm{Tr}(2)[1 - S_0(1,2,t - t_0)S_2(1,2,t_0 - t)]F_2(1,2,t_0).$$

After using the molecular chaos assumption for F_2 we find a result which in terms of F_A takes the form

$$F_A(1,t_0) = S_1(1,t - t_0)F_A(1,t) - n\mathrm{Tr}(2)[S_0(1,2,t - t_0)S_2(1,2,t_0 - t) - 1]\mathcal{P}(1,2)F_A(1,t_0)F_A(2,t_0).$$

The preceding equation is to be substituted into (13.41) for F_2, and the limit $t_0 \to -\infty$ taken. After a tedious calculation whose details will be omitted, the time-independent functional is found to be

$$\begin{aligned}F_2(1,2) = {} & D(1,2) + \Omega(1,2)F_A(1)F_A(2)\Omega^\dagger(1,2) \\ & + n\mathrm{Tr}(3)\{\mathcal{P}(1,2)\Upsilon_0 G(1,2,3)\Upsilon_0^\dagger - \Omega(1,2)[\Omega(1,3)G(1,2,3)\Omega^\dagger(1,3) \\ & + \Omega(2,3)G(1,2,3)\Omega^\dagger(2,3) - G(1,2,3)]\Omega^\dagger(1,2) \\ & + \sum_\alpha \Upsilon_\alpha F_A(\alpha)D(\alpha)\Upsilon_\alpha^\dagger\}.\end{aligned} \tag{13.42}$$

When the functional (13.42) is inserted into the first equation of the hierarchy, the kinetic equation for F_1, then F_A, is obtained. We write the result in the form

$$\partial F_A/\partial t + (1/i\hbar)[F_A, H_1] = nC_2 + n^2C_3 + n^2C_A.$$

Here C_2 is the binary collision operator (13.12) while C_A is given in terms of the operator defined in (13.40) by

$$C_A(1) = -\mathrm{Tr}(2)\,C_M(1,2). \tag{13.43}$$

Finally C_3 is a triple collision operator given by

$$\begin{aligned}C_3(1) = (1/i\hbar)\,\mathrm{Tr}(2,3)\Big[U(1,2),\ \Big\{\mathcal{P}(1,2)\Big(\Upsilon_0 G\Upsilon_0^\dagger + \sum_\alpha \Upsilon_\alpha F_A(\alpha)D(\alpha)\Upsilon_\alpha^\dagger\Big) \\ - \Omega(1,2)[\Omega(1,3)G\Omega^\dagger(1,3) + \Omega(2,3)G\Omega^\dagger(2,3) - G]\Omega^\dagger(1,2)\Big\}\Big].\end{aligned}$$

In the absence of bound states, $\mathcal{P}$ reduces to 1 and the term containing $D(\alpha)$ drops out. Then C_3 reduces to the quantum mechanical version of the triple collision operator obtained in Chapter 7.

13.8. DESCRIPTION ON INTERNAL SPACE

The collision operator C_M contains the effects of the molecular processes of elastic and inelastic scattering, molecular formation and dissociation, and rearrangement collisions. In the general form these are described on the three-particle Hilbert space. However, it is more convenient to describe the molecule by a density matrix on the internal space, and collisions in terms of transitions between channels. In reducing the kinetic equation to this form, we also neglect the spatial extent of collisions.

We first define a set of momentum variables which is useful in the three-body problem. As before $\mathbf{P}$, p will denote the total and relative momenta for a pair, for example,

$$\mathbf{P}_1 = \mathbf{P}_{2,3} = \mathbf{p}_2 + \mathbf{p}_3, \quad \mathsf{p}_1 = \mathsf{p}_{2,3} = \frac{1}{2}(\mathbf{p}_2 - \mathbf{p}_3).$$

The total momentum of the three particles will be written $\mathbf{K}$

$$\mathbf{K} = \mathbf{p}_1 + \mathbf{p}_2 + \mathbf{p}_3,$$

while $\mathbf{k}$ will denote the relative momentum of a particle with respect to the remaining pair, for example,

$$\mathbf{k}_1 = \mathbf{k}_{2,3} = \mathbf{p}_1 - \frac{1}{3}\mathbf{K} = \frac{2}{3}\mathbf{p}_1 - \frac{1}{3}(\mathbf{p}_2 + \mathbf{p}_3).$$

In place of the particle momenta, we may use the total and relative momenta $\mathbf{K},\mathbf{k},\mathsf{p}$ where $\mathbf{k},\mathsf{p}$ denotes any of the three sets $\mathbf{k}_\alpha,\mathsf{p}_\alpha$, $\alpha = 1,2,3$. The volume element in momentum space can be written in several forms,

$$d^3p_1 d^3p_2 d^3p_3 = d^3K\, d^3k d^3\mathsf{p} = d^3P_3 d^3p_3 d^3\mathsf{p}_3, \text{ and so on.}$$

The kinetic energies associated with $\mathsf{p},\mathbf{k}$ are

$$u = \mathsf{p}^2/m, \; w = k^2/2\mu$$

where $\mu = 2m/3$ is the reduced mass for an atom plus a diatomic molecule. The sum $u_\alpha + w_\alpha$ is the same for all α and so for brevity can be written as $u + w$.

Because of conservation of momentum, the wave operators in momentum space have the forms

$$\Upsilon_0(\mathbf{p}_1,\mathbf{p}_2,\mathbf{p}_3|\mathbf{p}_1',\mathbf{p}_2',\mathbf{p}_3') = \Upsilon_0(\mathbf{k},\mathsf{p}|\mathbf{k}',\mathsf{p}')\,\delta(\mathbf{K} - \mathbf{K}'),$$

$$\Upsilon_\alpha(\mathbf{p}_1,\mathbf{p}_2,\mathbf{p}_3|\mathbf{p}_1',\mathbf{p}_2',\mathbf{p}_3') = \Upsilon_\alpha(\mathbf{k},\mathsf{p}|\mathbf{k}',\mathsf{p}')\,\delta(\mathbf{K} - \mathbf{K}').$$

The Wigner function for atoms is defined by an equation similar to (13.4):

$$W_A(\mathbf{x},\mathbf{p}) = (n/h^3)\int d^3p'\, F_A(\mathbf{p} + \frac{1}{2}\mathbf{p}'|\mathbf{p} - \frac{1}{2}\mathbf{p}')\exp(i\mathbf{p}'\cdot\mathbf{x}/\hbar).$$

To construct the Wigner function for molecules, let $D(\mathbf{P},\mathsf{p}|\mathbf{P}',\mathsf{p}')$ denote matrix elements of D but considered as a function of the total and relative momenta. The transformation (13.4) is to be applied to the total momentum and the result is then an operator on the space of relative momentum as well as a Wigner function of position and mass-center momentum:

$$W_M(\mathbf{x},\mathbf{P},\mathsf{p},\mathsf{p}') = n^2\int d^3P'\,D(\mathbf{P} + \frac{1}{2}\mathbf{P}',\mathsf{p}|\mathbf{P} - \frac{1}{2}\mathbf{P}',\mathsf{p}')\exp(i\mathbf{P}'\cdot\mathbf{x}/\hbar).$$

With the normalization as given, the densities n_A and n_M of atoms and molecules are

$$n_A(\mathbf{x}) = \int d^3p\,W_A(\mathbf{x},\mathbf{p}), \quad n_M(\mathbf{x}) = \int d^3P\,d^3\mathsf{p}\,W_M(\mathbf{x},\mathbf{P},\mathsf{p},\mathsf{p}).$$

The kinetic equation for W_M can be written

$$\frac{\partial}{\partial t}W_M + \mathbf{V}\cdot\nabla W_M + (1/i\hbar)[H_r,W_M] = J_M$$

where $\mathbf{V} = \mathbf{P}/2m$ and $H_r = \mathsf{p}^2/m + U$ is the Hamiltonian for the relative motion of the pair. A description on the internal space is provided by the matrix elements of W_M with respect to the bound-state eigenfunctions, which we denote by W_{ij}:

$$W_{ij}(\mathbf{x},\mathbf{P}) = \int d^3\mathsf{p}\,d^3\mathsf{p}'\,\bar{\xi}_i(\mathsf{p})\,W_M(\mathbf{x},\mathbf{P},\mathsf{p},\mathsf{p}')\,\xi_j(\mathsf{p}').$$

These satisfy

$$\frac{\partial}{\partial t}W_{ij} + \mathbf{V}\cdot\nabla W_{ij} + i\omega_{ij}W_{ij} = J_{ij}$$

where $\omega_{ij} = (\epsilon_i - \epsilon_j)/\hbar$ are the transition frequencies, ϵ_i being the eigenvalues of the Hamiltonian for the internal motion, and J_{ij} denotes the matrix elements of J_M.

When the spatial extent of a collision is neglected, F_A and D can be replaced by

$$nF_A(\mathbf{p}|\mathbf{p}') = h^3W_A(\mathbf{p})\,\delta(\mathbf{p} - \mathbf{p}'),$$

$$n^2D(\mathbf{P},\mathsf{p}|\mathbf{P}',\mathsf{p}') = h^3W_M(\mathbf{P},\mathsf{p},\mathsf{p}')\,\delta(\mathbf{P} - \mathbf{P}').$$

The collision operator J_M is then found to be

$$J_M = -i\pi h^5[1 - \mathcal{P}(1,2)]\,\mathrm{tr}\,[U^{(3)},\Upsilon_0W_A(1)W_A(2)W_A(3)\Upsilon_0^\dagger + (1/h^3)\sum_\alpha\Upsilon_\alpha W_A(\alpha)W_M(\alpha)\Upsilon_\alpha^\dagger]. \qquad (13.44)$$

As in Equation (13.15), tr means a trace with the delta function for overall conservation of momentum omitted. Since $\mathcal{P}$ projects on scattering states, it drops out when the matrix elements J_{ij} are calculated. Let J_{ij} be separated into parts corresponding to the terms in Equation (13.44):

$$J_{ij} = J_{ij}^{0} + \sum_{\alpha} J_{ij}^{\alpha}.$$

Here

$$\begin{aligned} J_{ij}^{0} &= -2\pi i h^5 \int d^3\mathrm{p}_3 d^3\mathrm{p}_3' \,\bar{\xi}_i(\mathbf{p}_3) \\ &\quad \times \operatorname{tr}[U^{(3)}, \Upsilon_0 W_A(1) W_A(2) W_A(3) \Upsilon_0^{\dagger}]\, \xi_j(\mathbf{p}_3'), \end{aligned} \tag{13.45}$$

$$\begin{aligned} J_{ij}^{\alpha} &= -2\pi i h^2 \int d^3\mathrm{p}_3 d^3\mathrm{p}_3' \,\bar{\xi}_i(\mathbf{p}_3) \\ &\quad \times \operatorname{tr}[U^{(3)}, \Upsilon_\alpha W_A(\alpha) W_M(\alpha) \Upsilon_\alpha^{\dagger}]\, \xi_j(\mathbf{p}_3'). \end{aligned}$$

The wave operators can be eliminated in favor of T-matrices which transform between channels. Consider first Υ_0 which satisfies

$$\int d^3\mathrm{p}_3 \bar{\xi}_i(\mathbf{p}_3)\Upsilon_0(\mathbf{k},\mathbf{p}|\mathbf{k}',\mathbf{p}') = -2\pi i \delta(w_3 + \epsilon_i - u' - w')\, T_F(\mathbf{k}_3, i|\mathbf{k}',\mathbf{p}'), \tag{13.46}$$

$$\int d^3\mathrm{p}_3 \bar{\xi}_i(\mathbf{p}_3)(U^{(3)}\Upsilon_0)(\mathbf{k},\mathbf{p}|\mathbf{k}',\mathbf{p}') = T_F(\mathbf{k}_3, i|\mathbf{k}',\mathbf{p}').$$

where $T_F(\mathbf{k}_3, i|\mathbf{k}', \mathbf{p}')$ is the T-matrix for a three-body collision leading to formation of the i'th bound state of particles 1 and 2. The wave operators $\Upsilon_{\alpha j}$ satisfy

$$\begin{aligned} &\int d^3\mathrm{p}_3\, \bar{\xi}_i(\mathbf{p}_3)\Upsilon_{3j}(\mathbf{k},\mathbf{p}|\mathbf{k}',\mathbf{p}') \\ &\quad = [\delta(\mathbf{k}_3 - \mathbf{k}_3')\delta_{ij} - 2\pi i \delta(w_3 + \epsilon_i - w_3' - \epsilon_j)\, T_S(\mathbf{k}_3, i|\mathbf{k}_3', j)]\bar{\xi}_j(\mathbf{p}_3'), \end{aligned}$$

$$\int d^3\mathrm{p}_3 \bar{\xi}_i(\mathbf{p}_3)\, (U^{(3)}\Upsilon_{3j})(\mathbf{k},\mathbf{p}|\mathbf{k}',\mathbf{p}') = T_S(\mathbf{k}_3, i|\mathbf{k}_3', j)\, \bar{\xi}_j(\mathbf{p}_3'), \tag{13.47}$$

where T_S is the T-matrix for scattering; in this case particle 3 is scattered by the 1,2 bound pair, with the bound-state eigenfunction changing from ξ_j to ξ_i. The corresponding equations for a rearrangement collision are

$$\begin{aligned} &\int d^3\mathrm{p}_3 \bar{\xi}_i(\mathbf{p}_3)\Upsilon_{1j}(\mathbf{k},\mathbf{p}|\mathbf{k}',\mathbf{p}') \\ &\qquad = -2\pi i \delta(w_3 + \epsilon_i - w_1' - \epsilon_j)\, T_R(\mathbf{k}_3, i|\mathbf{k}_1', j)\, \bar{\xi}_j(\mathbf{p}_1'), \end{aligned}$$

$$\int d^3\mathrm{p}_3 \bar{\xi}_i(\mathbf{p}_3)\, (U^{(3)}\Upsilon_{1j})(\mathbf{k},\mathbf{p}|\mathbf{k}',\mathbf{p}') = T_R(\mathbf{k}_3, i|\mathbf{k}_1', j)\, \bar{\xi}_j(\mathbf{p}_1'). \tag{13.48}$$

Here $\mathrm{T}_R(\mathbf{k}_3, i|\mathbf{k}_1', j)$ is the T-matrix for a rearrangement collision in which the 2,3 bound pair goes into a 1,2 bound pair. Similar equations hold for other arrangements of the particles.

When Equation (13.46) is used in Equation (13.45) we find for J^0_{ij}

$$J^0_{ij} = 4\pi^2 h^5 \int d^3p_3 d^3k'\, d^3\mathrm{p}'\, W_A(\mathbf{p}_1')\, W_A(\mathbf{p}_2')\, W_A(\mathbf{p}_3')\, T_F(\mathbf{k}_3, i|\mathbf{k}', \mathbf{p}')\, \overline{T}_F(\mathbf{k}_3, j|\mathbf{k}', \mathbf{p}') \times [\delta_+(u' + w' - w_3 - \epsilon_j) + \delta_-(u' + w' - w_3 - \epsilon_i)].$$

Here and in the following conservation of momentum, $\mathbf{K} = \mathbf{K}'$ is left implicit. We next use Equations (13.47) and (13.48) to express J^α_{ij} in terms of T-matrices. The terms with $\alpha = 1$ and $\alpha = 2$ give the same contribution, which is

$$J^1_{ij} = (2\pi h)^2 \sum_{rs} \int d^3p_3 d^3k_1'\, W_A(\mathbf{p}_1')\, W_{rs}(\mathbf{P}_1')\, T_R(\mathbf{k}_3, i|\mathbf{k}_1', r)\, \overline{T}_R(\mathbf{k}_3, j|\mathbf{k}_1', s) \times [\delta_+(w_1' + \epsilon_s - w_3 - \epsilon_j) + \delta_-(w_1' + \epsilon_r - w_3 - \epsilon_i)].$$

Similarly, we get for J^3_{ij}

$$\begin{aligned} J^3_{ij} = (2\pi h)^2 \sum_{rs} \int & d^3p_3 d^3k_3'\, W_A(\mathbf{p}_3')\, W_{rs}(\mathbf{P}_3') \\ & \times \{\delta(\mathbf{k}_3 - \mathbf{k}_3')[\delta_{js} T_S(\mathbf{k}_3, i|\mathbf{k}_3, r) - \delta_{ir} \overline{T}_S(\mathbf{k}_3, j|\mathbf{k}_3, s)] \\ & + [\delta_+(w_3' + \epsilon_s - w_3 - \epsilon_j) + \delta_-(w_3' + \epsilon_r - w_3 - \epsilon_i)] \\ & \times T_S(\mathbf{k}_3, i|\mathbf{k}_3', r)\, \overline{T}_S(\mathbf{k}_3, j|\mathbf{k}_3', s)\}. \end{aligned}$$

13.9. DIAGONAL DENSITY MATRIX

As was the case in the derivation of the Waldmann-Snider equation, the collision operator at this point depends on T-matrix elements which are off the energy shell and so cannot be expressed in terms of scattering and reaction amplitudes. Kinetic equations at the Waldmann-Snider level can be obtained by assuming W_{ij} to be diagonal in the internal energy. For simplicity we will make the more stringent assumption of complete diagonality, in which case the collision operator can be expressed in terms of cross sections. The kinetic equation then corresponds to that of Wang Chang, Uhlenbeck, and de Boer but with reactions included. Thus, we write

$$W_{ij} = \delta_{ij} W_i$$

and consider only diagonal elements of the kinetic equation,

$$\partial W_i/\partial t + \mathbf{V}\cdot\nabla W_i = J_{ii}.$$

We obtain immediately for J^0_{ii}

$$J^0_{ii} = 4\pi^2 h^5 \int d^3p_3 d^3k'\, d^3\text{p}'\, W_A(\mathbf{p}_1') W_A(\mathbf{p}_2') W_A(\mathbf{p}_3') \times \delta(u' + w' - w_3 - \epsilon_i)|T_F(\mathbf{k}_i, i|\mathbf{k}', \mathbf{p}')|^2.$$

With the diagonality condition, J^1_{ii} becomes

$$J^1_{ii} = (2\pi h)^2 \int d^3p_3 d^3k_1' W_A(\mathbf{p}_1') \sum_j \delta(w_1' + \epsilon_j - w_3 - \epsilon_i) W_j(\mathbf{P}_1')|T_R(\mathbf{k}_3, i|\mathbf{k}_1', j)|^2.$$

Similarly, J^3_{ii} is

$$J^3_{ii} = (2\pi h)^2 \int d^3p_3 d^3k_3' W_A(\mathbf{p}_3') \sum_j W_j(\mathbf{P}_3') \times \{(1/2\pi i)\delta_{ij}\delta(\mathbf{k}_3 - \mathbf{k}_3')[T_S(\mathbf{k}_3, i|\mathbf{k}_3, i) - \overline{T}_S(\mathbf{k}_3, i|\mathbf{k}_3, i)] + \delta(w_3' + \epsilon_j - w_3 - \epsilon_i)|T_S(\mathbf{k}_3, i|\mathbf{k}_3', j)|^2\}. \qquad (13.49)$$

We next make use of the optical theorem, which in the present case takes the form

$$(-1/2\pi i)[T_S(\mathbf{k}_3, i|\mathbf{k}_3, i) - \overline{T}_S(\mathbf{k}_3, i|\mathbf{k}_3, i)] = \int d^3k'\, d^3\text{p}'\, \delta(w_3 + \epsilon_i - w' - u')|\,T_F(\mathbf{k}_3, i|\mathbf{k}', \mathbf{p}')|^2 + \sum_j \int d^3k_3' \delta(w_3 + \epsilon_i - w_3' - \epsilon_j)|T_S(\mathbf{k}_3, i|\mathbf{k}_3', j)|^2 + \sum_j \sum_{\alpha=1,2} \int d^3k_\alpha' \delta(w_3 + \epsilon_i - w_\alpha' - \epsilon_j)|T_R(\mathbf{k}_3, i|\mathbf{k}_\alpha', j)|^2.$$

This is to be substituted into Equation (13.49). We combine the terms containing T_F in the optical theorem with J^1_{ii}, the terms containing T_R with J^1_{ii}, and the terms containing T_S with the rest of J^3_{ii}. This yields a separation of J_{ii} into parts corresponding to direct scattering, rearrangement collisions, and molecular formation or dissociation,

$$J_{ii} = S_i + R_i + F_i.$$

A straightforward calculation yields for S_i

$$S_i = (2\pi h)^2 \int d^3p_3 dk_3' \sum_j \delta(w_3' + \epsilon_j - w_3 - \epsilon_i)\, |T_S(\mathbf{k}_3, i|\mathbf{k}_3', j)|^2 \times [W_A(\mathbf{p}_3')W_j(\mathbf{P}_3') - W_A(\mathbf{p}_3)W_i(\mathbf{P}_3)].$$

(In deriving this equation it is to be noted that $\mathbf{K}' = \mathbf{K}, \mathbf{k}_3' = \mathbf{k}_3$ implies $\mathbf{p}_3' = \mathbf{p}_3, \mathbf{P}_3' = \mathbf{P}_3$.) The cross section for scattering from the state $\mathbf{k}_3', j$ to the state $\mathbf{k}_3, i$ is given by

$$\sigma_S(\mathbf{k}_3', j \rightarrow \mathbf{k}_3, i) = (g_3/g_3')(2\pi\,\mu h)^2\,|T_S(\mathbf{k}_3, i|\mathbf{k}_3', j)|^2$$

where $g_3 = k_3/\mu$ is the relative velocity. Hence, we get

$$S_i = \mu^{-2}\int d^3p_3 d^3k_3' \sum_j \delta(w_3' + \epsilon_j - w_3 - \epsilon_i)\,(g_3'/g_3)$$
$$\times\ \sigma_S(\mathbf{k}_3', j \rightarrow \mathbf{k}_3, i)\,[W_A(\mathbf{p}_3')\,W_j(\mathbf{P}_3') - W_A(\mathbf{p}_3)\,W_i(\mathbf{P}_3)].$$

A more convenient form can be obtained by using detailed balance. It follows from time-reversal invariance that on the energy shell

$$T_S(\mathbf{k}_3, i|\mathbf{k}_3', j) = T_S(\mathbf{k}_3', j^*|\mathbf{k}_3, i^*)$$

where i^* denotes the time-reversed state to i. We may choose the eigenfunctions to satisfy $\xi_{i^*} = \bar{\xi}_i$ and detailed balance then holds in the form

$$|T_S(\mathbf{k}_3, i|\mathbf{k}_3', j)|^2 = |T_S(\mathbf{k}_3', j|\mathbf{k}_3, i)|^2.$$

This is equivalent to

$$(g_3'/g_3)\,\sigma_S(\mathbf{k}_3', j \rightarrow \mathbf{k}_3, i) = (g_3/g_3')\,\sigma_S(\mathbf{k}_3, i \rightarrow \mathbf{k}_3', j).$$

The delta function makes it possible to carry out the integration over the magnitude of $\mathbf{k}_3'$ immediately, and we get

$$S_i = \sum_j \int d^3p_3\, g_3\, d\Omega\,(\mathbf{k}_3')\,\sigma_S(\mathbf{k}_3, i \rightarrow \mathbf{k}_3', j)\,[W_A(\mathbf{p}_3')\,W_j(\mathbf{P}_3') - W_A(\mathbf{p}_3)\,W_i(\mathbf{P}_3)] \tag{13.50}$$

where $d\Omega(\mathbf{k}_3')$ is the element of solid angle about $\mathbf{k}_3'$. Conservation of energy, which has been left implicit, is expressed by

$$(1/2\mu)k_3'^2 + \epsilon_j = (1/2\mu)k_3^2 + \epsilon_i$$

which is equivalent to

$$(1/2m)p_3'^2 + (1/4m)\,P_3'^2 + \epsilon_j = (1/2m)p_3^2 + (1/4m)\,P_3^2 + \epsilon_i. \tag{13.51}$$

If $\epsilon_i < \epsilon_j$, then for fixed $\mathbf{P}_3$ the region of integration for $\mathbf{p}_3$ in Equation (13.50) is restricted by

$$p_3^2/2m + P_3^2/4m \geq \epsilon_j - \epsilon_i.$$

The collision operator S_i is the same as obtained by Wang Chang, Uhlenbeck, and de Boer.

In a similar way we find for R_i

$$R_i = 2\sum_j \int d^3p_3\, g_3\, d\Omega\,(\mathbf{k}_1')\,\sigma_R(\mathbf{k}_3, i \rightarrow \mathbf{k}_1', j)$$
$$\times\ [W_A(\mathbf{p}_1')\,W_j(\mathbf{P}_1') - W_A(\mathbf{p}_3)\,W_i(\mathbf{P}_3)].$$

Here σ_R is the rearrangement cross section, given by

$$\sigma_R(\mathbf{k}_3, i \to \mathbf{k}_1', j) = (g_1'/g_3)(2\pi\,\mu h)^2 |T_R(\mathbf{k}_1', j|\mathbf{k}_3, i)|^2.$$

Conservation of energy is expressed by

$$(1/2\mu)k_1'^2 + \epsilon_j = (1/2\mu)k_3^2 + \epsilon_i$$

or equivalently by

$$(1/2m)p_1'^2 + (1/4m)P_1'^2 + \epsilon_j = (1/2m)p_3^2 + (1/4m)P_3^2 + \epsilon_i. \tag{13.52}$$

The corresponding expression for F_i is found to be

$$F_i = \int d^3p_3 g_3 dw_3' d\Omega(\mathbf{k}_3') \, d\Omega(\mathbf{p}_3') \, \sigma_D(\mathbf{k}_3, i \to \mathbf{k}', \mathbf{p}') \times [h^3 W_A(\mathbf{p}_1') \, W_A(\mathbf{p}_2') \, W_A(\mathbf{p}_3') - W_A(\mathbf{p}_3) \, W_i(\mathbf{P}_3)].$$

The cross section for dissociation is given by

$$\begin{aligned}\sigma_D(\mathbf{k}_3, i \to \mathbf{k}', p') &= d\sigma_D(\mathbf{k}_3, i \to \mathbf{k}', p')/d\Omega(\mathbf{k}_3') \, d\Omega(\mathbf{p}_3') \, dw_3' \\ &= 2m(\pi\,\mu h)^2 \, (k_3' \mathrm{p}_3'/k_3) \, |T_D(\mathbf{k}', \mathbf{p}'|\mathbf{k}_3, i)|^2.\end{aligned}$$

The condition for conservation of energy can be written as

$$(1/2m)(p_1'^2 + p_2'^2 + p_3'^2) = (1/2m)p_3^2 + (1/4m)P_3^2 + \epsilon_i. \tag{13.53}$$

In equilibrium the molecular density matrix has the Maxwellian form

$$W_i(\mathbf{P}) = n_M \frac{1}{(4\pi m\kappa T)^{3/2}} \exp\left[-\frac{1}{\kappa T}\left(\frac{P^2}{4m} + \epsilon_i\right)\right].$$

Using Equations (13.51) and (13.52), one can readily confirm that the collision operators R_i and S_i then vanish. The Maxwellian for atoms is

$$W_A(\mathbf{p}) = n_A(2\pi m\kappa T)^{-3/2} \exp(-p^2/2m\kappa T).$$

Requiring that F_i vanish, we obtain a relation between the densities of atoms and molecules

$$n_M = \frac{h^3}{(\pi m\kappa T)^{3/2}} n_A^2 Z$$

which is the law of mass action.

Notes on Chapter 13

13.1 For the hierarchy and kinetic equation in quantum mechanics, see Hoffman, Mueller, and Curtiss [1965], Imam-Rahajoe and Curtiss [1967], van Weert and de Boer [1978], Boercker and Dufty [1979, 1981], and Kraeft, Schlanges, and Kremp [1986].

13.3 See Wang Chang, Uhlenbeck, and de Boer [1964]. The Waldmann-Snider equation was derived in Waldmann [1957, 1973], and Snider [1960a]. For a review of the Waldmann-Snider equation and its applications, see Moraal [1975]. Radiative effects, particularly line broadening, have been treated on the basis of the Waldmann-Snider equation; for a recent reference, see Blackmore [1987].

13.4 For calculations of the transport coefficients on the basis of the equation of Wang Chang, Uhlenbeck, and de Boer, see Monchick and Mason [1961], Mason and Monchick [1962a, b], Monchick, Yun, and Mason [1963a, b], Monchick, Pereira, and Mason [1965], and Velasco and Uribe [1986]. There are many papers dealing with the classical approach to the effects of internal structure; see Dahler and Theodosopulu [1975] and Cole and Evans [1986] for reviews.

13.5 See Lhuillier and Laloe [1982a, b, 1983].

13.6 The effect of bound states on the binary collision operator in classical mechanics is discussed in Marchetti and Dufty [1980, 1981].

13.7 See Snider and Sanctuary [1971], Snider and Lowry [1974], Lowry and Snider [1974], Eu [1975], Olmsted and Snider [1976], Hoffman, Kouri, and Top [1979], Evans, Hoffman, and Kouri [1983], and Evans [1986]. The particular approach used here is developed in Klimontovich and Kremp [1981], McLennan [1981, 1982a, 1982b], and Lagan and McLennan [1984].

Three-body collisions in the presence of bound states are discussed in Faddeev [1965] and Newton [1982].

APPENDIX:

SCATTERING THEORY

Scattering theory had its historical beginnings in kinetic theory but has far outgrown its roots. A general theory and a variety of approximation methods have been developed for applications to particle physics, nuclear physics, and atomic and molecular physics. This appendix provides a brief summary of the formalism of scattering theory with emphasis on those aspects which are needed in kinetic theory; it is of course not a substitute for a good book on scattering theory such as Newton [1982].

The general situation to which scattering theory addresses itself can be described as follows. Initially some number of particles are spatially well separated and so move without interaction. They come together in a scattering region and interact. Finally, there is a set of particles which has emerged from the scattering region and moved far enough apart that the interaction between them can again be neglected.

The particles need not be elementary but may be composite systems of other particles which are bound together. The constituent particles will then be called elementary, but what they are depends on the point of view. For example, a molecule may be treated as a bound state of atoms, or from a more detailed point of view, as a bound state of nuclei and electrons. Because of reactions, the final set of particles may be different in number and in kind from those present initially. The essential point is that in both the initial and final states the interaction between particles can be neglected (but of course the interactions which hold a composite particle together are not to be neglected).

The initial and final states are specified by the momenta of the various particles, and the quantum numbers which label the internal states (if any). The transition from the initial to the final state is described by the scattering operator, $\mathcal{S}$. This is a unitary operator which transforms the initial state into the state after the collision. Its matrix elements (the S-matrix) are functions of the initial and final momenta and internal quantum numbers, from which the scattering cross section can be obtained.

Consider the case of a single spinless elementary particle scattering from a fixed central potential, or of two particles described in their center-of-mass frame of reference. Let H, K be the Hamiltonians for the motion under the influence of a potential energy U and for the free motion respectively,

$$H = K + U, \quad K = p^2/2m.$$

Here m is the particle mass if scattering from a force center is considered, or the reduced mass in the two-particle problem. The corresponding time development operators are

$$V(t) = \exp(-iHt), \quad V_0(t) = \exp(-iKt)$$

where units have been used such that $\hbar = 1$.

In the idealized situation the particle has initially a definite momentum $\mathbf{p}$ and its wave function is a plane wave. However, in order to deal with square-integrable functions we will suppose its initial state to be a wave packet ϕ with a small spread in momentum about the value $\mathbf{p}$; the limit to a plane wave will be taken later. We want the wave packet initially to be far enough from the force center that it moves freely according to the time development operator V_0. However, with the passage to a plane wave in mind, we want to use a wave packet with a large spatial extent. To insure that it not overlap with the force center, we first move it backward without interaction for a long period of time. After this preparation of a noninteracting state, we then move the wave packet forward with the interaction included. This two-step process is described by the wave operator

$$\Omega = \lim_{t \to \infty} V(t)\, V_0(-t). \tag{A.1}$$

Actually two different wave operators which differ in the sign of the time can be defined:

$$\Omega_{\pm} = \lim_{t \to \pm\infty} V(t)\, V_0(-t). \tag{A.2}$$

The intention here is that the strong limit be taken, that is, the meaning of this equation is

$$\lim_{t \to \pm\infty} \|V(t)\, V_0(-t) - \Omega_{\pm})\psi\| = 0$$

for any ψ. Conditions on K, H which insure convergence of the limit are known but will not be discussed here. The conditions are general enough to include cases of practical interest, and we assume them to be satisfied.

Of course in practice one does not move a wave packet backwards in order to prepare the initial scattering state, but the result imitates the turning on of an incident beam far from the target. The time dependence of the initial wave packet is given approximately by $\exp(-iEt)$ where $E = p^2/2m$. This is simply a phase factor which drops out when probabilities, or the cross section, are calculated. The operator $V_0(t)$ cancels out the irrelevant phase factor and then gives a result which has a convergent long time limit.

We proceed to discuss some of the important properties of the wave operators. Let Ω stand for either $\Omega_\pm$. Since Ω is a limit of unitary operators, it must be isometric:

$$\Omega^\dagger \Omega = 1. \tag{A.3}$$

Indeed it follows from the existence of the limit in (A.2) that

$$(\Omega\psi, \Omega\psi) = \lim\,(V(t)\,V_0(-t)\,\psi, V(t)\,V_0(-t))\,\psi) = (\psi, \psi).$$

It need not be the case that $\Omega\,\Omega^\dagger = 1$, so Ω may not be unitary. Indeed it is not if the particle has a bound state about the force center, as will be discussed later. (In this case the limit of the product $V_0(t)\,V(-t)$ does not exist.)

The intertwining property of Ω is obtained by the calculation

$$\begin{aligned} V(t')\Omega &= \lim V(t')\,V(t)\,V_0(-t) = \lim V(t'+t)\,V_0(-t) \\ &= \lim V(t'+t)\,V_0(-t'-t)\,V_0(t') = \Omega\,V_0(t'). \end{aligned}$$

By differentiating with respect to t' then setting $t' = 0$, we get

$$H\Omega = \Omega\,K. \tag{A.4}$$

Because of the intertwining property, Ω "turns on" the interaction; it transforms a state of free motion into one with interaction. Thus, if $\phi(t)$ satisfies

$$K\,\phi = i\partial\phi/\partial t$$

then $\psi(t) = \Omega\,\psi(t)$ satisfies

$$H\,\psi = i\partial\psi/\partial t.$$

If ϕ is a plane wave satisfying $K\,\phi = E\,\phi$, then $\psi = \Omega\,\phi$ is a solution to $H\,\psi = E\,\psi$. In momentum representation, a plane wave with momentum $\mathbf{p}_0$ is represented by $\phi = \delta(\mathbf{p} - \mathbf{p}_0)$ so ψ is the matrix element of Ω, $\psi = \Omega(\mathbf{p}, \mathbf{p}_0)$. In particular $\psi_+ = \Omega_+\,\phi$ contains an outgoing scattered wave, while $\psi_- = \Omega_-\,\phi$ has an incoming "scattered" wave.

It follows from Equations (A.3) and (A.4) that

$$\Omega^\dagger\,H\Omega = K.$$

If Ω is unitary, then H and K are equivalent. This is not so if the potential in H is such that bound states exist; then K and H have different spectra and so cannot be equivalent. In this case $\Omega\Omega^\dagger \neq 1$ but instead

$$\Omega\,\Omega^\dagger = 1 - P$$

where P projects onto the bound states. Using Equation (A.3) we find

$$\Omega^\dagger P = 0, \quad P\Omega = 0.$$

The product $V_0(t)\,V(-t)$ does not have a limit on the full Hilbert space but does so on the subspace orthogonal to the bound states and

$$\lim V_0(t)\,V(-t)(1 - P) = \Omega^\dagger.$$

(Of course $\Omega^\dagger$ is defined as the adjoint of Ω and not by a limit such as (A.1).)

The scattering operator $\mathcal{S}$ is defined by

$$\mathcal{S} = \Omega_-^\dagger\,\Omega_+.$$

By using Equation (A.4) and its adjoint

$$\Omega^\dagger H = K\Omega^\dagger.$$

we find that

$$\mathcal{S}K = K\mathcal{S}.$$

Thus, $\mathcal{S}$ transforms a state of free motion into another such state; the initial state of the scattering problem is transformed into the state after the collision.

The proof that $\mathcal{S}$ is unitary proceeds by

$$\mathcal{S}^\dagger\mathcal{S} = \Omega_+^\dagger\,\Omega_-\,\Omega_-^\dagger\,\Omega_+ = \Omega_+^\dagger\,(1 - P)\Omega_+ = \Omega_+^\dagger\,\Omega_+ = 1$$

and in addition

$$\mathcal{S}\mathcal{S}^\dagger = \Omega_-^\dagger\,\Omega_+\,\Omega_+^\dagger\,\Omega_- = \Omega_-^\dagger\,(1 - P)\Omega_- = \Omega_-^\dagger\,\Omega_- = 1.$$

The preceding definitions of the wave operators and scattering operator do not provide a practical basis for calculations. Equations with a greater utility can be obtained as follows. By Abel's theorem

$$\lim_{\epsilon\to 0} \epsilon\int_0^\infty dt\,e^{-\epsilon t}f(t) = \lim_{t\to\infty} f(t)$$

provided that the limit on the right exists. This allows us to write

$$\Omega_\pm = \lim_{\epsilon\to 0} \epsilon\int_0^\infty dt\,e^{-\epsilon t}V(\pm t)\,V_0(\mp t).$$

Let this operate on a plane wave ϕ with energy E, $K\phi = E\phi$. Then

$$\Omega_{\pm}\phi = \lim_{\epsilon\to 0} \epsilon \int_0^{\infty} dt \exp[-\epsilon \pm i(E - H)t]$$
$$= \lim_{\epsilon\to 0} \epsilon[\epsilon \pm i(E - H)]^{-1}.$$

We write this in terms of the resolvent operator $R(z) = (H - z)^{-1}$ where z is a complex number. Then

$$\Omega_{\pm}\phi = \lim_{\epsilon\to 0} (\mp i\epsilon) R(E \pm i\epsilon)\phi.$$

Frequently this formula is abbreviated as

$$\Omega_{\pm} = \lim_{\epsilon\to 0} (\mp i\epsilon) R(E \pm i\epsilon) \tag{A.5}$$

but it must be kept in mind that E is the energy of the plane wave on which $\Omega_{\pm}$ operates. In terms of matrix elements in momentum representation, we have

$$\Omega_{\pm}(\mathbf{p},\mathbf{p}') = \lim_{\epsilon\to 0} (\mp i\epsilon) R(E_{p'} \pm i\epsilon, \mathbf{p}, \mathbf{p}')$$

where $E_p = p^2/2m$.

As a function of z, R has a branch cut along the positive real axis, which corresponds to the continuous spectrum of H. In order to take the limit in Equation (A.5) it is convenient to exhibit this branch cut in the following way. Let R_0 denote the resolvent for K,

$$R_0(z) = (K - z)^{-1}.$$

An algebraic manipulation shows that

$$R(z) = R_0(z) - R_0(z)\,U R(z). \tag{A.6}$$

In momentum representation R_0 is given by

$$R_0(z,\mathbf{p},\mathbf{p}') = (E_p - z)^{-1}\delta(\mathbf{p} - \mathbf{p}').$$

From Equations (A.5) and (A.6) we obtain

$$\begin{aligned}\Omega_{\pm}(\mathbf{p},\mathbf{p}') &= -\lim_{\epsilon\to 0} (\pm i\epsilon)(E_p - E_{p'} \mp i\epsilon)^{-1}[\delta(\mathbf{p} - \mathbf{p}') - (U R(E_{p'} \pm i\epsilon))(\mathbf{p},\mathbf{p}')] \\ &= \delta(\mathbf{p} - \mathbf{p}') - (E_p - E_{p'} \mp i0)^{-1}(U\Omega_{\pm})(\mathbf{p},\mathbf{p}').\end{aligned} \tag{A.7}$$

This can be abbreviated as

$$\Omega_{\pm} = 1 - R_0(E \pm i0)U\Omega_{\pm}$$

where again E is the energy of the plane wave on which $\Omega_\pm$ operates. The notation $\mp i0$ means that the singularity along the positive real axis is to be approached from above or below respectively; this prescription yields a well-defined result.

The quantities $T_\pm = U\Omega_\pm$ are transition operators, or T-matrices. (Frequently in the literature $T^\dagger = \Omega_-^\dagger U$ is denoted by T_-. Usually by "the" T-matrix one means T_+.) The wave operators and T-matrices are related by

$$\Omega_\pm = 1 - R_0(E \pm i0)T_\pm. \tag{A.8}$$

In addition

$$T_\pm = U - UR_0(E \pm i0)T_\pm.$$

The wave functions $\psi_\pm = \Omega_\pm \phi$ where ϕ is a plane wave satisfy

$$\psi_\pm = \phi - (K - E \mp i0)^{-1}U\psi_\pm.$$

These equations are called the Lippman-Schwinger equations. They determine scattering wave functions which consist of the incident plane wave plus an outgoing or incoming scattered wave, and which satisfy $H\psi_\pm = E\psi_\pm$.

Let $\mathbf{k}$ be the momentum of the incident plane wave, which is customarily normalized in coordinate representation so that $\phi(\mathbf{x}) = \exp(i\mathbf{k}\cdot\mathbf{x})$. Denoting the momentum representation by a hat, we find

$$\hat{\phi}(\mathbf{p}) = (2\pi)^{3/2}\,\delta(\mathbf{p} - \mathbf{k}).$$

Then $\hat{\psi}_\pm(\mathbf{p})$ is related to the wave operator by

$$\hat{\psi}_\pm(\mathbf{p}) = (2\pi)^{3/2}\,\Omega_\pm(\mathbf{p},\mathbf{k}).$$

The Lippman-Schwinger equation in momentum representation is

$$\hat{\psi}_\pm(\mathbf{p}) = (2\pi)^{3/2}\,\delta(\mathbf{p} - \mathbf{k}) - 2m(p^2 - k^2 \mp i0)^{-1}(U\hat{\psi}_\pm)(\mathbf{p}). \tag{A.9}$$

Equation (A.9) is converted back to coordinate representation by taking the Fourier transform which yields

$$\psi_\pm(\mathbf{x}) = \exp(i\mathbf{k}\cdot\mathbf{x}) - (2\pi)^{3/2}\int d^3x\,G_\pm(\mathbf{x} - \mathbf{x}')\,U(\mathbf{x}')\,\psi_\pm(\mathbf{x}')$$

where

$$\begin{aligned} G_\pm(\mathbf{x}) &= (2\pi)^{-3/2}\int d^3p\,\exp(i\mathbf{p}\cdot\mathbf{x})\,2m(p^2 - k^2 \mp i0)^{-1} \\ &= \sqrt{2\pi}\,m\frac{e^{\pm ikr}}{r}. \end{aligned}$$

Hence the equation for $\psi_\pm$ is

$$\psi_\pm(\mathbf{x}) = \exp(i\mathbf{k}\cdot\mathbf{x}) - \frac{m}{2\pi}\int d^3x'\frac{\exp(\pm ik|\mathbf{x} - \mathbf{x}'|)}{|\mathbf{x} - \mathbf{x}'|}U(\mathbf{x}')\,\psi_\pm(\mathbf{x}').$$

This shows that ψ_+ and ψ_- contain respectively an outgoing or incoming scattered wave.

The scattering amplitude f is extracted from the asymptotic form of ψ_+ for large r according to

$$\psi_+(\mathbf{x}) \to e^{i\mathbf{k}\cdot\mathbf{x}} + f\,\frac{e^{ikr}}{r}$$

where

$$f = -(m/2\pi)\int d^3x' \exp(i\mathbf{k}'\cdot\mathbf{x}')\,U(\mathbf{x}')\,\psi_+(\mathbf{x}')$$

Here $\mathbf{k}'$ has the same magnitude as $\mathbf{k}$ but points in the direction of observation (that is, the direction of $\mathbf{x}$). Since $U\psi_+ = U\Omega_+\phi = T_+\phi$, the scattering amplitude is essentially the same as a matrix element of T_+,

$$f(\mathbf{k}',\mathbf{k}) = -\frac{m}{2\pi}\int d^3x' \exp(-i\mathbf{k}'\cdot\mathbf{x}')\,T_+\exp(i\mathbf{k}\cdot\mathbf{x}') = -\frac{m}{4\pi^2}T_+(\mathbf{k}',\mathbf{k}).$$

In SI units this reads

$$f(\mathbf{k}',\mathbf{k}) = -2\pi m h T_+(\mathbf{k}',\mathbf{k}).$$

Since the magnitudes of $\mathbf{k}$ and $\mathbf{k}'$ are equal, the energies $E = k^2/2m$ and $E' = k'^2/2m$ are also equal, so T_+ is evaluated "on the energy shell." The cross section is related to the scattering amplitude by

$$\sigma(\mathbf{k}\to\mathbf{k}') = |f(\mathbf{k}',\mathbf{k})|^2. \tag{A.10}$$

Several useful relations can be obtained from Equation (A.6). Define an operator $T(z)$ by

$$U R(z) = T(z)\,R_0(z). \tag{A.11}$$

Then

$$R(z) = R_0(z) - R_0(z)\,T(z)\,R_0(z).$$

Taking a matrix element of this equation, we get

$$R(z,\mathbf{p},\mathbf{p}') = (E_p - z)^{-1}[\delta(\mathbf{p}-\mathbf{p}') - T(z,\mathbf{p},\mathbf{p}')\,(E_{p'} - z)^{-1}].$$

Equation (A.5) then shows that

$$\Omega_\pm(\mathbf{p},\mathbf{p}') = \delta(\mathbf{p}-\mathbf{p}') - (E_p - E_{p'} \mp i0)^{-1}T(E_{p'} \pm i0,\mathbf{p},\mathbf{p}').$$

Comparing this with Equation (A.8) we see that

$$T_\pm(\mathbf{p},\mathbf{p}') = T(E_{p'} \pm i0,\mathbf{p},\mathbf{p}').$$

The term transition operator (or T-matrix) is applied to $T(z)$ as well as to $T_\pm$. If z does not have either of the values $E_{p'} \pm i0$, then $T(z)$ is said to be off the energy shell; $T_{\pm}(\mathbf{p},\mathbf{p}')$ has two of the three energies equal ($z = E_{p'} \pm i0$) and so is said to be half on shell to distinguish it from the fully off shell $T(z)$. The main advantage of the T-matrices is that they have finite on-shell values. This is not the case for the wave operators or S-matrix, which are singular when the initial and final energies are equal.

Since H and K are self-adjoint operators, the resolvents satisfy

$$R^{\dagger}(z) = R(\bar{z}), \quad R_0^{\dagger}(z) = R_0(\bar{z}).$$

It follows that T satisfies a similar relation,

$$T^{\dagger}(z) = T(\bar{z}),$$

or in terms of its matrix elements

$$T^{\dagger}(z,\mathbf{p},\mathbf{p}') = \overline{T}(z,\mathbf{p}',\mathbf{p}) = T(\bar{z},\mathbf{p},\mathbf{p}').$$

By setting $z = E_{p'} - i0$ we obtain

$$T_{+}(\mathbf{p},\mathbf{p}') = \overline{T}(E_{p'} - i0,\mathbf{p}',\mathbf{p}).$$

Hence on the energy shell

$$T_{+}(\mathbf{p},\mathbf{p}') = \overline{T}_{-}(\mathbf{p},\mathbf{p}') = T_{-}^{\dagger}(\mathbf{p},\mathbf{p}'), \quad \text{for} \quad E_p = E_{p'}. \tag{A.12}$$

Since the potential energy in coordinate representation is real, its Fourier transform satisfies $\overline{U}(\mathbf{p}) = U(-\mathbf{p})$ and it follows that

$$\overline{R}(z,\mathbf{p},\mathbf{p}') = R(\bar{z},-\mathbf{p},\mathbf{p}').$$

As a consequence,

$$\overline{T}(z,\mathbf{p},\mathbf{p}') = T(\bar{z},-\mathbf{p},-\mathbf{p}').$$

(This argument is essentially an application of time-reversal invariance.) Inversion symmetry of the potential, $U(\mathbf{x}) = U(-\mathbf{x})$, or $\hat{U}(\mathbf{p}) = \hat{U}(-\mathbf{p})$, leads to $R(z,\mathbf{p},\mathbf{p}') = R(z,-\mathbf{p},-\mathbf{p}')$ which implies

$$T(z,\mathbf{p},\mathbf{p}') = T(z,-\mathbf{p},-\mathbf{p}').$$

These relations can be summarized as

$$T(z,\mathbf{p},\mathbf{p}') = T(z,\mathbf{p}',\mathbf{p}) = T(z,-\mathbf{p},-\mathbf{p}') = \overline{T}(\bar{z},\mathbf{p},\mathbf{p}').$$

Taking $z = E_p + i0 = \mathrm{E}_{p'} + i0$ we find for the scattering amplitude

$$f(\mathbf{p},\mathbf{p}') = f(\mathbf{p}',\mathbf{p}) = f(-\mathbf{p},-\mathbf{p}')$$

and so the cross section satisfies

$$\sigma(\mathbf{p}' \to \mathbf{p}) = \sigma(-\mathbf{p} \to -\mathbf{p}') = \sigma(\mathbf{p} \to \mathbf{p}').$$

The first of these equations follows from time-reversal invariance and is usually referred to as reciprocity; the second equality then follows from inversion symmetry.

The resolvent $R(z)$ satisfies

$$R(z_1) - R(z_2) = (z_1 - z_2)R(z_1)R(z_2)$$

and a similar relation holds for $R_0(z)$. By writing Equation (A.10) for two different values of z and subtracting the two equations, we obtain after some manipulation

$$\begin{aligned} T(z_1) - T(z_2) &= -(z_1 - z_2)\, T(z_1)\, R_0(z_1)\, R_0(z_2)\, T(z_2) \\ &= -(z_1 - z_2)\, T(z_2)\, R_0(z_1)\, R_0(z_2)\, T(z_1). \end{aligned} \tag{A.13}$$

The second form is obtained from the first by interchanging z_1 and z_2. Take a matrix element of this equation and put $z_1 = E_{p'} + i\epsilon$, $z_2 = E_{p'} - i\epsilon$. This yields

$$\begin{aligned} &T(E_{p_{''}} + i\epsilon, \mathbf{p}, \mathbf{p}') - T(E_{p'} - i\epsilon, \mathbf{p}, \mathbf{p}') \\ &\qquad = -2i\epsilon \int d^3p''\, T(E_{p'} - i\epsilon, \mathbf{p}, \mathbf{p}'')\, T(E_{p'} + i\epsilon, \mathbf{p}'', \mathbf{p}') \\ &\qquad\qquad \times [(E_{p'} - E_{p''})^2 + \epsilon^2]^{-1}. \end{aligned} \tag{A.14}$$

In the limit $\epsilon \to 0$ we get

$$\begin{aligned} &T_+(\mathbf{p}, \mathbf{p}') - \overline{T}_+(\mathbf{p}', \mathbf{p}) \\ &\qquad = -2\pi i \int d^3p''\, \delta(E_{p'} - E_{p''})\, T_+(\mathbf{p}'', \mathbf{p}')\, \overline{T}_+(\mathbf{p}'', \mathbf{p}). \end{aligned} \tag{A.15}$$

Putting $E_p = E_{p'}$, we obtain an equation for the scattering amplitude,

$$\begin{aligned} &f(\mathbf{p}, \mathbf{p}') - \bar{f}(\mathbf{p}', \mathbf{p}) \\ &\qquad = (i/2\pi m) \int d^3p''\, \delta(E_{p'} - E_{p''})\, f(\mathbf{p}'', \mathbf{p}')\, \bar{f}(\mathbf{p}'', \mathbf{p}). \end{aligned} \tag{A.16}$$

The volume element can be written as

$$d^3p'' = p''^2\, dp''\, d\Omega'' = mp''\, dE_{p''}\, d\Omega''$$

which yields

$$f(\mathbf{p}, \mathbf{p}') - \bar{f}(\mathbf{p}', \mathbf{p}) = (ip/2\pi) \int d\Omega''\, f(\mathbf{p}'', \mathbf{p}')\, \bar{f}(\mathbf{p}'', \mathbf{p}).$$

In the forward direction ($\mathbf{p} = \mathbf{p}'$) this becomes

$$\operatorname{Im} f(\mathbf{p}, \mathbf{p}) = (p/4\pi) \int d\Omega'' |f(\mathbf{p}'', \mathbf{p})|^2 = (p/4\pi)\sigma_T$$

where σ_T is the total cross section. This is the optical theorem, and Equation

(A.14) or (A.15) is the generalized optical theorem. In SI units, the optical theorem reads

$$\operatorname{Im} f(\mathbf{p},\mathbf{p}) = (p/4\pi\hbar)\sigma_T.$$

If z_1 and z_2 are interchanged in the derivation of the optical theorem, the result is

$$T_+(\mathbf{p},\mathbf{p}') - \overline{T}_+(\mathbf{p}',\mathbf{p}) = -2\pi i \int d^3p''\,\delta(E_{p'} - E_{p''})\,T_+(\mathbf{p},\mathbf{p}'')\,\overline{T}_+(\mathbf{p}',\mathbf{p}'').$$

The relation for the scattering amplitude becomes

$$f(\mathbf{p},\mathbf{p}') - \bar{f}(\mathbf{p}',\mathbf{p})$$

$$= (i/2\pi m)\int d^3p''\,\delta(E_p - E_{p''})\,f(\mathbf{p},\mathbf{p}'')\bar{f}(\mathbf{p}',\mathbf{p}'') \qquad \text{(A.17)}$$

or in the forward direction

$$\operatorname{Im} f(\mathbf{p},\mathbf{p}) = (p/4\pi)\int d\Omega''\,|f(\mathbf{p},\mathbf{p}'')|^2.$$

Thus

$$\int d\Omega'\,\sigma(\mathbf{p}\to\mathbf{p}') = \int d\Omega'\,\sigma(\mathbf{p}'\to\mathbf{p}).$$

Reciprocity and inversion symmetry imply the stronger relation,

$$\sigma(\mathbf{p}\to\mathbf{p}') = \sigma(\mathbf{p}'\to\mathbf{p}).$$

The kinetic energy K has a continuous spectrum along the positive real axis, and this coincides (under reasonable conditions on the potential) with the continuous spectrum of H. If the potential supports bound states, then H also has eigenvalues at a discrete set of points on the negative real axis. (Except for long-range potentials, of which the Coulomb potential is the main example, the number of bound states is finite.) The corresponding properties of $T(z)$ can be described as follows. Let (in momentum representation) $\psi_i(\mathbf{p})$ denote the eigenfunctions of H,

$$H\psi_i = \epsilon_i\psi_i.$$

(The eigenvalues are negative numbers.) The ψ_i may be assumed to form an orthonormal set,

$$\int d^3p\,\overline{\psi}_i(\mathbf{p})\,\psi_j(\mathbf{p}) = \delta_{ij}.$$

Then T can be shown to have the representation

$$T(z,\mathbf{p},\mathbf{p}') = \sum (E_p - \epsilon_i)(E_{p'} - \epsilon_i)(z - \epsilon_i)^{-1}\,\psi_i(\mathbf{p})\overline{\psi}_i(\mathbf{p}') + \tau(z,\mathbf{p},\mathbf{p}').$$

The first term displays the effect of the bound states, which are seen to contribute simple poles to $T(z)$. The remaining function, $\tau(z)$, is analytic in

z except for a branch cut along the positive real axis; furthermore, it has finite limiting values as z approaches this branch cut from above or below. (This is not the case for $R_0(z)$ and $R(z)$ which diverge as z approaches the positive real axis.) The preceding formula yields for the T-matrices on the half shell

$$T_{\pm}(\mathbf{p},\mathbf{p}') = \sum (E_p - \epsilon_i)\, \psi_i(\mathbf{p})\, \overline{\psi}_i(\mathbf{p}') + \tau(E_{p'} \pm i0, \mathbf{p}, \mathbf{p}').$$

One of the basic results of scattering theory is a formula for collision rate. Consider a general collision, in which m beams of particles of various kinds are incident on the scattering region, and n counters are set to observe emitted particles. Let $f_i(\mathbf{p})d^3p$ be the number of incident particles of the i'th kind, per unit volume. Then the rate of collisions, per unit volume, such that emitted particles (of various kinds) have momenta in the ranges $d^3p_1'..d^3p_n'$ is given by

$$d\mathcal{R} = (2\pi)^2\ h^{3m-4} \prod_{i=1}^{m} f_i(\mathbf{p}_i)\, d^3p_i \prod_{j=1}^{n} d^3p_j'\, |T|^2\ \delta(\mathbf{K} - \mathbf{K}')\ \delta(E - E')$$

where T is the T-matrix for the transition (reduced by removal of the δ-function for overall conservation of momentum).

The reactive collision operators derived from the hierarchy in Section 13.9 can be obtained more simply from the collision rate. Consider for example a dissociative collision in which atom number 1 collides with a bound pair of particles 2, 3; for simplicity we assume there to be only a single bound state. The collision rate for this process, expressed in terms of the Wigner functions W_A, W_M for atoms and molecules, is

$$\begin{aligned} d\mathcal{R}_D = (2\pi h)^2\, W_A(\mathbf{p}_1)\, d^3p_1 W_M(\mathbf{P}_{23}) d^3P_{23} \\ \times\ d^3p_1' d^3p_2' d^3p_3' |T_D(\mathbf{k}_1', \mathbf{q}_{23}'|\mathbf{k}_1)|^2\ \delta(\mathbf{K} - \mathbf{K}')\ \delta(E - E') \end{aligned}$$

where T_D denotes the T-matrix for dissociation. The δ-functions impose four restrictions on the nine components of the final momenta, so five scattering parameters are needed to specify the collision. We take these to be the directions of $\mathbf{k}_1'$, $\mathbf{q}_{23}'$ plus the energy w_1'. The volume element for the final momenta can be written

$$d^3p_1' d^3p_2'\ d^3p_3'\ =\ d^3K'\ \mu_3 k_1' dw_1'\ d\Omega(\mathbf{k}_1')\ \mu_2 q_{23}' du_{23}' d\Omega(\mathbf{q}_{23}')$$

where $\mu_2\ =\ m/2$ and $\mu_3\ =\ 2m/3$ are reduced masses for the two- and three-particle systems. The energy difference is

$$E\ -\ E'\ =\ w_1\ +\ \epsilon\ -\ w_1'\ -\ u_{23}'.$$

The product of incident flux times target density is

$$(k_1/\mu_3) W_A(\mathbf{p}_1) d^3p_1 W_M(\mathbf{P}_{23}) d^3P_{23}.$$

To identify the cross section, we divide by this product and integrate over

$\mathbf{K}'$, u'_{23}. This yields

$$\sigma_D(\mathbf{k}_1 \to \mathbf{k}'_1, \mathbf{q}'_{23}) = \frac{d\sigma}{d\Omega(\mathbf{k}'_1)\, d\Omega(\mathbf{q}'_{23})\, dw'_1}$$

$$= (2\pi h)^2\, \mu_2\, \mu_3^2(k'_1\, q'_{23}/k_1)\, |T_D(\mathbf{k}'_1, \mathbf{q}'_{23}|\mathbf{k}_1)|^2.$$

Molecular formation corresponds to a collision in which three atomic beams are incident and both atoms and diatomic molecules are emitted. If the emitted molecule is a bound pair of particles 2, 3, the collision rate is

$$d\mathcal{R}_F = (2\pi)^2\, h^5 W_A(\mathbf{p}_1) d^3p_1 W_A(\mathbf{p}_2)\, d^3p_2 W_A(\mathbf{p}_3)\, d^3p_3$$
$$\times\, d^3p'_1\, d^3P'_{23}|T_F(\mathbf{k}'_1|\mathbf{k}_1, \mathbf{q}_{23})|^2\, \delta(\mathbf{K} - \mathbf{K}')\, \delta(E - E')$$

where T_F is the T-matrix for molecular formation. It is a consequence of microscopic reversibility and invariance to spatial inversions that

$$\overline{T}_D(\mathbf{k}'_1, \mathbf{q}'_{23}|\mathbf{k}_1) = T_F(\mathbf{k}_1|\mathbf{k}'_1, \mathbf{q}'_{23})$$

so the rates of molecular formation and dissociation can be written in terms of either T-matrix.

If a bound pair of particles 2, 3 collides with atom 1, the net rate of loss to molecules with momentum in d^3P_{23} is

$$\text{rate of loss} = \Big\{(2\pi h)^2 \int d^3p_1\, W_A(\mathbf{p}_1) W_M(\mathbf{P}_{23})$$
$$\times \int d^3p'_1 d^3p'_2 d^3p'_3 |T_D(\mathbf{k}'_1, \mathbf{q}'_{23}|\mathbf{k}_1)|^2\, \delta(\mathbf{K} - \mathbf{K}')\, \delta(E - E')\Big\}\, d^3P_{23}.$$

To obtain the corresponding rate of gain, consider a collision leading to a bound pair of particles 2, 3. We want the rate of gain to molecules with momentum in the range d^3P_{23}, and so now denote the initial momenta by a prime. The net rate of gain is

$$\text{rate of gain} = \Big\{(2\pi)^2 h^5 \int d^3p'_1\, d^3p'_2\, d^3p'_3 W_A(\mathbf{p}'_1)\, W_A(\mathbf{p}'_2)\, W_A(\mathbf{p}'_3)$$
$$\times \int d^3p_1 |T_F(\mathbf{k}_1|\mathbf{k}'_1, \mathbf{q}'_{23})|^2\, \delta(\mathbf{K} - \mathbf{K}')\, \delta(E - E')\Big\}\, d^3P_{23}.$$

Combining the gain and loss rates, we obtain the collision operator

$$J_M(\mathbf{P}_{23}) = (2\pi h)^2 \int d^3p'_1\, d^3p'_2\, d^3p'_3\, d^3p_1\, |T_D(\mathbf{k}'_1, \mathbf{q}'_{23}|\mathbf{k}_1)|^2\, \delta(\mathbf{K} - \mathbf{K}')$$
$$\times\, \delta(E - E')[h^3 W_A(\mathbf{p}'_1)\, W_A(\mathbf{p}'_2)\, W_A(\mathbf{p}'_3) - W_A(\mathbf{p}_1)\, W_M(\mathbf{P}_{23})].$$

This can be expressed in terms of the cross section, and the integration carried out in part with the aid of the delta functions. The result is then the same as the collision operator denoted by F in Section 13.9.

By a similar argument, the corresponding collision operator in the kinetic equation for atoms is found to be

$$J_A(\mathbf{p}_1) = 2\,(2\pi h)^2 \int d^3p_1'\, d^3P_{23}'\, d^3p_2\, d^3p_3\, |T_D(\mathbf{k}_1, \mathbf{q}_{23}|\mathbf{k}_1')|^2\, \delta(E - E')\,\delta(\mathbf{K} - \mathbf{K}') \times [W_A(\mathbf{p}_1')W_M(\mathbf{P}_{23}') - h^3\, W_A(\mathbf{p}_1)\, W_A(\mathbf{p}_2)\, W_A(\mathbf{p}_3)].$$

The argument based on collision rate does not apply to the coherent effects associated with off-diagonal elements of the molecular density matrix. It is also inadequate for the derivation of the triple collision operator. Indeed the total collision rate for a collision with three particles in both initial and final states is infinite because of contributions from completed two-body collisions. Just these contributions are removed by the binary-collision terms in the triple collision operator.

BIBLIOGRAPHY

The numbers in parentheses at the end of a reference are the chapter and section in which it is cited.

ALDER, B. J., AND T. WAINWRIGHT (1969): "Velocity Autocorrelations for Hard Spheres," *Phys. Rev. Letters* **18**, 988. (2.12, 12.11)

——— AND ——— (1970): "Decay of the Velocity Autocorrelation Function," *Phys. Rev.* **A1**, 18. (2.12, 12.11)

———, D. M. GASS, AND T. WAINWRIGHT (1971): "Decay of Time Correlations in Two Dimensions," *Phys. Rev.* **A4**, 233. (2.12, 12.11)

———, AND W. E. ALLEY (1978): "Long-Time Correlation Effects on Displacement Distributions," *J. Stat. Phys.* **19**, 341. (2.10)

———, AND ——— (1981): "The Lorentz Gas," in *Perspectives in Statistical Physics*, ed. H. J. Raveche, vol. IX of *Studies in Statistical Mechanics*, eds. E. W. Montroll and J. L. Lebowitz. Amsterdam: North-Holland Publishing Company. (2.10)

ALLE, C., AND Y. POMEAU (1973): "Developpements d'Enskog et du Viriel en Theorie des Gaz Denses." *Physica* **66**, 145. (12.11)

ALLEY, W. E., AND B. J. ALDER (1979): "Modification of Fick's Law," *Phys. Rev. Letters* **43**, 653. (2.10)

———, AND ——— (1983a): "Generalized Transport Coefficients for Hard Spheres," *Phys. Rev.* **A 27**, 3158. (12.11)

———, ———, AND S. YIP (1983b): "The Neutron Scattering Function for Hard Spheres," *Phys. Rev.* **A 27**, 3174. (12.11)

ALTERMAN, Z., K. FRANKOWSKI, AND C. L. PEKERIS (1962): "Eigenvalues and Eigenfunctions of the Linearized Boltzmann Collision Operator for Maxwell Molecules and for Rigid-Sphere Molecules," *Astrophys. J. Suppl. Series* **7**, 291. (5.7, 6.7)

ARKERYD, L. (1972a): "On the Boltzmann Equation. Part I: Existence." *Arch. Rational Mech. Anal.* **45**, 1. (3)

______ (1972b): "On the Boltzmann Equation. Part II: The Full Initial Value Problem," *Arch. Rational Mech. Anal.* **45**, 17. (3)

______ (1982): "Asymptotic Behavior of the Boltzmann Equation with Infinite Range Forces," *Commun. Math. Phys.* **86**, 475. (3)

ARSEN'EV, A. A. (1968): "The Cauchy Problem for the Linearized Boltzmann Equation," *Sov. Phys. Doklady* **10**, 725. (6.8)

ASMOLOV, E. S., N. K. MAKASHEV, AND V. I. NOSIK (1980): "Heat Transfer Between Plane Parallel Plates in a Gas of Maxwellian Molecules," *Sov. Phys. Doklady* **24**, 892. (5.3)

BALESCU, R. (1975): *Equilibrium and Nonequilibrium Statistical Mechanics*. New York: John Wiley and Sons. (11)

BARAJAS, L., L. S. GARCIA-COLIN, AND E. PINA (1973): "On the Enskog-Thorne Theory for a Binary Mixture of Dissimilar Rigid Spheres," *J. Stat. Phys.* **7**, 161. (7.8)

BEDEAUX, D., R. KAPRAL, AND P. MAZUR (1977): "The Effective Shear Viscosity of a Uniform Suspension of Spheres," *Physica* **A88**, 88. (2)

BEENAKKER, J. J. M. (1973): "Nonequilibrium Angular Momentum Polarization in Rotating Molecules," *Acta Physica Austriaca Supp.* **X**, 267. (8.3)

BELINFANTE, F. J. (1939): "On the Spin Angular Momentum of Mesons," *Physica* **6**, 887. (8.3)

BENIN, D. (1970): "Improved Variational Principles for Transport Coefficients," *Phys. Rev.* **B6**, 2777. (4.6)

BERGMANN, P. G., AND J. L. LEBOWITZ (1955): "New Approach to Nonequilibrium Processes," *Phys. Rev.* **99**, 578. (9.12)

BERNARD, W., AND H. B. CALLEN (1959): "Irreversible Thermodynamics of Nonlinear Processes and Noise in Driven Systems," *Revs. Modern Phys.* **31**, 1017. (9.6)

BERNE, B. J., AND D. FORSTER (1971): "Topics in Time-Dependent Statistical Mechanics," *Ann. Rev. Phys. Chem.* **22**, 563. (9.10)

BHATNAGAR, P. L., E. P. GROSS, AND M. KROOK (1954): "A Model for Collision Processes in Gases. I. Small Amplitude Processes in Charged and Neutral One-Component Systems," *Phys. Rev.* **94**, 511. (5.5)

BIXON, M., AND R. ZWANZIG (1969): "The Boltzmann-Langevin Equation and Hydrodynamic Fluctuations," *Phys. Rev.* **187**, 267. (9.11)

______, J. R. DORFMAN, AND K. C. MO (1971): "General Hydrodynamic Equations for the Linearized Boltzmann Equation," *Phys. Fluids* **14**, 1049. (6.1)

BLACKMORE, R. (1987): "Modified Boltzmann Kinetic Equation for Line Shape Functions," *J. Chem. Phys.* **87**, 791. (13.3)

BLUM, L., AND J. L. LEBOWITZ (1969): "Binary Collision Expansion for the Kinetic Equation of Time-Displaced Correlation Functions," *Phys. Rev.* **185**, 273. (12.4)

BOBYLEV, A. V. (1975a): "Fourier Transform Method in the Theory of the Boltzmann Equation for Maxwellian Molecules," *Sov. Phys. Doklady* **20**, 820. (5.4, 5.5)

——— (1975b): "Exact Solutions of the Boltzmann Equation," *Sov. Phys. Doklady* **20**, 822. (5.8)

——— (1976): "A Class of Invariant Solutions of the Boltzmann Equation," *Sov. Phys. Doklady* **21**, 632. (5.8)

——— (1979): "On the Structure of Spatially Homogeneous Normal Solutions of the Nonlinear Boltzmann Equation for a Mixture of Gases," *Sov. Phys. Doklady* **25**, 30. (5.8)

——— (1980): "The Structure of the General Solution and the Classification of Particular Solutions of the Nonlinear Boltzmann Equation for Maxwellian Molecules," *Sov. Phys. Doklady* **25**, 257. (5.8)

——— (1981): "Asymptotic Properties of Solutions of the Boltzmann Equation," *Sov. Phys. Doklady* **26**, 1127. (3)

——— (1982): "The Chapman-Enskog and Grad Methods for Solving the Boltzmann Equation," *Sov. Phys. Doklady* **27**, 29. (4.10)

BOERCKER, D. B., AND J. W. DUFTY (1979): "Degenerate Quantum Gases in the Binary Collision Approximation," *Annals of Phys.* **119**, 43. (13.1)

——— AND ——— (1981): "Quantum Kinetic Theory of Time Correlation Functions," *Phys. Rev.* **A23**, 1952. (13.1)

BOGOLIUBOV, N. N. (1946): "Problems of a Dynamical Theory in Statistical Physics," English trans. by E. K. Gora, in *Studies in Statistical Mechanics* **I**, eds. G. E. Uhlenbeck and J. de Boer. Amsterdam: North-Holland Publishing Co., 1962. (7.1, 7.7)

BOLEY, C. D., AND R. C. DESAI (1973a): "Kinetic Theory of a Dense Gas: Triple-Collision Memory Function," *Phys. Rev.* **A7**, 1700. (7.7)

——— AND ——— (1973b): "Kinetic Theory of Dense Gas: Properties of Collision Kernels of the Bogoliubov Type," *Phys. Rev.* **A7**, 2192. (7.7)

BOON, J. P., AND A. BOUILLER (1976): "Experimental Observation of 'Long Time Tails'?," *Phys. Letters* **55A**, 391. (2.12)

———, AND S. YIP (1980): *Molecular Hydrodynamics*. New York: McGraw-Hill. (9.10)

BORN, M., AND H. S. GREEN (1949): *A General Kinetic Theory of Liquids*. Cambridge: Cambridge University Press. (7.1)

BREY, J. J., R. ZWANZIG, AND J. R. DORFMAN (1981): "Nonlinear Transport Equations in Statistical Mechanics," *Physica* **109A**, 425. (9.13)

——— (1983): "Long Time Behavior of the Burnett Transport Coefficients," *J. Chem. Phys.* **79**, 4585. (9.13, 12.11)

BRINSER, G. B., AND D. W. CONDIFF (1973): "Kinetic Theory of Moderately Dense Gases—Additional Computations of the Triple Collision Contributions to the Transport Coefficients for the Rigid Sphere Model," *J. Chem. Phys.* **59**, 2754. (7.7)

BRUSH, S. G. (1972): *Kinetic Theory*. New York: Pergamon Press. (3, 9.10, 12.10)

BUCKNER, J. K., AND J. H. FERZIGER (1966a): "Linearized Initial Value Problem for a Gas," *Phys. Fluids* **9**, 2309. (5.9)

——— AND ——— (1966b): "Linearized Boundary Value Problem for a Gas and Sound Propagation," *Phys. Fluids* **9**, 2315. (6.5)

BURSCHKA, M. A., AND U. M. TITULAER (1981a): "The Kinetic Boundary Layer for the Fokker-Planck Equation with Absorbing Boundary," *J. Stat. Phys.* **25**, 569. (2.5)

_____ AND _____ (1981b): "The Kinetic Boundary Layer for the Fokker-Planck Equation: Selectively Absorbing Boundaries," *J. Stat. Phys.* **26**, 59. (2.5)

CAFLISCH, R. E. (1980): "The Fluid Dynamic Limit of the Nonlinear Boltzmann Equation," *Comm. Pure and Appl. Math.* **33**, 651. (4.2)

_____ (1983): "Fluid Dynamics and the Boltzmann Equation," in *Studies in Statistical Mechanics* **X**, eds. E. W. Montroll and J. L. Lebowitz. Amsterdam: North-Holland Publishing Company. (4.2)

CALLEN, H. B., AND T. A. WELTON (1951): "Irreversibility and Generalized Noise," *Phys. Rev.* **83**, 34. (9.6)

_____ (1985): *Thermodynamics and an Introduction to Thermostatistics*. New York: John Wiley and Sons. (9.3)

CARLEMAN, T. (1933): "Sur la Theorie de l'Equation Integro-Differentielle de Boltzmann," *Acta Math.* **60**, 91. (3)

_____ (1957): *Problemes Mathematiques dans la Theorie Cinetique des Gas*. Uppsala: Almqvist and Wiksells. (3, 6.7)

CASE, K. M. (1969): "On the Boundary Value Problems of Linear Transport Theory," in *Transport Theory*, SIAM-AMS Proceedings Vol. I, American Mathematical Society, Providence. (6.10)

CASIMIR, H. B. G. (1945): "On Onsager's Principle of Microscopic Reversibility," *Revs. Modern Phys.* **17**, 343. (1.9)

CERCIGNANI, C. (1969): *Mathematical Methods in Kinetic Theory*. New York: Plenum Press. (3, 4.2, 5.5, 6)

_____, C. W. GREENBERG, AND P. F. ZWEIFEL (1979): "Global Solutions of the Boltzmann Equation on a Lattice," *J. Stat. Phys.* **20**, 149. (3)

_____ (1983): "Solution of the Boltzmann Equation," In *Studies in Statistical Mechanics* **X**, eds. E. W. Montroll and J. L. Lebowitz. Amsterdam: North-Holland Publishing Company. (3)

CHANDRASEKHAR, S. (1943): "Stochastic Problems in Physics and Astronomy," *Revs. Modern Phys.* **15**, 1. Reprinted in *Noise and Stochastic Processes*, ed. N. Wax. New York: Dover, 1954. (2.5, 2.8)

CHAPMAN, S., AND T. G. COWLING (1970): *The Mathematical Theory of Non-Uniform Gases*, 3rd ed. Cambridge: Cambridge University Press. (3)

CHAPPELL, W. R. (1970): "Kinetic Equations for Autocorrelation Functions in Dilute Gases," *J. Stat. Phys.* **2**, 267. (11.3)

CHESTER, G. V., AND A. THELLUNG (1959): "On the Electrical Conductivity of Metals," *Proc. Phys. Soc.* **73**, 745. (11.6)

CHOH, S. T. (1958): "The Kinetic Theory of Phenomena in Dense Gases," Ph.D. Dissertation, University of Michigan. Available from University Microfilms International, Ann Arbor, Michigan. (7.7)

CHOQUARD, PH. (1963): "Lattice Energy Currents in Solids and Lattice Thermal Conductivity," *Helv. Phys. Acta* **36**, 415. (8.2)

CHOW, T. S., AND J. J. HERMANS (1972): "Effect of Inertia on the Brownian Motion of Rigid Particles in a Viscous Fluid," *J. Chem. Phys.* **56**, 3150. (2.12)

COHEN, E. G. D. (1962a): "On the Generalization of the Boltzmann Equation to General Order in the Density," *Physica* **28**, 1025. (7.3, 7.9)

——— (1962b): "Cluster Expansions and the Hierarchy. I. Non-Equilibrium Distribution Functions," *Physica* **28**, 1045. (7.3)

——— (1962c): "Cluster Expansions and the Hierarchy. II. Equilibrium Distribution Functions," *Physica* **28**, 1060. (7.3, 11.2)

——— (1963): "On the Kinetic Theory of Dense Gases," *J. Math. Phys.* **4**, 183. (7.3, 7.10)

——— (1973): "The Generalization of the Boltzmann Equation to Higher Densities," *Acta Physica Austriaca Supp.* **X**, 157. (7.3)

COLE, R. G., AND G. T. EVANS (1986): "Dynamics in Polyatomic Fluids: A Kinetic Theory Approach," *Ann. Rev. Phys. Chem.* **37**, 105. (13.4)

CONDIFF, D. W., AND J. S. DAHLER (1964): "Fluid Mechanical Aspects of Antisymmetric Stress," *Phys. Fluids* **7**, 842. (8.3)

COOKE, R. G. (1953): *Linear Operators; Spectral Theory and Some Other Applications*. London: Macmillan. (10.1)

CORNGOLD, N. (1964): "Some Transient Phenomena in Thermalization I. Theory," *Nucl. Sci. Eng.* **19**, 80. (6.8)

——— (1968): "Some Recent Results in the Theory of the Transport of Thermal Neutrons," in *Transport Theory*, SIAM-AMS Proceedings Vol. I, American Mathematical Society, Providence. (6.8)

CUKIER, R. I., R. KAPRAL, J. R. LEBENHAFT, AND J. R. MEHAFFEY (1980): "On the Microscopic Origin of Stokes' Law," *J. Chem. Phys.* **73**, 5244. (2.3)

CURTISS, C. F. (1967): "Transport Phenomena in Gases," *Ann. Rev. Phys. Chem.* **18**, 125. (9.10)

DAHLER, J. S., AND M. THEODOSOPULU (1975): "The Kinetic Theory of Dense Polyatomic Fluids," *Adv. Chem. Phys.* **31**, 155. (13.4)

DE BOER, J. (1948): "Molecular Distributions and Equations of State of Gases," *Repts. Progr. Phys.* **12**, 305. (11.2)

DEKKER, H. (1986): "Kramers' Activation Rate for a Sharp Edged Potential Barrier: The Double Oscillator," *Physica* **136A**, 124. (2.9)

DE GROOT, S. R., AND P. MAZUR (1962): *Non-Equilibrium Thermodynamics*. Amsterdam: North-Holland. (1)

DELALE, C. F. (1982): "The Hilbert Expansion to the Boltzmann Equation for Steady Flow," *J. Stat. Phys.* **28**, 589. (4.1)

DE SCHEPPER, I., H. VAN BEIJEREN AND M. H. ERNST (1974): "The Nonexistence of the Linear Diffusion Equation Beyond Fick's Law," *Physica* **75**, 1. (2.10, 12.11)

———, AND E. G. D. COHEN (1978): "Long Time Behavior of Four Point Velocity Correlations in Hard Disk Systems," *Physica* **93A**, 611. (9.13)

———, AND ——— (1979): "Self-diffusion beyond Fick's law," *Physica* **98A**, 189. (2.10)

———, AND E. G. D. COHEN (1982): "Very-Short-Wavelength Collective Modes in Fluids," *J. Stat. Phys.* **27**, 223. (12.11)

_____ , J. C. van Rijs, and E. G. D. Cohen (1985): "Sound Propagation Gaps from the Navier-Stokes Equations," *Physica* **134A**, 1. (6.2)

DeVault, G. P., and J. A. McLennan (1965a): "Theory of Phonon Contribution to Internal Friction of Solids," *Phys. Rev.* **138**, A856. (11.6)

_____ , and _____ (1965b): "Statistical Mechanics of Viscoelasticity," *Phys. Rev.* **137**, A724. (11.6)

_____ (1966): "Anharmonic Contribution to Momentum-Flux Operator for a Lattice," *Phys. Rev.* **149**, 624. (8.2)

_____ , and R. J. Hardy (1967a): "Lowest-Order Contributions to the Lattice Viscosity," *Phys. Rev.* **155**, 869. (11.6)

_____ (1967b): "Lowest-Order Nonvanishing Contribution to Lattice Viscosity," *Phys. Rev.* **155**, 875. (11.6)

Domaradzki, J. A. (1977): "The Liouville Operator for the Step-Type Particle Interaction," *Physica* **86A**, 169. (12.3)

Donoghue, W. F. (1969): *Distributions and Fourier Transforms*. New York: Academic Press. (10.1)

Dorfman, J. R. (1963): "Note on the Linearized Boltzmann Integral Equation for Rigid Sphere Molecules," *Proc. Nat. Acad. Sci.* **50**, 804. (6.7)

_____ (1967): "The Binary Collision Expansion Method in Kinetic Theory," in *Lectures in Theoretical Physics* **9C**, ed. W. E. Brittin. New York: Gordon and Breach. (12.4)

_____ and E. G. D. Cohen (1972): "Velocity Correlation Functions in Two and Three Dimensions: Low Density," *Phys. Rev.* **A6**, 776. (11.5, 12.10)

_____ (1973): "Velocity Correlation Functions for Moderately Dense Gases," *Acta Physica Austriaca Supp* **X**, 209. (11.5)

_____ and _____ (1975a): "Velocity Correlation Functions in Two and Three Dimensions: II. Higher Density," *Phys. Rev.* **A12**, 292. (11.5, 12.10)

_____ (1975b): "Kinetic and Hydrodynamic Theory of Time Correlation Functions," in *Fundamental Problems in Statistical Mechanics* **III**, ed. E. G. D. Cohen. Amsterdam: North-Holland Publishing Co. (11.5)

_____ , and H. van Beijeren (1977): "The Kinetic Theory of Gases," in *Statistical Mechanics* Pt. B, ed. B. J. Berne. New York: Plenum Press. (7.3)

_____ (1981): "Some Recent Developments in the Kinetic Theory of Gases," in *Perspectives in Statistical Physics*, ed. H. J. Raveche, *Studies in Statistical Mechanics* **IX**, eds. E. W. Montroll and J. L. Lebowitz. Amsterdam: North-Holland Publishing Company. (2.10, 12.10)

Duderstadt, J. J., and W. R. Martin (1979): *Transport Theory*. New York: John Wiley and Sons. (3)

J. W. Dufty (1968a): "Hydrodynamics and Time-Correlation Functions," *Phys. Rev.* **176**, 398. (9.8)

_____ , and J. A. McLennan (1968b): "Correlation Functions as Hydrodynamic Green's Functions," *Phys. Rev.* **172**, 176. (9.8)

_____ (1972): "Long-Time Behaviour of Space- and Time-Dependent Correlation Functions," *Phys. Rev.* **A5**, 2247. (2.10)

——— AND ——— (1974): "Persistent Correlations in Diffusion," *Phys. Rev.* **A9**, 1266. (2.10)

——— (1976a): "Nonlinear Burnett Order Thermal Stress Coefficient," *Phys. Letters* **56A**, 247. (9.13)

——— (1976b): "Kinetic Theory of Multitime Correlation Functions," *Phys. Rev.* **A13**, 2299. (9.13)

———, AND M. J. LINDENFELD (1979): "Nonlinear Transport in the Boltzmann Limit," *J. Stat. Phys.* **20**, 259. (9.13)

———, AND M. C. MARCHETTI (1981): "Viscometric Functions for a Simple Fluid," *J. Chem. Phys.* **75**, 422. (9.13)

——— (1983): "Nonlinear Diffusion and Mode-Coupling Anomalies," *Phys. Rev.Letters* **51**, 2159. (12.11)

——— (1984a): "Viscoelastic and Non-Newtonian Effects in Shear Flow," *Phys. Rev.* **A30**, 622. (12.11)

——— (1984b): "Diffusion in Shear Flow," *Phys. Rev.* **A30**, 1465. (12.11)

———, AND J. M. RUBI (1987): "Generalized Onsager Symmetry," *Phys. Rev.* **A36**, 222. (1.9, 10.2)

EGELSTAFF, P. A. (1967): *An Introduction to the Liquid State*. New York: Academic Press. (9.10)

ELLIS, R., AND M. PINSKY (1975): "The First and Second Fluid Approximations to the Linearized Boltzmann Equation," *J. Math. Pures Appl.* **54**, 125. (6.9)

ERNST, M. H. (1966a): "Formal Theory of Transport Coefficients to General Order in the Density," *Physica* **32**, 209. (9.10, 11.5)

——— (1966b): "Hard Sphere Transport Coefficients from Time Correlation Functions," *Physica* **32**, 273. (11.5)

———, L. K. HAINES AND J. R. DORFMAN (1969a): "Theory of Transport Coefficients for Moderately Dense Gases," *Revs. Modern Phys.* **41**, 296. (11.7, 12.10)

———, J. R. DORFMAN, W. R. HOEGY, AND J. M. J. VAN LEEUWEN (1969b): "Hard Sphere Dynamics and Binary Collision Operators," *Physica* **45**, 127. (12.3, 12.4)

——— (1970): "The Linearized Boltzmann Equation. Navier-Stokes and Burnett Transport Coefficients," *Amer. J. Phys.* **38**, 908. (6.1)

———, AND J. R. DORFMAN (1972): "Non-Analytic Dispersion Relations in Classical Fluids. I. The Hard Sphere Gas," *Physica* **61**, 157. (12.10)

——— AND ——— (1975): "Nonanalytic Dispersion Relations for Classical Fluids. II. The General Fluid," *J. Stat. Phys.* **12**, 311. (12.10)

———, B. CICHOCKI, J. R. DORFMAN, J. SHARMA, AND H. VAN BEIJEREN (1978): "Kinetic Theory of Nonlinear Viscous Flow in Two and Three Dimensions," *J. Stat. Phys.* **18**, 237. (12.11)

——— (1979): "Exact Solution of the Non-Linear Boltzmann Equation for Maxwell Models," *Phys. Letters* **69A**, 390. (5.7)

——— (1981a): "Nonlinear Model-Boltzmann Equations and Exact Solutions," *Phys. Reports* **78**, 1. (5.8)

———, AND H. VAN BEIJEREN (1981b): "On a Conjecture of Alley and Alder for Fluids and Lorentz Models," *J. Stat. Phys.* **26**, 1. (2.10)

——— (1983): "Exact Solutions of the Nonlinear Boltzmann Equation and Related Kinetic Equations," in *Studies in Statistical Mechanics* **X**, eds. E. W. Montroll and J. L. Lebowitz. Amsterdam: North-Holland Publishing Company. (5.8)

ERPENBECK, J. J., AND W. W. WOOD (1981): "Molecular Dynamics Calculations of Shear Viscosity Correlation Functions for Hard Spheres," *J. Stat. Phys.* **24**, 455. (12.11)

——— (1984): "Shear Viscosity of the Hard-Sphere Fluid via Nonequilibrium Molecular Dynamics," *Phys. Rev. Letters* **52**, 1333. (12.11)

EU, B. C. (1971): "Cluster Expansion and Autocorrelation Functions: Self-Diffusion Coefficient," *J. Chem. Phys.* **55**, 4613. (11.5)

——— (1974): "On the Estimate of the Three-Body Contribution to Time-Correlation Functions for Transport Coefficients," *J. Chem. Phys.* **60**, 1914. (7.7)

——— (1975): "Kinetic Equations for Reacting Chemical Systems: Exchange Reaction," *J. Chem. Phys.* **63**, 303. (13.6)

——— (1983): "Shear-Rate Dependence of Viscosity for Simple Fluids," *Phys. Letters* **96A**, 29. (12.11)

EVANS, D. J. (1980): "Enhanced $t^{-3/2}$ Long-Time Tail for the Stress-Stress Time Correlation Function," *J. Stat. Phys.* **22**, 81. (12.11)

EVANS, J. W., D. K. HOFFMAN, AND D. J. KOURI (1983): "The Reactive Quantum Boltzmann Equations: A Derivation from an Arrangement Channel Space Representation and BBGKY Hierarchy," *J. Chem. Phys.* **78**, 2665. (13.6)

——— (1986): "Two-Hilbert Space Formulations of the Quantum Statistical Mechanics of Reactive Fluids: Dimer Formation and Decay," *J. Chem. Phys.* **85**, 5991. (13.6)

FADDEEV, L. D. (1965): *Mathematical Aspects of the Three-Body Problem in the Quantum Scattering Theory*. New York: Daniel Davey and Co. (13.7)

FEDELE, P. D. AND Y. W. KIM (1980): "Direct Measurement of the Velocity Autocorrelation Function for a Brownian Test Particle," *Phys. Rev. Letters* **44**, 691. (2.12)

FELDERHOF, B. U. (1966): "Onsager Relations and the Spectrum of Critical Opalescence," *J. Chem. Phys.* **44**, 602. (1.9)

——— (1978): "Diffusion of Interacting Brownian Particles," *J. Phys.* **A11**, 929. (2)

———, AND U. M. TITULAER (1980): "Normal Modes in Irreversible Thermodynamics," *Physica* **100A**, 563. (6.2)

FELLER, W. (1957): *An Introduction to Probability Theory and its Applications*. New York: John Wiley and Sons. (10.1)

FERZIGER, J. H., AND H. G. KAPER (1972): *Mathematical Theory of Transport Processes in Gases*. Amsterdam: North-Holland Publishing Co. (3)

FILIPPOV, B. V., AND G. B. IL'INSKAYA (1966): "Existence and Uniqueness of the Solution to the Boltzmann Equation," *Sov. Phys. Doklady* **10**, 915. (3)

FOCH, J. D. (1967a): "On the Theory of Sound Propagation in Monatomic Gases and Binary Mixtures," Ph.D. Dissertation, Rockefeller University. Available from University Microfilms International, Ann Arbor, Michigan. (4.10, 6.5)

———, AND G. E. UHLENBECK (1967b): "Propagation of Sound in Monatomic Gases," *Phys. Rev. Letters* **19**, 1025. (4.10, 6.5)

———, AND G. W. FORD (1970): "The Dispersion of Sound in Monoatomic Gases," in *Studies in Statistical Mechanics* **II**, eds. J. de Boer and G. E. Uhlenbeck. Amsterdam: North-Holland Publishing Co. (4.10, 6.5)

FORD, G. W. (1968): "Matrix Elements of the Linearized Collision Operator," *Phys. Fluids* **11**, 515. (5.6)

——— (1973): "Sound from the Boltzmann Equation," *Acta Phys. Austriaca Supp* **X**, 141. (6.5)

FOX, R. F. (1978): "Gaussian Stochastic Processes in Physics," *Phys. Reports* **48**, 179. (2)

——— (1983): "Long-Time Tails and Diffusion," *Phys. Rev.* **A27**, 3216. (12.11)

FURTH, R. (ED.) (1956): *Investigations on the Theory of the Brownian Movement*, trans. of the 1926 edition by A. D. Cowper. New York: Dover. (2)

GALKIN, V. S. (1958): "On a Class of Solutions of Grad's Moment Equations," *J. Appl. Math. Mech.* **22**, 532. (5.3)

——— (1963): "One-Dimensional Unsteady Solution of the Equation for the Kinetic Moments of a Monatomic Gas," *J. Appl. Math. Mech.* **28**, 186. (5.3)

——— (1966): "Exact Solutions of the Kinetic-Moment Equations of a Mixture of Monatomic Gases," *Izv. Akad. Nauk. Fluid Dynamics* **1**, 29. (5.3)

GALLAVOTTI, G. (1969): "Divergences and the Approach to Equilibrium in the Lorentz and Wind-Tree Models," *Phys. Rev.* **185**, 308. (11.5)

———, O. E. LANFORD, AND J. L. LEBOWITZ (1970): "Thermodynamic Limit of Time-Dependent Correlation Functions for One-Dimensional Systems," *J. Math. Phys.* **11**, 2898. (11.5)

GARCIA-COLIN, L. S., AND A. FLORES (1966a): "The Generalization of Choh-Uhlenbeck's Method in the Kinetic Theory of Dense Gases," *J. Math. Phys.* **7**, 254. (7.7)

———, M. S. GREEN, AND F. CHAOS (1966b): "The Chapman-Enskog Solution of the Generalized Boltzmann Equations," *Physica* **32**, 450. (7.7)

———, L. BARAJAS, AND E. PINA (1971): "Corrections to Thorne's Equations for Binary Mixtures," *Phys. Letters* **A37**, 395. (7.8)

GEIGENMULLER, U., U. M. TITULAER, AND B. U. FELDERHOF (1983): "The Approximate Nature of the Onsager-Casimir Reciprocal Relations," *Physica* **119A**, 53. (1.9)

GLIKSON, A. (1972): "On the Existence of General Solutions of the Initial-Value Problem for the Nonlinear Boltzmann Equation," *Arch. Rational Mech. Anal.* **45**, 35. (3)

GOLDHIRSCH, I., AND I. PROCACCIA (1981): "The Number of Propagating Modes in Hydrodynamic Systems," *Physica* **105A**, 330. (10.3)

GOTZE, W., E. LEUTHEUSSER, AND S. YIP (1981): "Correlation Functions of the Hard-Sphere Lorentz Model," *Phys. Rev.* **A24**, 1008. (11.3)

GOYAL, V. K., AND P. K. SHARMA (1980): "Shear Viscosity of Isotopically Disordered Insulating Solids," *Physica* **B101**, 376. (11.6)

GRAD, H. (1949): "On the Kinetic Theory of Rarefied Gases," *Commun. Pure Appl. Math.* **2**, 331. (3.8, 3.9)

_____ (1958): "Principles of the Kinetic Theory of Gases," in *Encyclopedia of Physics*, ed. S. Flugge. New York: Springer-Verlag. (3, 4.2, 12.8)

_____ (1963a): "Asymptotic Theory of the Boltzmann Equation," *Phys. Fluids* **6**, 147 (1.6, 3, 4.2, 6.6, 9.11)

_____ (1963b): "Asymptotic Theory of the Boltzmann Equation, II," in *Proc. Third International Rarefied Gas Dynamics Symposium* **I**, ed. J. A. Laurmann. New York: Academic Press. (3, 4, 6)

_____ (1965): "Asymptotic Equivalence of the Navier-Stokes and Nonlinear Boltzmann Equations," in *Proc. Symp. Appl. Math.* **17**, American Mathematical Society, Providence. (4.1)

_____ (1966): "High Frequency Sound According to the Boltzmann Equation," J. *SIAM Appl. Math.* **14**, 935. (6.5)

GREEN, H. S. (1961): "Theories of Transport in Fluids," *J. Math. Phys.* **2**, 344. (9.10)

GREEN, M. S. (1951): "Brownian Motion in a Gas of Non-Interacting Molecules," *J. Chem. Phys.* **19**, 1036. (3.13)

_____ (1952): "Markoff Random Processes and the Statistical Mechanics of Time-Dependent Phenomena," *J. Chem. Phys.* **20**, 1281. (9.10)

_____ (1954): "Markoff Random Processes and the Statistical Mechanics of Time-Dependent Phenomena. II. Irreversible Processes in Fluids," *J. Chem. Phys.* **22**, 398. (1.3, 9.10, 11.4)

_____ (1956): "Boltzmann Equation from the Statistical Mechanical Point of View," *J. Chem. Phys.* **25**, 836. (7.9, 7.10)

_____ (1958): "The Non-Equilibrium Pair Distribution Function at Low Densities," *Physica* **24**, 393. (7.10)

_____ (1960): "Comment on a Paper of Mori on Time-Correlation Expressions for Transport Properties," *Phys. Rev.* **119**, 829.

_____ , AND R. A. PICCIRELLI (1963): "Basis of the Functional Assumption in the Theory of the Boltzmann Equation," *Phys. Rev.* **132**, 1388. (7.10)

_____ (1964): "Surface Integral Form for Three-Body Collision in the Boltzmann Equation," *Phys. Rev.* **136A**, 905. (7.7)

GREENBERG, W., J. POLEWCZAK, AND P. F. ZWEIFEL (1983): "Global Existence Proofs for the Boltzmann Equation," in *Studies in Statistical Mechanics* **X**, eds. E. W. Montroll and J. L. Lebowitz. Amsterdam: North-Holland Publishing Company. (3)

GRMELA, M. (1977): "Onsager's Symmetry in Higher Order Fluid Dynamics," *Helv. Phys. Acta* **50**, 393. (1.9)

_____ , AND I. ISCOE (178): "Reductions in a Class of Dissipative Dynamical Systems of Macroscopic Physics," *Ann. Inst. Henri Poincare* **28**, 111. (1.9, 10.3)

_____ , R. ROSEN, AND L. S. GARCIA-COLIN (1981): "Compatibility of the Enskog Kinetic Theory with Hydrodynamics," *J. Chem. Phys.* **75**, 5474. (7.8)

GROSS, E. P. AND E. A. JACKSON (1959): "Kinetic Models and the Linearized Boltzmann Equation," *Phys. Fluids* **2**, 432. (5.5)

_____ , AND D. WISNIVESKY (1968): "Space-Time Correlation Functions for the Enskog Dense Gas Theory," *Phys. Fluids* **11**, 1387. (6)

GUIRAUD, J. P. (1968): "Probleme aux Limites Interieur pour l'Equation de Boltzmann Lineaire," *C. R. Acad. Sci.* **266A**, 671. (6.9)

——— (1972): "Probleme aux Limites Interieur pour l'Equation de Boltzmann en Regime Stationnaire, Faiblement Non Lineaire," *J. Mecanique* **11**, 183. (6.9)

GUNTON, J. D. (1979): "Mode Coupling Theory in Relation to the Dynamical Renormalization Group Method," in *Dynamical Critical Phenomena and Related Topics,* ed. C. P. Enz, *Lecture Notes in Physics* **104**. Berlin: Springer-Verlag. (9.14)

HAINES, L. K., J. R. DORFMAN AND M. H. ERNST (1966): "Divergent Transport Coefficients and the Binary Collision Expansion," *Phys. Rev.* **144**, 207. (12.4)

HANLEY, H. J. M., R. D. MCCARTY, AND E. G. D. COHEN (1972): "Analysis of the Transport Coefficients for Simple Dense Fluids: Application of the Modified Enskog Theory," *Physica* **60**, 322. (7.8)

HARDY, R. J. (1963): "Energy-Flux Operator for a Lattice," *Phys. Rev.* **132**, 168. (8.3)

———, R. J. SWENSON, AND W. C. SCHIEVE (1965a): "Perturbation Expansion for Lattice Thermal Conductivity," *J. Math. Phys.* **6**, 1741. (11.6)

——— (1965b): "Lowest-Order Contribution to the Lattice Thermal Conductivity," *J. Math. Phys.* **6**, 1749. (11.6)

HARRIS, S. (1980): "Brownian Motion with an Absorbing Boundary," *J. Chem. Phys.* **72**, 2659. (2.5)

HARRISON, A. K., AND R. ZWANZIG (1986): "On the Relation Between a Fluctuating Diffusion Equation and Long Time Tails in Stationary Random Media," *J. Stat. Phys.* **42**, 935. (12.11)

HAUGE, E. H., AND E. G. D. COHEN (1969): "Normal and Abnormal Diffusion in Ehrenfest's Wind-Tree Model," *J. Math. Phys.* **10**, 397. (12.10)

——— (1970): "Exact and Chapman-Enskog Solutions of the Boltzmann Equation for the Lorentz Model," *Phys. Fluids* **13**, 1201. (6.1)

HECKE, E. (1922): "Uber die Integralgleichung der Kinetische Gastheorie," *Math Z.* **12**, 274. (6.7)

HENDRIKS, E. M., AND T. M. NIEUWENHUIZEN (1982): "Solution to the Nonlinear Boltzmann Equation for Maxwell Models for Nonisotropic Initial Conditions," *J. Stat. Phys.* **29**, 591. (5.7)

HERCZYNSKI, R., AND I. PIENKOWSKA (1980): "Toward a Statistical Theory of Suspension," *Ann. Rev. Fluid Mech.* **12**, 237. (2)

HERMAN, P. T., AND J. A. MCLENNAN (1969): "Quantum Corrections to the Thermal Conductivity and Shear Viscosity of a Moderately Dense Gas," *J. Chem. Phys.* **50**, 3880. (12.6)

HIDY, G. M., AND J. R. BROCK (1970): *The Dynamics of Aerocolloidal Systems.* New York: Pergamon Press. (2)

HILL, T. L. (1956): *Statistical Mechanics: Principles and Selected Applications.* New York: McGraw-Hill. (11.2)

HIRSCHFELDER, J. O., C. F. CURTISS, AND R. B. BIRD (1954): *The Molecular Theory of Gases and Liquids.* New York: John Wiley and Sons. (3)

HOARE, M. R. (1984): "Quadratic Transport and Soluble Boltzmann Equations," *Adv. Chem. Phys.* **56**, 1. (5.8)

HOEGY, W. (1969): "Convergent Generalization of the Boltzmann Equation for a Hard-Sphere Lorentz Gas," *Phys. Rev.* **185**, 210. (12.10)

______ , AND J. V. SENGERS (1970): "Three-Particle Collisions in a Gas of Hard Spheres," *Phys. Rev.* **A2**, 2461. (7.7)

HOFFMAN, D. K., J. J. MUELLER, AND C. F. CURTISS (1965): "Quantum Mechanical Boltzmann Equation," *J. Chem. Phys.* **43**, 2878. (13.1)

______ , D. J. KOURI AND Z. H. TOP (1979): "Kinetic Theory of Reacting Fluid Mixtures. I. The BBGKY Hierarchy in Arrangement Channel Quantum Mechanics," *J. Chem. Phys.* **70**, 4640. (13.6)

HUBBARD, P. S. (1981): "Rotational Brownian Motion II," *Phys. Rev.* **A8**, 1429. (2)

HYNES, J. T., R. KAPRAL, AND M. WEINBERG (1975): "Microscopic Theory of Brownian Motion: Mori Friction Kernel and Langevin Equation Derivation," *Physica* **81A**, 485. (2.3)

______ (1977): "Statistical Mechanics of Molecular Motion in Dense Fluids," *Ann. Rev. Phys. Chem.* **28**, 301. (9.10)

______ , ______ , AND ______ (1979): "Molecular Theory of Translational Diffusion: Microscopic Generalization of the Normal Velocity Boundary Condition," *J. Chem. Phys.* **70**, 1456. (2.3)

IKENBERRY, E., AND C. TRUESDELL (1956): "On the Pressures and the Flux of Energy in a Gas according to Maxwell's Kinetic Theory, I," *J. Rat. Mech. Anal.* **5**, 1. (5.3)

ILLNER, R., AND I. KUSCER (1979): "Collision Operators as Generators of Markov Processes and Their Spectra," *J. Stat. Phys.* **20**, 303. (6.8)

IMAM-RAHAJOE, S., AND C. F. CURTISS (1967): "Collisonal Transfer Contributions in the Quantum Theory of Transport Coefficients," *J. Chem. Phys.* **47**, 5269. (12.6)

IRVING, J. H., AND J. G. KIRKWOOD (1950): "The Statistical Mechanical Theory of Transport Processes. IV. The Equations of Hydrodynamics," *J. Chem. Phys.* **18**, 817. (8.1)

______ , AND R. W. ZWANZIG (1951): "The Statistical Mechanical Theory of Transport Processes, V. Quantum Hydrodynamics." *J. Chem. Phys.* **19**, 1173. (8.2)

JAYNES, E. T. (1963): "Information Theory and Statistical Mechanics," in *Statistical Physics*, ed. K. W. Ford. New York: W. A. Benjamin, Inc. (3.7, 9.4)

______ (1979): "Where Do We Stand on Maximum Entropy?" in *The Maximum Entropy Formalism*, eds. R. D. Levine and M. Tribus. Cambridge: The MIT Press. (3.7, 9.4)

KAMGAR-PARSI, B., AND J. V. SENGERS (1983): "Logarithmic Density Dependence of the Transport Properties of Gases," *Phys. Rev. Letters* **51**, 2163. (12.10)

______ , E. G. D. COHEN, AND I. M. DE SCHEPPER (1987): "Dynamical Processes in Hard-Sphere Fluids," *Phys. Rev.* **A35**, 4781. (6)

KADANOFF, L. P., AND P. C. MARTIN (1963): "Hydrodynamic Equations and Correlation Functions," *Ann. Phys.* **24**, 419. (9.11)

KAN, Y., J. R. DORFMAN, AND J. V. SENGERS (1977): "Logarithmic Terms in the Density Expansion of the Transport Coefficients of Dense Gases: A Survey," in *Proceedings of the Seventh Symposium on Thermophysical Properties*, ed. A. Cezairliyan. The American Society of Mechanical Engineers, New York. (12.10)

KANIEL, S. (1977): "On the Derivation of Thermohydrodynamic Equations from the Boltzmann Equation," *J. Stat. Phys.* **16**, 415. (4.1)

———, AND M. SHINBROT (1978): "The Boltzmann Equation. I. Uniqueness and Local Existence," *Commun. Math. Phys.* **58**, 65. (3)

KAPER, H. G., C. G. LEKKERKERKER, AND J. HEJTMANEK (1982): *Spectral Methods in Linear Transport Theory*. Basel: Birkhauser Verlag. (3)

KARKHECK, J. P., AND G. STELL (1979a): "Transport Properties of the Widom-Rowlinson Hard-Sphere Mixture Model," *J. Chem. Phys.* **71**, 3620. (7.8)

——— AND ——— (1979b): "Bulk Viscosity of Dense Simple Fluid Mixtures," *J. Chem. Phys.* **71**, 3636. (7.8)

KATO, T. (1980): *Perturbation Theory for Linear Operators*. Berlin: Springer-Verlag. (6.8)

KAWASAKI, K., AND I. OPPENHEIM (1964): "Correlation-Function Method for the Transport Coefficients of Dense Gases. I. First Density Correction to the Shear Viscosity," *Phys. Rev.* **136**, A1519. (12.6)

——— AND ——— (1965): "Logarithmic Term in the Density Expansion of Transport Coefficients," *Phys. Rev.* **139**, A1763. (12.10)

———, AND J. D. GUNTON (1973): "Theory of Nonlinear Transport Processes: Nonlinear Shear Viscosity and Normal Stress Effects," *Phys. Rev.* **A8**, 2048. (12.11)

KAWASHIMA, S., A. MATSUMURA, AND T. NISHIDA (1979): "On the Fluid-Dynamical Approximation to the Boltzmann Equation at the Level of the Navier-Stokes Equation," *Commun. Math. Phys.* **70**, 97. (4.1)

KESTIN, J., AND E. A. MASON (1973): "Transport Properties in Gases (Comparison between Theory and Experiment)," in *Transport Phenomena*, ed. J. Kestin, AIP Conference Proceedings Number 11, American Institute of Physics, New York. (4.7)

——— (1977): "Transport Properties of Mixtures," in *Proceedings of the Seventh Symposium on Thermophysical Properties*, ed. A. Cezairliyan. The American Society of Mechanical Engineers, New York. (4.7)

KEYES, T., AND I. OPPENHEIM (1973): "On the Moments of the Displacement of a Tagged Particle," *Physica* **70**, 100. (2.10)

——— (1977): "Principles of Mode-Mode Coupling Theory," in *Statistical Mechanics*, Pt. B, ed. B. J. Berne. New York: Academic Press. (12.10)

———, AND A. J. MASTERS (1985): "Tagged-Particle Motion in Dense Media: Dynamics Beyond the Boltzmann Equation," *Adv. Chem. Phys.* **58**, 1. (2.12, 12.10)

KIM, Y. W., AND J. E. MATTA (1973): "Long-Time Behavior of the Velocity Autocorrelation: A Measurement," *Phys. Rev. Letters* **31**, 208. (2.12)

KINCAID, J. M. (1978): "Mutual (Isothermal) Diffusion in Binary Hard-Sphere Mixtures: The Effects of the Onsager Reciprocity Relations," *Phys. Letters* **64A**, 429. (7.8)

KIRKPATRICK, T. R. (1984): "Large Long-Time Tails and Shear Waves in Dense Classical Liquids," *Phys. Rev. Letters* **53**, 1735. (12.11)

——— (1985): "Short-Wavelength Collective Modes and Generalized Hydrodynamic Equations for Hard-Sphere Particles," *Phys. Rev.* **A32**, 3130. (12.11)

______, AND J. C. NIEUWOUDT (1986a): "Stability Analysis of a Dense Hard-Sphere Fluid Subjected to Large Shear: Shear-Induced Ordering," *Phys. Rev. Letters* **56**, 885. (12.11)

______, AND ______ (1986b): "Mode-Coupling Theory of the Large Long-Time Tails in the Stress-Tensor Autocorrelation Function," *Phys. Rev.* **B33**, 2651. (12.11)

KIRKWOOD, J. G. (1946): "The Statistical Mechanical Theory of Transport Processes. I. General Theory," *J. Chem. Phys.* **14**, 180; errata **14**, 347. (2.11, 7.1)

______ (1947): "The Statistical Mechanical Theory of Transport Processes. II. Transport in Gases," *J. Chem. Phys.* **15**, 72. (7.1)

KLIMONTOVICH, IU. L., AND D. KREMP (1981): "Quantum Kinetic Equations in Systems with Bound States," *Physica* **109A**, 517. (13.7)

KOGAN, M. K. (1969): *Rarefied Gas Dynamics*. New York: Plenum Press. (3)

KOHLER, M. (1948): "Behandlung von Nichtgleichgewichtsvorgangen mit Hilfe eines Extremalprinzipes," *Z. Physik* **124**, 772. (4.6)

KONIJNENDIJK, H. H. U., AND J. M. J. VAN LEEUWEN (1973): "Kinetic Equation for Hard-Sphere Correlation Functions," *Physica* **64**, 342. (12.3)

KRAEFT, W. D., M. SCHLANGES, AND D. KREMP (1986): "A Quantum Kinetic Equation for Non-Ideal Gases in the Three-Particle Collision Approximation," *J. Phys.* **A19**, 3251. (13.1)

KRAMERS, H. A. (1940): "Brownian Motion in a Field of Force and the Diffusion Model of Chemical Reactions," *Physica* **7**, 284. (2.9)

KROOK, M., AND T. T. WU (1976): "Formation of Maxwellian Tails," *Phys. Rev. Letters* **36**, 1107. (5.8)

______, AND ______ (1977): "Exact Solutions of the Boltzmann Equation," *Phys. Fluids* **20**, 1589. (5.8)

KUBO, R. (1957a): "Statistical-Mechanical Theory of Irreversible Processes I. General Theory and Simple Applications to Magnetic and Conduction Problems," *J. Phys. Soc. Japan* **12**, 570. (9.6)

______, M. YOKOTA, AND S. NAKAJIMA (1957b): "Statistical-Mechanical Theory of Irreversible Processes. II. Response to Thermal Disturbance," *J. Phys. Soc. Japan* **12**, 1203. (9.10)

______ (1966): "The Fluctuation-Dissipation Theorem," *Repts. Progr. Phys.* **29**, 255. (9.6)

KUSCER, I., AND M. M. R. WILLIAMS (1967): "Relaxation Constants of a Hard Sphere Gas," *Phys. Fluids* **10**, 1922. (6.7)

______, AND R. ILLNER (1980): "Eigenvalues of Fokker-Planck Operators," *J. Phys.* **A13**, 2599. (2.5)

LAGAN, S., AND J. A. MCLENNAN (1984): "Kinetic Theory for a Weakly-Associated Diatomic Gas with Reactions," *Physica* **128A**, 178. (13.7)

LAMB, H. (1932): *Hydrodynamics*. Cambridge: Cambridge University Press. (2.12)

LANDAU, L. D., AND E. M. LIFSHITZ (1959): *Fluid Mechanics*. London: Pergamon Press. (1, 2.12, 9.11)

LAX, M. (1960): "Fluctuations from the Nonequilibrium Steady State," *Revs. Modern Phys.* **32**, 25. (9.6)

LEBOWITZ, J. L., AND P. G. BERGMANN (1957): "Irreversible Gibbsian Ensembles," *Annals of Phys.* **1**, 1. (9.12)

———, AND A. SHIMONY (1962): "Statistical Mechanics of Open Systems," *Phys. Rev.* **128**, 1945. (9.12)

———, AND J. PERCUS (1967): "Kinetic Equations and Density Expansions: Exactly Solvable One-Dimensional System," *Phys. Rev.* **155**, 122. (7.3)

———, ———, AND J. SYKES (1968): "Time Evolution of the Total Distribution Function of a One-Dimensional System of Hard Rods," *Phys. Rev.* **171**, 224. (7.3)

———, ———, AND——— (1969): "Kinetic-Equation Approach to Time-Dependent Correlation Functions," *Phys. Rev.* **188**, 487. (11.3, 11.7)

LEKKERKERKER, N. N. W., AND W. G. LAIDLAW (1974): "Onsager Symmetry Relations and the Maximum Number of Propagating Modes," *Phys. Rev.* **A9**, 431. (10.3)

LEUTHEUSSER, E., S. YIP, B. J. ALDER, AND W. E. ALLEY (1983): "Dynamical Correlations in a Hard-Disk Fluid: Generalized Enskog Theory," *J. Stat. Phys.* **32**, 503. (7.8)

LEWIS, R. M. (1961): "Solution of the Equations of Statistical Mechanics," *J. Math. Phys.* **2**, 222. (7.9)

LHUILLIER, C., AND F. LALOE (1982a): "Transport Properties in a Spin Polarized Gas, I," *J. Physique* **43**, 197. (13.5)

———, AND ——— (1982b): "Transport Properties in a Spin Polarized Gas, II," *J. Physique* **43**, 225. (13.5)

———, AND ——— (1983): "Transport Properties in a Spin Polarized Gas, III," *J. Physique* **44**, 1. (13.5)

LINDENFELD, M. (1977): "Identity for Memory Operators in Classical Kinetic Theory," *Phys. Rev.* **A15**, 1801. (11.7)

LOYALKA, S. K., AND J. F. FERZIGER (1967): "Model Dependence of the Slip Coefficient," *Phys. Fluids* **10**, 1833. (4.2)

LOWRY, J. T., AND R. F. SNIDER (1974): "Kinetic Theory of Dimer Formation and Decay," *J. Chem. Phys.* **61**, 2320. (13.7)

LUTSKO, J. F., AND J. W. DUFTY (1986): "Possible Instability for Shear-Induced Order-Disorder Transition," *Phys. Rev. Letters* **57**, 2775. (12.11)

MACHTA, J., M. H. ERNST, H. VAN BEIJEREN, AND J. R. DORFMAN (1984): "Long Time Tails in Stationary Random Media II: Applications," *J. Stat. Phys.* **35**, 413. (12.11)

——— (1986a): "Anomalous Long Time Tails Due to Trapping," *J. Stat. Phys.* **42**, 941. (12.11)

———, AND B. REINHOLD (1986b): "Decay of Correlations in the Regular Lorentz Gas," *J. Stat. Phys.* **42**, 949. (12.11)

MAKASHEV, N. K. (1979): "On the Boundary Conditions for the Equations of Gas Dynamics Corresponding to the Higher Approximations in the Chapman-Enskog Method of Solution of the Boltzmann Equation," *Fluid Dynamics* **14**, 385. (4.2)

MARCHETTI, M. C., AND J. W. DUFTY (1980): "Memory Function for the Velocity Autocorrelation Function at Moderate Densities," *Chem. Phys. Letters* **70**, 539. (13.6)

_____ AND _____ (1981): "Bound State and Finite Collision Time Effects in the Binary Collision Approximation," *Phys. Rev.* **A24**, 2117. (13.6)

MASON, E. A., AND L. MONCHICK (1962a): "Heat Conductivity of Polyatomic and Polar Gases," *J. Chem. Phys.* **36**, 1622. (13.4)

_____ AND _____ (1962b): "Transport Properties of Polar-Gas Mixtures," *J. Chem. Phys.* **36**, 2746. (13.4)

MASTERS, A. J., AND T. KEYES (1985): "Translational and Rotational Diffusion of an Anisotropic Particle in a Molecular Liquid: Long-Time Tails and Brownian Limit," *J. Stat. Phys.* **39**, 215. (2)

MATKOWSKI, B. J., Z. SCHUSS, AND C. TIER (1984): "Uniform Expansion of the Transition Rate in Kramers' Problem," *J. Stat. Phys.* **35**, 443. (2.9)

MAYER, J. E., AND M. G. MAYER (1977): *Statistical Mechanics.* New York: John Wiley and Sons. (11.2)

MAZENKO, G. F. (1972a): "Properties of the Low-Density Memory Function," *Phys. Rev.* **A5**, 2545. (11.7)

_____, T. Y. C. WEI, AND S. YIP (1972b): "Thermal Fluctuations in a Hard-Sphere Gas," *Phys. Rev.* **A6**, 1981. (11.7)

_____ (1974): "Fully Renormalized Kinetic Theory. III Density Fluctuations," *Phys. Rev.* **A9**, 360. (11.7)

_____, AND S. YIP (1977): "Renormalized Kinetic Theory of Dense Fluids," in *Statistical Mechanics*, Pt. B, ed. B. J. Berne. New York: Academic Press. (11.7)

MAZUR, R., AND W. VAN SAARLOOS (1982): "Many-Sphere Hydrodynamic Interactions and Mobilities in a Suspension," *Physica* **115A**, 21. (2)

MCLENNAN, J. A. (1959): "Statistical Mechanics of the Steady State," *Phys. Rev.* **115**, 1405. (9.12)

_____ (1960): "Statistical Mechanics of Transport in Fluids," *Phys. Fluids* **3**, 493. (9.12)

_____ (1963a): "The Formal Statistical Theory of Transport Processes," *Adv. Chem. Phys.* **5**, 261. (9.12)

_____, AND J. R. SWENSON (1963b): Theory of Transport Coefficients in Low Density Gases. *J. Math. Phys.* **4**, 1527. (11.6)

_____ (1963c): "Theory of Transport in Moderately Dense Gases," *Phys. Letters* **7**, 332. (12.3)

_____ (1965): "Convergence of the Chapman-Enskog Expansion for the Linearized Boltzmann Equation," *Phys. Fluids* **8**, 1580. (6.6, 6.8)

_____ (1966a): "Symmetry of the Stress Tensor," *Physica* **32**, 689. (8.3)

_____ (1966b): "Correlation Functions for Dilute Systems," *Phys. Fluids* **9**, 1581. (2.5, 6.5, 6.8)

_____ (1967): "Space and Time Dependent Fluctuations in a Continuous Medium," *Helv. Phys. Acta* **40**, 645. (9.8)

_____ (1973): "Burnett Coefficients and Correlation Functions," *Phys. Rev.* **A8**, 1479. (2.10, 9.13)

_____ (1974): "Onsager's Theorem and Higher Order Hydrodynamic Equations," *Phys. Rev.* **A10**, 1272. (1.9, 10.3)

——— (1981): "Bound-State Contributions to the Triple-Collision Operator," *Physica* **106A**, 278. (13.7)

——— (1982a): "Triple-Collision Operator for a Quantum Gas with Bound States," *J. Stat. Phys.* **28**, 257. (13.7)

——— (1982b): "Kinetic Equations for a Quantum Gas with Bound States," *J. Stat. Phys.* **28**, 521. (13.7)

MEAD, C. A. (1982): "Generalizations of the Entropy Concept: Principles, Limitations, and Prospects for Applications," *Adv. Chem. Phys.* **51**, 113. (3.7)

MEIXNER, J., AND H. G. REIK (1959): "Thermodynamik der Irreversiblen Prozesse," in *Encyclopedia of Physics*, Vol. III/2, ed. S. Flugge. Berlin: Springer-Verlag. (9.14)

MICHAELS, I. A., AND I. OPPENHEIM (1975): "Long-Time Tails and Brownian Motion," *Physica* **81A**, 221. (2.12)

MILLER, D. G. (1973): "The Onsager Relations: Experimental Evidence," in *Foundations of Continuum Thermodynamics*, eds. J. J. Delgado Domingos, M. N. R. Nina, and J. H. Whitelaw. New York: John Wiley and Sons. (1.9)

MILLIKAN, R. (1917): *The Electron*. Chicago: The University of Chicago Press. 1935 revision titled *Electrons + and −*. (2)

MONCHICK, L., AND E. A. MASON (1961): "Transport Properties of Polar Gases," *J. Chem. Phys.* **35**, 1676. (13.4)

———, K. S. YUN, AND E. A. MASON (1963a): "Relaxation Effects in the Transport Properties of a Gas of Rough Spheres," *J. Chem. Phys.* **38**, 1282. (13.4)

———, ———, AND ——— (1963b): "Formal Kinetic Theory of Transport Phenomena in Polyatomic Gas Mixtures," *J. Chem. Phys.* **39**, 654. (13.4)

———, A. N. G. PEREIRA AND E. A. MASON (1965): "Heat Conductivity of Polyatomic and Polar Gases and Gas Mixtures," *J. Chem. Phys.* **42**, 3241. (13.3)

MORAAL, H. (1975): "Quantum Kinetic Theory of Polyatomic Gases," *Phys. Reports* **17**, 225. (4.10, 13.3)

MORGENSTERN, D. (1954): "General Existence and Uniqueness Proof for Spatially Homogeneous Solutions of the Maxwell-Boltzmann Equation in the Case of Maxwellian Molecules," *Proc. Nat. Acad. Sci.* **40**, 719. (3)

——— (1955): "Analytical Studies Related to the Maxwell-Boltzmann Equation," *J. Rat. Mech. Anal.* **4**, 533. (3)

MORI, H. (1958a): "Time Correlation Functions in the Statistical Mechanics of Transport Processes," *Phys. Rev.* **111**, 694. (9.7, 9.10)

——— (1958b): "Statistical-Mechanical Theory of Transport in Fluids," *Phys. Rev.* **112**, 1829. (9.7, 9.10)

——— (1962): "Collective Motion of Particles at Finite Temperatures," *Prog. Theor. Phys.* **28**, 763. (9.10)

——— (1965a): "Transport, Collective Motion and Brownian Motion," *Prog. Theor. Phys.* **33**, 423. (2.3, 9.8, 11.7)

——— (1965b): "A Continued-Fraction Representation of the Time-Correlation Function," *Prog. Theor. Phys.* **34**, 399. (11.7)

MOTT-SMITH, H. M. (1954): *A New Approach in the Kinetic Theory of Gases*. Lincoln

Laboratory Group Report V-2, MIT, available from Defense Documentation Center, Arlington, VA (4.2, 5.6, 5.7, 6.8)

MOUNTAIN, R. D. (1966): "Spectral Distribution of Scattered Light in a Simple Fluid," *Revs. Modern Phys.* **38**, 205. (9.11)

NELKIN, M., AND A. GHATAK. (1964): "Simple Binary Collision Model for van Hove's $G_s(r, t)$," *Phys. Rev.* **135A**, 4. (11.3)

______, AND S. YIP (1966): "Brillouin Scattering by Gases as a Test of the Boltzmann Equation," *Phys. Fluids* **9**, 380. (11.3)

NEWTON, R. G. (1982): *Scattering Theory of Waves and Particles.* New York: Springer-Verlag. (13.7, 14)

NIKOL'SKII, A. A. (1964a): "The Simplest Exact Solutions of the Boltzmann Equation for the Motion of a Rarefied Gas," *Sov. Phys. Doklady* **8**, 633. (5.8)

______ (1964b): "The Three-Dimensional Expansion-Contraction of a Rarefied Gas with Power Interaction Functions," *Sov. Phys. Doklady* **8**, 639. (5.8)

NISHIDA, T. (1978): "Fluid Dynamical Limit of the Nonlinear Boltzmann Equation to the Level of the Compressible Euler Equation," *Commun. Math. Phys.* **61**, 119. (4.1)

NYQUIST, H. (1928): "Thermal Agitation of Electric Charge in Conductors," *Phys. Rev.* **32**, 110. (9.6)

OLMSTED, R. D., AND R. F. SNIDER (1976a): "Differences in Fluid Dynamics Associated wth an Atomic Versus a Molecular Description of the Same System," *J. Chem. Phys.* **65**, 3407. (8.3, 13.7)

______, AND ______ (1976b): "Symmetry of the Pressure Tensor in Macromolecular Fluids," *J. Chem. Phys.* **65**, 3423. (8.3, 13.7)

ONO, S., AND T. SHIZUME (1963): "Statistical Mechanics of Transport Phenomena in Gases at Moderate Density," *J. Phys. Soc. Japan* **18**, 29. (11.5)

ONSAGER, L. (1931a): "Reciprocal Relations in Irreversible Processes. I," *Phys. Rev.* **37**, 405. (1.9)

______ (1931b): "Reciprocal Relations in Irreversible Processes. II," *Phys. Rev.* **38**, 2265. (1.9)

PASTERNAK, V. E., A. A. SENKEVICH, A. A. IUSHKANOV, AND IU. I. IALOMOV (1980): "Thermal Slip of a Moderately Dense Gas along a Flat Surface," *J. Eng. Physics* **37**, 933. (4.2)

PAO, Y. (1974a): "Boltzmann Collision Operator with Inverse-Power Intermolecular Potentials, I," *Commun. Pure Appl. Math.* **27**, 407. (6.7)

______ (1974b): "Boltzmann Collision Operator with Inverse-Power Intermolecular Potentials, II," *Commun. Pure Appl. Math.* **27**, 559. (6.7)

PAUL, G. L., AND P. N. PUSEY (1981): "Observation of a Long-Time Tail in Brownian Motion," *J. Phys. A* **14**, 3301. (2.12)

PAZY, A. (1983): *Semigroups of Linear Operators and Applications to Partial Differential Equations.* New York: Springer-Verlag. (6.9)

PEKERIS, C. L. (1955): "Solution of the Boltzmann-Hilbert Integral Equation," *Proc. Nat. Acad. Sci.* **41**, 661. (4.6)

———, Z. ALTERMAN, L. FINKELSTEIN, AND K. FRANKOWSKI (1962): "Propagation of Sound in a Gas of Rigid Spheres," *Phys. Fluids* **5**, 1608. (6.5)

PERRIN, J. (1914): *Les Atomes.* Paris: Librairie Felix Alcan. (2)

PETERSON, R. L. (1967): "Formal Theory of Nonlinear Response," *Revs. Modern Phys.* **39**, 69. (9.6)

PICCIRELLI, R. A. (1966): "Some Properties of the Long-Time Values of the Probability Densities for Moderately Dense Gases," *J. Math. Phys.* **7**, 922. (7.10)

PINSKY, M. (1976): "On the Navier-Stokes Approximation to the Linearized Boltzmann Equation," *J. Math. Pures Appl.* **55**, 217. (6.9)

POMEAU, Y. (1968): "A New Kinetic Theory for a Dense Classical Gas," *Phys. Letters* **27A**, 601. (12.10)

——— (1971a): "Phenomenes de Transport pour des Molecules Optiquement Actives," *Phys. Lett.* **A34**, 143. (10.3)

——— (1971b): "Correlation Length in a Nonequilibrium Gas," *J. Math. Phys.* **12**, 2286. (12.10)

——— (1971c): "Transport Theory for a Two-Dimensional Dense Gas," *Phys. Rev.* **A3**, 1174. (12.10)

———, AND P. RESIBOIS (1975): "Time Dependent Correlation Functions and Mode-Mode Coupling Theories," *Phys. Reports* **19C**, 63. (12.10, 12.11)

PRIGOGINE, I. (1949): "Le Domaine de Validite de la Thermodynamique des Phenomenes Irreversibles," *Physica* **15**, 272. (3.8)

——— (1962): *Nonequilibrium Statistical Mechanics.* New York: Interscience Publishers. (11)

——— (1967): "*Introduction to Thermodynamics of Irreversible Processes*. New York: Interscience Publishers. (1)

RANGANATHAN, S., AND S. YIP (1966): "Time-Dependent Correlations in a Maxwell Gas," *Phys. Fluids* **9**, 372. (11.3)

RANNINGER, J. (1965): "Lattice Thermal Conductivity," *Phys. Rev.* **140**, A2031. (11.6)

REICHL, L. E. (1981): "Rotational Brownian Motion in a Fluid with Internal Degrees of Freedom," *Phys. Rev.* **A24**, 1617. (2)

——— (1982): "Effect of Fluid Rotational Degrees of Freedom on Brownian Motion: Estimates for Water," *Phys. Rev. Letters* **49**, 85. (2)

RESIBOIS, P. (1963): "Structure of the Three-Particle Scattering Operator in Classical Gases," *J. Math. Phys.* **4**, 166. (7.7)

——— (1965): "The Three Particle Collision Operator in Quantum Mechanics," *Physica* **31**, 645. (7.7)

——— (1966): "The Three-Body Collision Operator in Quantum Mechanics. II. Renormalization and Quasiparticles," *Physica* **32**, 1473. (7.7)

RISKEN, H. (1984): *The Fokker-Planck Equation: Methods of Solution and Applications*. New York: Springer-Verlag. (2.4)

RONIS, D. (1979): "Statistical Mechanics of Systems Nonlinearly Displaced from Equilibrium," *Physica* **99A**, 403. (9.13)

______ (1984a): "Theory of Fluctuations in Colloidal Suspensions Undergoing Steady Shear Flow," *Phys. Rev.* **A29**, 1453. (2)

______ (1984b): "Microscopic Theory of Photoacoustic Pulse Generation," *Phys. Rev.* **A29**, 2125. (9.6)

RYTOV, S. M. (1958a): "Correlation Theory of Thermal Fluctuations in an Isotropic Medium," *Sov. Phys. JETP* **6**, 130. (9.11)

______ (1958b): "Correlation Theory for Rayleigh Scattering of Light I," *Sov. Phys. JETP* **6**, 401. (9.11)

______ (1958c): "Correlation Theory for Rayleigh Scattering of Light II," *Sov. Phys. JETP* **6**, 513. (9.11)

SAHU, D. N., AND P. K. SHARMA (1983): "Thermal Conductivity of an Imperfect Anharmonic Crystal," *Phys. Rev.* **B28**, 3200. (11.6)

SATHER, N. F., AND J. S. DAHLER (1962): "Approximate Theory of Viscosity and Thermal Conductivity in Dense Polyatomic Fluids," *Phys. Fluids* **5**, 754. (8.3)

SCHARF, G. (1967): "Functional-Analytic Discussion of the Linearized Boltzmann Equation," *Helv. Phys. Acta* **40**, 929. (6.9)

______ (1969): "Normal Solutions of the Linearized Boltzmann Equation," *Helv. Phys. Acta* **42**, 5. (6.9)

SENGERS, J. V. (1966): "Triple Collision Contribution to the Transport Coefficients of a Rigid Sphere Gas," *Phys. Fluids* **9**, 1333. (7.7)

______ (1967): "Triple Collision Contribution to the Transport Coefficients of Gases," in *Lectures in Theoretical Physics* **9C**, ed. W. E. Brittin. New York: Gordon and Breach. (7.7)

______ (1971): "Triple Collision Effects in the Transport Properties for a Gas of Hard Spheres," in *Kinetic Equations*, eds. R. L. Liboff and N. Rostoker. New York: Gordon and Breach. (7.7)

______, M. H. ERNST AND D. T. GILLESPIE (1972): "Three-Particle Collision Integrals for a Gas of Hard Spheres," *J. Chem. Phys.* **56**, 5583. (7.7)

______ (1973): "The Three-Particle Collision Term in the Generalized Boltzman Equation," *Acta Phys. Austriaca Supp.* **X**, 177. (7.7)

______, D. T. GILLESPIE, AND J. J. PEREZ-ESANDI (1978): "Three-Particle Collision Effects in the Transport Properties of a Gas of Hard Spheres," *Physica* **90A**, 365. (7.7)

SIEGERT, A. J. F., AND E. TERAMOTO (1958): "Simplified Derivation of the Binary Collision Expansion and It's Connection with the Scattering Operator Expansion," *Phys. Rev.* **110**, 1232. (12.4)

SIMONS, S. (1978): "Is the Solution of the Boltzmann Equation Positive?" *Physics Letters* **69A**, 239. (3)

SIROVICH, L., AND J. K. THURBER (1963): "Sound Propagation According to the Kinetic Theory of Gases," in *Proc. Third International Rarefied Gas Dynamics Symposium*, Vol. I, ed. J. A. Laurmann. New York: Academic Press. (6.5)

______ (1966): "Mixtures of Maxwell Molecules," *Phys. Fluids* **9**, 2323. (5.5)

______, AND J. K. THURBER (1969): "Wave Propagation and Other Spectral Problems in Kinetic Theory," *J. Math. Phys.* **10**, 239. (6.5, 6.8)

SKVORTSOV, G. E. (1966): "Propagation Solutions of the Boltzmann Equation," *Sov. Phys. JETP* **22**, 864. (6.8)

——— (1967): "Spectrum and Evolution of Boltzmann Systems," *Sov. Phys. JETP* **25**, 853. (6.8)

SNIDER, R. F., AND C. F. CURTISS (1958): "Kinetic Theory of Moderately Dense Gases," *Phys. Fluids* **1**, 122. (7.6)

——— (1960): "Quantum-Mechanical Modified Boltzmann Equation for Degenerate Internal States," *J. Chem. Phys.* **32**, 1051. (13.3)

———, AND ——— (1963): "Kinetic Theory of Moderately Dense Gases: Inverse Power Potentials," *Phys. Fluids* **6**, 1020. (7.6)

——— (1964a): "Variational Methods for Solving the Boltzmann Equation," *J. Chem. Phys.* **41**, 591. (4.6)

——— (1964b): "Perturbation Variation Methods for a Quantum Boltzmann Equation," *J. Math. Phys.* **5**, 1580. (4.6)

———, AND B. C. SANCTUARY (1971): "Generalized Boltzmann Equation for Molecules with Internal States," *J. Chem. Phys.* **55**, 1555. (13.7)

———, AND J. T. LOWRY (1974): "Quantum Kinetic Theory of Chemical Recombination," *J. Chem. Phys.* **61**, 2330. (13.7)

SPOHN, H. (1979): "Boltzmann Equation on a Lattice: Existence and Uniqueness of Solutions," *J. Stat. Phys.* **20**, 463. (2)

STRUMINSKII, V. V. (1965): "Hilbert's Method of Solving the Boltzmann Kinetic Equation," *Sov. Phys. Doklady* **9**, 733. (4.1)

SWENSON, R. J. (1962): "Derivation of Generalized Master Equations," *J. Math. Phys.* **3**, 1017. (11.6)

——— (1963): "Generalized Master Equation and *t*-Matrix Expansion," *J. Math. Phys.* **4**, 544. (11.6)

SYKES, J. (1973): "Short-Time Kinetic Equations for Hard Spheres: Comparison with Other Theories," *J. Stat. Phys.* **8**, 279. (7.10)

TANAKA, T., K. MOURJANI, AND T. MORITA (1967): "Green's Function Theory of Nonlinear Transport Coefficients," *Phys. Rev.* **155**, 388. (9.6)

TAYLOR, G. I. (1920): "Diffusion by Continuous Movements," *Proc. Lond. Math. Soc.* **20**, 196. (2.11)

TESKE, A. (ED.) (1977): *Marian Smoluchowski: Leben und Werk*. Warsaw: Polskiej Akademii Nauk. (2)

THOMAS, M. W., AND R. F. SNIDER (1970): "Boltzmann Equation and Angular Momentum Conservation," *J. Stat. Phys.* **2**, 61. (8.3)

TITULAER, U. M. (1978): "A Systematic Solution Procedure for the Fokker-Planck Equation of a Brownian Particle in the High-Friction Case," *Physica* **91A**, 321. (2.5)

———, AND B. U. FELDERHOF (1980): "The Eigenvalue Problem for Linear Macroscopic Equations," *Physica* **100A**, 573. (6.2)

TREMBLAY, A. M. S. (1984): "Theories of Fluctuations in Nonequilibrium Systems," in *Recent Developments in Nonequilibrium Thermodynamics,* eds. J. Casa-Vazquez, D. Jou, and G. Lebon, *Lecture Notes in Physics* **199**. New York: Springer-Verlag. (9.13)

TRUESDELL, C. (1956): "On the Pressures and the Flux of Energy in a Gas According to Maxwell's Kinetic Theory, II," *J. Rat. Mech. Anal.* **5**, 55. (5.3)

______, AND R. G. MUNCASTER (1980): *Fundamentals of Maxwell's Kinetic Theory of a Simple Monatomic Gas: Treated as a Branch of Rational Mechanics.* New York: Academic Press. (5)

UHLENBECK, G. E., AND L. S. ORNSTEIN (1930): "On the Theory of the Brownian Motion," *Phys. Rev.* **36**, 823. Reprinted in *Noise and Stochastic Processes*, ed. N. Wax. New York: Dover, 1954. (2.3)

______, AND G. W. FORD (1963): "Lectures in Statistical Mechanics," vol. I of *Lectures in Applied Mathematics*, ed. M. Kac. American Mathematical Society, Providence. (3)

UKAI, S. (1976): "Les Solutions Globales de l'Equation de Boltzmann dans l'Espace tout Entier et dans le Demi-Espace," *C. R. Acad. Sc. Paris* **282**, 317. (3)

VAN BEIJEREN, H., AND M. H. ERNST (1973a): "The Modified Enskog Equation," *Physica* **68**, 437. (7.8)

______, AND ______ (1973b): "The Modified Enskog Equation for Mixtures," *Physica* **70**, 225. (7.8)

______, AND ______ (1979): "Kinetic Theory of Hard Spheres," *J. Stat. Phys.* **21**, 125. (12.7)

______, AND J. R. DORFMAN (1980a): "Kinetic Theory of Hydrodynamic Flows. I. The Generalized Normal Solution Method and its Application to the Drag on a Sphere," *J. Stat. Phys.* **23**, 335. (2.3)

______, AND ______ (1980b): "Kinetic Theory of Hydrodynamic Flows. II. The Drag on a Sphere and on a Cylinder," *J. Stat. Phys.* **23**, 443. (2.3)

______ (1982): "Transport Properties of Stochastic Lorentz Models," *Revs. Modern Phys.* **54**, 195. (12.10)

VAN DEN BROECK, C., F. LOSTAK, AND H. N. W. LEKKERKERKER (1981): "The Effect of Direct Interactions on Brownian Diffusion," *J. Chem. Phys.* **74**, 2006. (2)

VAN KAMPEN, N. G. (1971): "The Case Against Linear Response Theory," *Phys. Norvegica* **5**, 279. (9.6)

______ (1975): "Stochastic Differential Equations," in *Fundamental Problems in Statistical Mechanics* **III**, ed. E. G. D. Cohen. Amsterdam: North-Holland Publishing Co. (2)

______ (1981): *Stochastic Processes in Physics and Chemistry*. Amsterdam: North-Holland Publishing Co. (2)

VAN LEEUWEN, J. M. J., AND S. YIP (1965): "Derivation of Kinetic Equations for Slow-Neutron Scattering," *Phys. Rev.* **139**, A1138. (11.3)

______, AND A. WEYLAND (1967a): "Non-Analytic Density Behaviour of the Diffusion Coefficient of a Lorentz Gas. I. Divergences in the Expansion in Powers in the Density," *Physica* **36**, 457. (12.10)

______, AND ______ (1967b): "The Density Expansion of the Diffusion Coefficients of a Lorentz Gas," in *Statistical Mechanics, Foundations and Applications*, ed. T. A. Bak. New York: Benjamin. (12.10)

VAN OOST, E., AND H. N. W. LEKKERKERKER (1978): "Geometry of the Roots of Matrices with Onsager-Casimir Symmetry," *J. Math. Phys.* **19**, 2223. (10.3)

VAN WEERT, CH. G., AND W. P. H. DE BOER (1978): "On the Derivation of Quantum Kinetic Equations. I. Collision Expansions," *J. Stat. Phys.* **18**, 271. (13.1)

VELASCO, R. M., AND F. J. URIBE (1986): "Transport Coefficients for Polyatomic Gases," *Physica* **A134**, 339. (13.4)

VERBOVEN, E. (1960): "On the Quantum Theory of Electrical Conductivity," *Physica* **26**, 1091. (11.6)

VESTNER, H., AND L. WALDMANN (1977): "Differential Equations and Boundary Conditions for Higher Gas Kinetic Moments. Heat Transfer between Parallel Plates," *Z. Naturforsch.* **32a**, 667. (5.7)

VISSCHER, P. B. (1976): "Escape Rate for a Brownian Particle in a Potential Well," *Phys. Rev.* **B13**, 3272. (2.9)

WALDMANN, L. (1957): "Die Boltzmann Gleichung fur Gase mit Rotierenden Molekulen," *Z. Naturforsch.* **12A**, 660. (13.3)

——— (1958): "Transporterscheinungen in Gasen von Mittlerem Druck," in *Encyclopedia of Physics* **12**, ed. S. Flugge. Berlin: Springer-Verlag. (3.3, 5.5)

WANG, M. C., AND G. E. UHLENBECK (1945): "On the Theory of the Brownian Motion II," *Revs. Modern Phys.* **17**, 323. Reprinted in *Noise and Stochastic Processes*, ed. N. Wax. New York: Dover, 1954. (2.4, 2.9)

WANG CHANG, C. S., G. E. UHLENBECK, AND J. DE BOER (1964): "The Heat Conductivity and Viscosity of Polyatomic Gases," in *Studies in Statistical Mechanics* **II**, eds. J. de Boer and G. E. Uhlenbeck. Amsterdam: North-Holland Publishing Company. (13.3)

——— AND ——— (1970): "The Kinetic Theory of Gases," in *Studies in Statistical Mechanics* **V**, eds. J. de Boer and G. E. Uhlenbeck. Amsterdam: North-Holland Publishing Co. This is a collection of University of Michigan Research Institute Reports written by the authors during the period 1948 to 1956. (2.5, 3.13, 5.5, 6.5)

WAX, N. ED. (1954): *Noise and Stochastic Processes*. New York: Dover. (2)

WEINSTOCK, J. (1962): "Binary Collision Expansion of the Classical N-Body Green's Function," *Phys. Rev.* **126**, 341. (12.4)

——— (1963a): "Cluster Formulation of the Exact Equation for the Evolution of a Classical Many-Body System," *Phys. Rev.* **132**, 454. (11.5)

——— (1963b): "Three-Body Scattering Operator in Nonequilibrium Statistical Mechanics," *Phys. Rev.* **132**, 470. (7.7)

WELANDER, P. (1954): "On the Temperature Jump in a Rarefied Gas," *Arkiv. Fysik* **7**, 507. (4.1, 5.5)

WIDOM, A. (1971): "Velocity Fluctuations of a Hard-Core Brownian Particle," *Phys. Rev.* **A3**, 1394. (2.12)

WILD, E. (1951): "On Boltzmann's Equation in the Kinetic Theory of Gases," *Proc. Camb. Phil. Soc.* **47**, 602. (3)

WOLYNES, P. G. (1976): "Hydrodynamic Boundary Conditions and Mode-Mode Coupling Theory," *Phys. Rev.* **A13**, 1235. (12.11)

WONG, C. K., J. A. MCLENNAN, M. LINDENFELD, AND J. W. DUFTY (1978): "Theory of Nonlinear Transport in Burnett Order," *J. Chem. Phys.* **68**, 1563. (9.13)

WOOD, W. W. (1975): "Computer Studies on Fluid Systems of Hard-Core Particles," in *Fundamental Problems in Statistical Mechanics* **III**, ed. E. G. D. Cohen. Amsterdam: North-Holland Publishing Company. (12.11)

———, AND J. J. ERPENBECK (1976): "Molecular Dynamics and Monte Carlo Calculations in Statistical Mechanics," *Ann. Rev. Phys. Chem.* **27**, 319. (12.11)

WYLLIE, G. (1980): "Random Motion and Brownian Motion," *Phys. Reports* **61**, 327. (2)

YAMADA, T., AND K. KAWASAKI (1975): "Application of Mode-Coupling Theory to Nonlinear Stress Tensor in Fluids," *Prog. Theor. Phys.* **53**, 111. (12.11)

YIP, S., AND M. NELKIN (1964): "Application of a Kinetic Model to Time-Dependent Density Correlations in Fluids," *Phys. Rev.* **135**, A1241. (5.9)

——— (1979): "Renormalized Kinetic Theory of Dense Fluids," *Ann. Rev. Phys. Chem.* **30**, 547. (11.7)

———, W. E. ALLEY, AND B. J. ALDER (1982): "Evaluation of Time Correlation Functions From a Generalized Enskog Equation," *J. Stat. Phys.* **27**, 201. (7.8)

YOSHIDA, K. (1980): *Functional Analysis*. New York: Springer-Verlag. (6.9)

YVON, J. (1935): *La Theorie Statistique des Fluides et l'Equation d'Etat*. Paris: Hermann et Cie. (7.1)

ZIMAN, J. (1956): "The General Variational Principle of Transport Theory," *Can. J. Phys.* **34**, 1256. (4.6)

ZUBAREV, D. N. (1962): "The Statistical Operator for Nonequilibrium Systems," *Sov. Phys. Doklady* **6**, 776. (9.12)

——— (1965): "Local Equilibrium of a Gibbs Ensemble and its Connection with the Theory of Fluctuations and Transport Phenomena," *Sov. Phys. Doklady* **10**, 452. (9.12)

——— (1966): "A Statistical Operator for Nonstationary Processes," *Sov. Phys. Doklady* **10**, 850. (9.12)

——— (1970): "The Method of the Non-Equilibrium Statistical Operator and its Applications," *Fortschritte der Physik* **18**, 125. (8.2, 9.12)

——— (1974): *Nonequilibrium Statistical Mechanics*. New York: Consultants Bureau. (9.12)

ZWANZIG, R. (1961): "Memory Effects in Irreversible Thermodynamics," *Phys. Rev.* **124**, 983. (11.7)

——— (1963): "Method for Finding the Density Expansion of Transport Coefficients of Gases," *Phys. Rev.* **129**, 486. (11.7, 12.4)

——— (1964): "Elementary Derivation of Time-Correlation Formulas for Transport Coefficients," *Phys. Rev.* **40**, 2527 (11.7)

——— (1965a): "Frequency-Dependent Transport Coefficients in Fluid Mechanics," *J. Chem. Phys.* **43**, 714. (9.14)

———, AND R. D. MOUNTAIN (1965b): "High-Frequency Elastic Moduli of Simple Fluids," *J. Chem. Phys.* **43**, 4464. (9.14)

——— (1965c): "Time-Correlation Functions and Transport Coefficients in Statistical Mechanics," *Ann. Rev. Phys. Chem.* **16**, 67. (9.10)

———, AND M. BIXON (1970): "Hydrodynamic Theory of the Velocity Correlation Function," *Phys. Rev.* **A2**, 2005. (2.12)

———, AND M. BIXON (1975): "Compressibility Effects in the Hydrodynamic Theory of Brownian Motion," *J. Fluid Mech.* **69**, 21. (2.12)

——— (1979): "Non-Linear Shear Viscosity of a Gas," *J. Chem. Phys.* **71**, 4416. (12.11)

——— (1981): "Nonlinear Shear Viscosity and Long Time Tails," *Proc. Nat. Acad. Sci.* **78**, 3296. (12.11)

INDEX